KU-166-643

Rainfall–Runoff Modelling

Rainfall–Runoff Modelling

The Primer

Keith J. Beven

Professor of Hydrology and Fluid Dynamics
Lancaster University, UK

JOHN WILEY & SONS, LTD

Chichester • New York • Weinheim • Brisbane • Singapore • Toronto

National 01243 779777
International (+44) 1243 779777
e-mail (for orders and customer service enquiries): cs-books@wiley.co.uk
Visit our Home Page on http://www.wiley.co.uk
or http://www.wiley.com

Other Wiley Editorial Offices

John Wiley & Sons, Inc., 605 Third Avenue,
New York, NY 10158-0012, USA

WILEY-VCH Verlag GmbH, Pappelallee 3,
D-69469 Weinheim, Germany

John Wiley & Sons Australia, Ltd, 33 Park Road, Milton,
Queensland 4064, Australia

John Wiley & Sons (Asia) Pte Ltd, Clementi Loop #02-01,
Jin Xing Distripark, Singapore 129809

John Wiley & Sons (Canada) Ltd, 22 Worcester Road,
Rexdale, Ontario M9W 1L1, Canada

Library of Congress Cataloguing-in-Publication Data

Beven, K. J.
Rainfall–runoff modelling : the primer / Keith J. Beven.
p. cm.
Includes bibliographical references (p.).
ISBN 0-471-98553-8 (alk. paper)
1. Runoff–Mathematical models. 2. Rain and rainfall – Mathematical models. I. Title.

GB980 .B48 2000
551.48′8–dc21

00-043340

British Library Cataloguing in Publication Data

A catalogue record for this book is available from the British Library

ISBN 0-471-98553-8

Typeset in 10/12pt Times from the author's disks by Laser Words, Madras, India
Printed and bound in Great Britain by Bookcraft (Bath) Ltd, Midsomer Norton

This book is printed on acid-free paper responsibly manufactured from sustainable forestry, in which at least two trees are planted for each one used for paper production.

Contents

Preface

Models are undeniably beautiful, and a man may justly be proud to be seen in their company. But they may have their hidden vices. The question is, after all, not only whether they are good to look at, but whether we can live happily with them.

A. Kaplan, 1964

One is left with the view that the state of water resources modelling is like an economy subject to inflation – that there are too many models chasing (as yet) too few applications; that there are too many modellers chasing too few ideas; and that the response is to print ever-increasing quantities of paper, thereby devaluing the currency, vast amounts of which must be tendered by water resource modellers in exchange for their continued employment.

Robin Clarke, 1974

There is one fundamental problem in studying hydrological systems: most of the action takes place underground. Despite all the advances that have been made in remote sensing, ground-probing radar and other techniques for exploring the subsurface, our knowledge of what goes on underground is still very limited. What we do know from detailed studies of water movement through soils and rocks in both the laboratory and small-scale field studies is that the patterns of water movement are very complex indeed and change in very nonlinear ways with flow rates and degree of wetness. From the point of view of rainfall–runoff modelling at scales of practical interest (generally medium to large catchments), this complexity means that we cannot hope to reproduce all the details of the flow processes that give rise to the stream hydrograph: too much of the complexity involved is simply unknowable with current measurement techniques. At one level, therefore, rainfall–runoff modelling is an impossible problem!

That has not, of course, stopped many different hydrologists, teams of hydrologists and collaborating hydrological institutes from developing rainfall–runoff models. The word 'plethora' springs to mind, despite the fact that at different times in my career I have been involved in the development and testing of many different models trying, in different ways, to do the job properly. This book cannot hope to enumerate all the different rainfall–runoff models currently available and I must apologise in advance to all the modellers whose models get only a brief mention or none at all. It is now virtually impossible for any one person to be aware of all the models that are reported in the literature, let alone know something of the historical framework of the different initiatives. I gave up making a list of available models when I reach a count of 100 more than 20 years ago (and that was before I too reported on another one in my PhD thesis).

Therefore I have attempted instead to outline the most interesting themes that are currently being explored in trying to improve our predictive capability in hydrology, while still trying to reflect the historical context of the subject. That means that the reader will find very little on the type of conceptual storage models that started the whole process in the 1960s and which are still widely used today. Nor will you find much on monthly water balance models. These represent the past; I hope the reader will find that this book looks to the future. Even so, such a book can only represent a snapshot of the progress that has been made, and even then one that is only sharply in focus in some places. The activity of rainfall–runoff modelling was already a major activity among hydrologists around the world when Robin Clarke made his analogy with an economy subject to inflation 25 years ago. Today, the number of trees being sacrificed to report on progress in the field is even greater. To provide a complete review of the literature in a primer such as this would be impossible, but I have tried to provide references to both recent and classic papers to allow the reader to explore more of the research literature as needed.

There remains a continuing and important need for rainfall–runoff modelling for very practical problems in water resource assessment, flood forecasting, design of engineered channels, assessing the impact of effluents on water quality, predicting pollution incidents, and many other purposes. Fortunately, the situation is not quite as dire as suggested by the impossibility of predicting the flow pathways of water in detail. For many practical purposes we do not need to include all the detail in developing a predictive model. Indeed, many successful rainfall–runoff models are essentially very simple. This book is intended as an introduction to recent developments in rainfall–runoff modelling that allow such practical predictions to be made. This will be done, however, in the context of an understanding that the impossibility of making detailed predictions of the flow processes involved necessarily means that all rainfall–runoff models can only be very approximate descriptions of the rainfall–runoff processes and, as such, must be considered to be uncertain in their predictions.

Thus, a whole section of this book is devoted to considerations of predictive uncertainty. This might normally be considered an 'advanced' topic. My own view is that it is essential to understand the uncertainties in any environmental modelling exercise and that, using modern computer-intensive Monte Carlo techniques, the estimation of uncertainty can be introduced in a conceptually very simple way. Such uncertainty estimation feeds directly into the assessment of risk in decision-making and, in most practical cases, rainfall–runoff modelling is carried out precisely to make decisions. Should a flood forecast be issued given forecasts of river stage over the next 6 hours? What should be the reservoir overspill channel capacity to cope with the flood expected once in 50 years? The limitations of our hydrological knowledge are such that these decisions should be taken within a risk assessment framework, recognizing the inherent uncertainty in our predictions.

It is hoped that an understanding of the material set out in the text of this book, including the section on uncertainty estimation, will be enhanced by the demonstration software provided. The software is primarily based on methods that have been developed at Lancaster University over the last decade, and many colleagues, research assistants and students have stimulated ideas or made direct contributions to that development. I would particularly like to mention Peter Young, whose idea that the data (rather than theory alone) may suggest an appropriate model structure has been an important influence. Andrew Binley, Kathy Bashford, David Cameron, James Fisher, Stewart Franks, Jim Freer, Rob Lamb, Matthew Lees, Paul Quinn, Renata Romanowicz, Karsten Schulz and Jonathon Tawn, all at, or formerly at, Lancaster have made important contributions to modelling projects. Collaborations with other groups has also been very important, particularly those with George Hornberger (Charlottesville), Bruno Ambroise (Strasbourg), Charles Obled and Georges-Marie Saulnier (Grenoble), Eric Wood (Princeton), Peter Germann (Bern), Sarka Blazkova (Prague) and Philippe Merot (Rennes). Parts of this book were written while on sabbatical in Santa Barbara with support from Tom Dunne and Jeff Dozier, in Lausanne with support from André Musy, and in Leuven with support from Jan Feyen and the Francqui Foundation.

I would also like to acknowledge my lasting debt to Mike Kirkby. A long while ago, his lectures while I was an undergraduate at Bristol University made me realize that it was possible to model geomorphological and hydrological systems in a thoughtful and insightful way, while the origins of TOPMODEL lie in his wealth of ideas during in my post-doctoral work with him at Leeds. His skill in capturing the essence of a problem in a set of relatively simple assumptions has been a lasting inspiration, even while I was struggling to understand what he was getting at! I hope he will recognize some of that influence in what follows.

Finally, this book is intended to introduce hydrological modelling concepts to a new generation of students and it is dedicated to one very special recent graduate in particular. If, by some chance of fate, Anna ever has to read this and try to understand it, I hope she will find it a clear and useful guide to both present and future uses of rainfall–runoff models. It has been written primarily for her generation.

Keith Beven
Pully, 1997
Lancaster, Outhgill and Leuven, 1999

The publishers gratefully acknowledge the help of Günter Blöschl in reviewing the final manuscript.

1

Down to Basics: Runoff Processes and the Modelling Process

As scientists we are intrigued by the possibility of assembling our knowledge into a neat package to show that we do, after all, understand our science and its complex interrelated phenomena.

W. M. Kohler, 1969

1.1 Why Model?

As noted in the Preface, there are many different reasons why we need to model the rainfall–runoff processes of hydrology. The main reason is, however, a result of the limitations of hydrological measurement techniques. We are not able to measure everything we would like to know about hydrological systems. We have, in fact, only a limited range of measurement techniques and a limited range of measurements in space and time. We therefore need a means of extrapolating from those available measurements in both space and time, particularly to ungauged catchments (where measurements are not available) and into the future (where measurements are not possible) to assess the likely impact of future hydrological change. Models of different types provide a means of quantitative extrapolation or prediction that will hopefully be helpful in decision-making.

There is much rainfall–runoff modelling that is carried out purely for research purposes as a means of formalizing knowledge about hydrological systems. The demonstration of such understanding is an important way of developing an area of science. We generally learn most when a model or theory is shown to be in conflict with reliable data so that some modification of the understanding on which the model is based must be sought. However, the ultimate aim of prediction using models must be to improve decision-making about a hydrological problem, whether that be in water resources planning, flood protection, mitigation of contamination, or licensing of abstractions, etc.

With increasing demands on water resources throughout the world, improved decision-making, within a context of fluctuating weather patterns from year to year, requires improved models. That is what this book is about.

Rainfall–runoff modelling can be carried out within a purely analytical framework based on observations of the inputs and outputs to a catchment area. The catchment is treated as a '*black box*', without any reference to the internal processes that control the rainfall to runoff transformation. Some models developed in this way are described in Chapter 4, where it is shown that it may also be possible to make some physical interpretations of the resulting models based on an understanding of the nature of catchment response. This understanding should be the starting point for any rainfall–runoff modelling study.

There are, of course, many hydrological texts that describe hydrological processes with varying degrees of mathematical analysis and numbers of equations, but the more mathematical descriptions do not always point out the important simplifications that are being made in their analyses; they present the equations as if they applied everywhere. However, it is only necessary to sprinkle a coloured dye solution onto the soil surface and then dig to see where the dye has stained the soil to realize the limitations of hydrological theory (see Figure 1.1). Whenever detailed studies of flow pathways are carried out in the field we find great complexity. We can perceive that complexity quite easily, but producing a mathematical description suitable for quantitative prediction is much more difficult and will always involve important simplication and approximation. This initial chapter will therefore be concerned with a perceptual model of catchment response as the first stage of the modelling process. This complexity is one reason why there is no commonly agreed modelling strategy for the rainfall–runoff process but a variety of options and approaches that will be discussed in the chapters that follow.

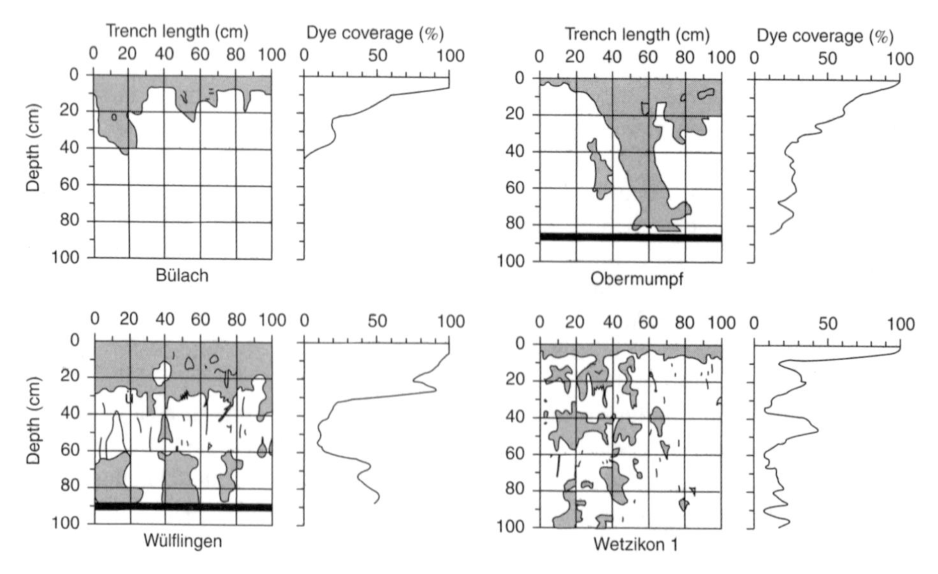

Figure 1.1 *Staining by dye in sections through different soil profiles in Switzerland after infiltration of 40 mm of water (after Flury et al. 1994). Reproduced from Water Resources Research 30(7): 1945–1954, 1994, copyright by the American Geophysical Union*

1.2 How to Use This Book

It should be made clear right at the beginning that this is not a book only about the theory that underlies the different types of rainfall–runoff model that are now available to the user. You will find, for example, that relatively few equations are used in the main text of the book. Where it has been necessary to show some theoretical development this has been printed in boxes at the ends of the chapters that can be skipped at a first reading. The theory can also be followed up in the many (but necessarily selected) references given.

This is much more a book about the concepts that underlie different modelling approaches and the critical analysis of the software packages that are now widely available for hydrological prediction. The presentation of models as software is becoming increasingly sophisticated, with links to geographical information systems and the display of impressive three-dimensional graphical outputs. It is easy to be seduced by these displays into thinking that the output of the model is a good simulation of the real catchment response, especially if few data are available to check on the predictions. However, even for the most sophisticated models currently available, this is not necessarily so, and evaluation of the model predictions will be necessary. It is hoped that the reader will learn from this book the concepts and techniques necessary to evaluate both the different modelling approaches and packages available and the simulations of a model in a particular application.

There are four software packages available to accompany this book: TFM, TOPMODEL, DTMAnalysis and GLUE, which are described and used in Chapters 4, 6 and 7. The first two of these represent examples of different generic types of models that make specific assumptions about the response of the rainfall–runoff system. One of the aims of this book is to train the reader to evaluate models, not only in terms of how well the model can reproduce any data that are available for testing, but also by critically assessing the assumptions made. Thus, wherever possible, models are presented together with a list of the assumptions made. The reader is encouraged to make a similar list whenever he or she encounters a model for the first time. The GLUE package is an uncertainty estimation methodology that can be used with any hydrological model (see Chapter 7). The latest versions of these packages, written to run under PC Windows, can be downloaded over the Internet (see Appendix A). A list of World Wide Web links to other pages concerned with different rainfall–runoff modelling packages is also available at the same site.

At the end of each chapter a review of the major points arising from that chapter has been provided. It is generally a good strategy to read that summary before reading the bulk of the chapter. A glossary of terms used in hydrological modelling is provided in Appendix B. These terms are highlighted when they first appear in the text.

1.3 The Modelling Process

Most books on modelling start with the choice of model to be used for a particular application. Here, we will start at an earlier stage in the modelling process: the *perceptual model* of the rainfall–runoff processes in a catchment (see Figure 1.2). The perceptual model is the summary of our perceptions of how the catchment responds

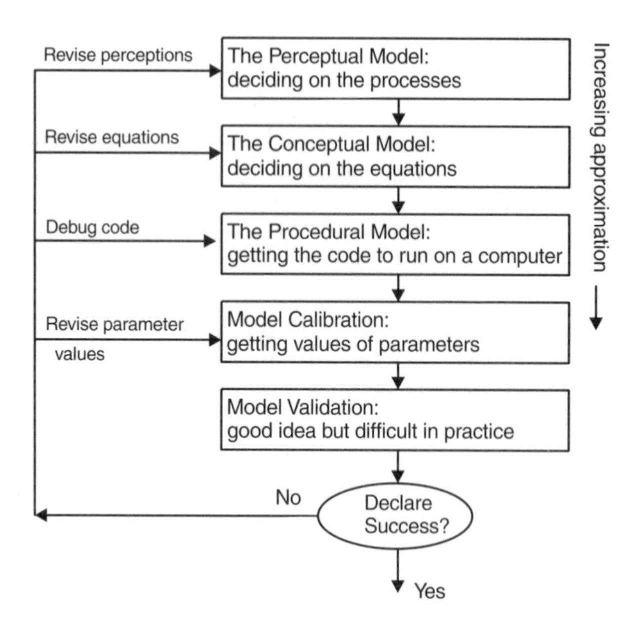

Figure 1.2 *A schematic outline of the different steps in the modelling process*

to rainfall under different conditions, or rather, *your* perceptions of that response. A perceptual model is necessarily personal. It will depend on the training that a hydrologist has had, the books and articles they have read, the data sets they have analysed, and particularly the field sites they have had experience of in different environments. Thus it is to be expected that one hydrologist's perceptual model will differ from that of another (for a typical personal example, see Section 1.4 below).

An appreciation of the perceptual model for a particular catchment is important because it must be remembered that all the mathematical descriptions that might be used for making predictions will inevitably be simplifications of the perceptual model, in some cases gross simplifications, but perhaps still sufficient to provide adequate predictions. There are good reasons for this. The perceptual model is not constrained by mathematical theory. It exists primarily in the head of each hydrologist and need not even be written down. We can *perceive* complexities of the flow processes in a purely qualitative way (see, for example, the flow visualization experiments of Flury *et al.* (1994) in Figure 1.1) that may be very difficult indeed to describe in the language of mathematics. However, a mathematical description is, traditionally, the first stage in the formulation of a model that will make *quantitative* predictions.

This mathematical description we will call here the *conceptual model* of the process or processes being considered. At this point the hypotheses and assumptions being made to simplify the description of the processes need to be made explicit. For example, many models have been based on using a description of flow through the soil based on Darcy's law which states that flow is proportional to a gradient of hydraulic potential (see Box 5.1). Measurements show that gradients of hydraulic potential in structured soils can vary significantly over small distances so that if Darcy's law is applied at the scale of a soil profile or greater, it is implicitly assumed that some average gradient can be used to characterize the flow and that the effects of preferential flow through macropores in the soil (one explanation of the observations of Figure 1.1) can be neglected. It is worth noting that in many articles and model user manuals, while the equations on which the model is based may be given, the underlying simplifying assumptions may not actually be stated explicitly. Usually, however, it is not difficult to list the assumptions, knowing something of the background to the equations. This should be the starting point for the evaluation of a particular model relative to the perceptual model in mind. Making a list of all the assumptions of a model is a useful practice that we will follow here in the presentation of different modelling approaches.

The conceptual model may be more or less complex, ranging from the use of simple mass balance equations for components representing storages in the catchment, to coupled nonlinear partial differential equations. Some equations may be easily translated directly into programming code for use on a digital computer. However, if the equations cannot be solved analytically given some *boundary conditions* for the real system (which is usually the case for the partial differential equations found in some hydrological models) then an additional stage of approximation is necessary using the techniques of numerical analysis to define a *procedural model* in the form of code that will run on the computer. An example is the replacement of the differentials of the original equations by finite-difference or finite-volume equivalents. Great care has been taken at this point: the transformation from the equations of the conceptual model to the code of the procedural model has the potential to add significant error relative to the true solution of the original equations. Because such models are often highly *nonlinear*, assessing that error may be difficult for all the conditions in which the model may be used.

With the procedural model we have a code that will run on the computer. Before we can apply the code to make quantitative predictions for a particular catchment, however, it is generally necessary to go through a stage of parameter *calibration*. All the models used in hydrology have equations that involve a variety of different input and state variables. There are inputs that define the geometry of the catchment which are normally considered constant during the duration of a particular simulation. There are variables that define the time variable boundary conditions during a simulation, such as the rainfalls and other meteorological variables at a given time step. There are the state variables, such as soil water storage or water table depth, which change during a simulation as a result of the model calculations. There are the initial values of the state variables that define the state of the catchment at the start of a simulation. Finally, there are the model *parameters* which define the characteristics of the catchment area or flow domain.

The model parameters may include characteristics such as the porosity and hydraulic conductivity of different soil horizons in a spatially distributed model, or the mean residence time in the saturated zone for a model that uses state variables at the catchment scale. They are usually considered constant during the period of a simulation (although for some parameters, such as the capacity of the interception storage of a developing vegetation canopy, there may be a strong time dependence that may be important for some applications). In all cases, even if they are considered as constant in time, it is not easy to specify the values of the parameters for a particular catchment *a priori*.

Indeed, the most commonly used method of parameter calibration is to use a technique of adjusting the values of the parameters to achieve the best match between the model predictions and any observations of the actual catchment response that may be available (see Section 1.8 and Chapter 7).

Once the model parameter values have been specified, a simulation may be made and quantitative predictions about the response obtained. The next stage is then the *validation* or *evaluation* of those predictions. This evaluation may also be carried out within a quantitative framework, calculating one or more indices of the performance of the model relative to the observations available (if any) about the runoff response. The problem at this point is usually not that it is difficult to find an acceptable model, particularly if it has been possible to calibrate the model parameters by a comparison with observed discharges; most model structures have a sufficient number of parameters that can be varied to allow reasonable fits to the data. The problem is more often that there are *many* different combinations of model structure and sets of parameter values that will give reasonable fits to the discharge data. Thus, in terms of discharge prediction alone, it may be difficult to differentiate between different feasible models and therefore to *validate* any individual model. This will be addressed in more detail in Chapter 7 in the context of assessing the uncertainty in model predictions.

On the other hand, the discharge predictions, together with any predictions of the internal responses of the catchment, may also be evaluated relative to the original perceptual model of the catchment of interest. Here, it is usually much more difficult to find a model that is totally acceptable. The differences may lead to a revision of the parameter values being used; to a reassessment of the conceptual model being used; or even in some cases to a revision of the perceptual model of the catchment as understanding is gained from the attempt to model the hydrological processes.

The remainder of this chapter will be concerned with the different stages in the modelling processes: Section 1.4 outlines an example of a perceptual model of catchment responses to rainfall; Section 1.5 discusses the additional information that might be gained from considering geochemical information; Section 1.5 gives the functional requirements of runoff production and runoff routing; Section 1.7 gives a definition of a conceptual model; and model calibration and validation issues are discussed in Section 1.8.

1.4 Perceptual Models of Catchment Hydrology

There are many outlines of the processes of catchment response available in the literature. Most general hydrological texts deal, in greater or less detail, with the processes of catchment response. The volumes edited by Kirkby (1978) and Anderson and Burt (1990) are of particular interest in that the different chapters reflect the experiences of a number of different hydrologists. Hydrological systems are sufficiently complex that each hydrologist will have his or her own impression or perceptual model of what is most important in the rainfall–runoff process, so different hydrologists might not necessarily agree about what are the most important processes or the best way of describing them. There are sure to be general themes in common, as reflected in hydrological texts, but our understanding of hydrological responses is still evolving and the details will depend on experience, in particular the type of hydrological environments that a hydrologist has experienced. Different processes may be dominant in different environments and in catchments with different characteristics of topography, soil, vegetation and bedrock.

One of the problems involved in having a complete understanding of hydrological systems is that most of the water flows take place underground in the soil or bedrock. Our ability to measure and assess subsurface flow processes is generally very limited. Most of the measurement techniques available reflect conditions only in the immediate area of the measurement probe. When the characteristics of the flow domain vary rapidly in space (and sometimes in time), the small-scale nature of such measurements can give only a very partial picture of the nature of the flow. Thus, there is much that remains unknown about the nature of subsurface flow processes, and is indeed unknowable given the limitations of current measurement techniques. It is necessary to make inferences about the processes from the available measurements. Such inferences add information to the perceptual model of hydrological response, but they are inferences.

One way of gaining further understanding is to examine a part of the system in much greater detail. Many studies have been made of the flow processes on particular hillslopes or plots, or columns of undisturbed soil brought back to the laboratory. It has generally been found in such studies that more detailed investigation reveals greater complexity and variability in the flow pathways. The same has generally been true of adding different types of information, such as the use of artificial or environmental tracers. Figure 1.1 is a good example of this (see also Section 1.5). Such complexity can be made part of the perceptual model. As noted above, it is not necessary that the perceptual model be anything more than a set of qualitative impressions, but complexity inevitably creates difficulty in the choice of assumptions in moving from the perceptual model to a set of equations defining a conceptual model. Choices must be made at this point to simplify the description, and as we will see, such choices have not always had a good foundation in hydrological reality.

Consider, briefly, one hydrologist's perceptual model. It is based on an outline set out in Beven (1991a), with some revision based on additional experience since then. In recession periods between storms, storage in the soil and rock gradually declines (Figure 1.3(a)). If there is a water table, the level and gradient will gradually fall. Storage will often be higher and water tables closer to the surface in the valley bottom *riparian areas*, partly because of downslope flow, particularly where there is convergence of flow in hillslope hollows. Storage in riparian areas may also be maintained by return flows from deeper layers (e.g. Huff *et al.* 1982; Genereux *et al.* 1993), but also because soils tend to be deeper in valley bottoms (e.g. Piñol *et al.* 1997). Loss of water by evapotranspiration will have a greater or less effect on the profile of storage depending on season, climate and vegetation type and rooting depth. Many plants, however, may extract water from considerable depths, with roots penetrating up to tens of metres into the soil and bedrock fractures and root channels also acting as pathways for infiltrating water (e.g. the Jarrah trees of Western Australia). Plants that are *phreatophytes* (such as the Cottonwoods of the western United States) will extract water directly from beneath the water table. These evapotranspiration and drainage processes will be important in controlling the *antecedent conditions* prior to a storm event.

(a)

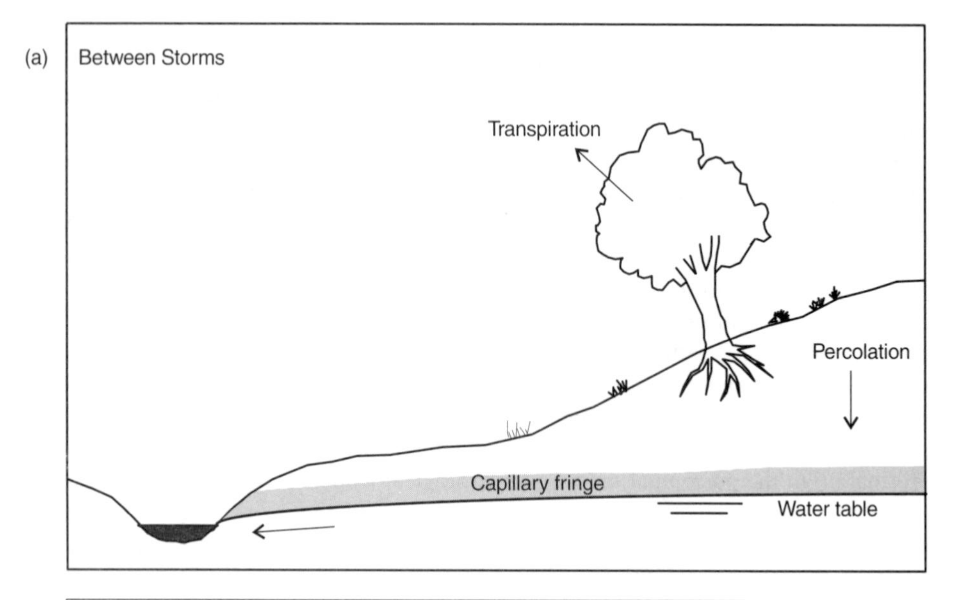

(b)

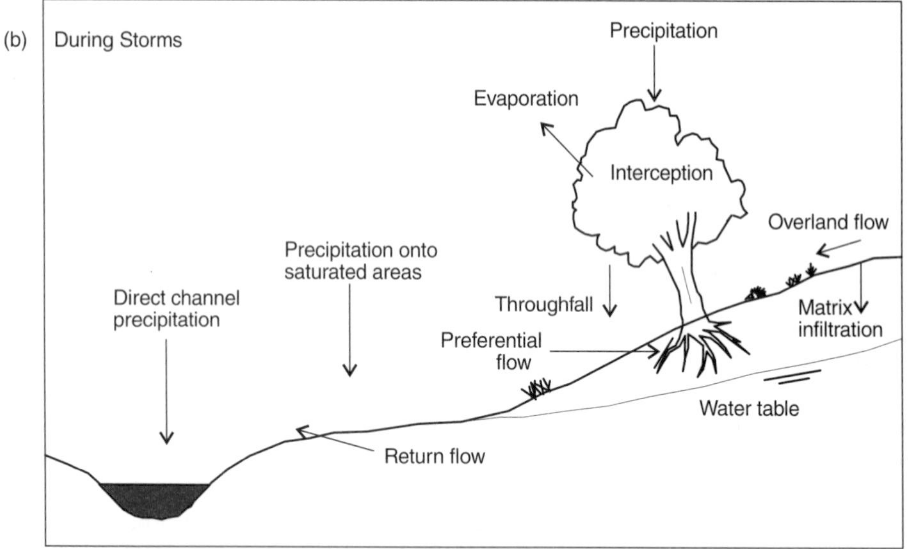

Figure 1.3 *A representation of the processes involved in one perceptual model of hillslope hydrology*

The antecedent conditions, as well as the volume and intensity of rainfall (or snowmelt), will be important in governing the processes by which a catchment responds to rainfall and the proportion of the input volume that appears in the stream as part of the hydrograph (Figure 1.3(b)). Unless the stream is ephemeral, there will always be a response from precipitation directly onto the channel and immediate riparian area. This area, although a relatively small area of the catchment (perhaps 1–3 percent), may be an important contributor to the hydrograph in catchments and storms with low runoff coefficients. Even in *ephemeral streams*, surface flow will often start first in the stream channels. The extent of the channel network will generally expand into headwater areas as a storm progresses and will be greater during wet seasons than dry (e.g. Hewlett 1974).

Rainfall and snowmelt inputs are not spatially uniform, but can show rapid changes in intensity and volume over relatively short distances, particularly in convective events (e.g. Newson 1980; Smith *et al.* 1996; Goodrich *et al.* 1997). The variability at ground level, after the pattern of intensities has been affected by the vegetation canopy, may be even greater. Some of the rainfall will fall directly to the ground as direct *throughfall*. Some of the rainfall will be intercepted and evaporated from the canopy back to the atmosphere. Some evaporation of intercepted water may occur even during events, especially from rough canopies, under windy conditions, when the air is not saturated with vapour. Differences of up to 30 percent between incident rainfall and throughfall have been measured in a Mediterranean catchment even for large storms (Llorens *et al.* 1997). The remaining rainfall will drip from the vegetation canopy as throughfall or run down the branches, trunks and stems as *stemflow*. The latter process may be important since, for some canopies, 10 percent or more of the incident rainfall may reach the ground as stemflow, resulting in local concentrations of water at much higher intensities than the incident rainfall. Some plants, such as maize, have a structure designed to channel water to their roots in this way.

Snowmelt rates will vary with elevation and aspect in that they affect the air temperature and radiation inputs to the snowpack. The water equivalent of the snowpack can vary dramatically in space, due particularly to the effects of wind drifting during snow events and after the snowpack has formed, as affected by the topography and vegetation cover. Much deeper snow will often be found in the lee of ridges, a feature that has been well documented in the Reynolds Creek catchment in Idaho and elsewhere (see Bathurst and Cooley 1996; Section 5.3). There can also be feedback effects in that deeper snow cover might lead to more water being available to the vegetation, leading to greater growth and, in the case of trees, to greater trapping of snow drifting in the wind.

Once the rain or snowmelt water has reached the ground it will start to infiltrate the soil surface, except on impermeable areas of bare rock, on areas of completely frozen soil, or some artificial surfaces where surface runoff will start almost immediately. The rate and amount of infiltration will be limited by the local ground level rainfall, throughfall or steamflow intensity and the *infiltration capacity* of the soil. Where the input rate exceeds the infiltration capacity of the soil, infiltration excess *overland flow* will be generated. Soils tend to be locally heterogeneous in their characteristics, so that infiltration capacities and rates of overland flow generation might vary greatly from place to place (e.g. Loague and Kyriakidis 1997). In many places, particularly on vegetated surfaces, rainfall will only very rarely exceed the infiltration capacity of the soil unless the soil becomes completely saturated. Elsewhere, where infiltration capacities are exceeded, this will start in areas where soil permeabilities are lowest or initial water contents are highest and, since infiltration capacities tend to decrease with increased wetting, will gradually expand to areas with higher permeability. Bare soil areas will be particularly vulnerable to such *infiltration excess runoff* generation since the energy of the raindrops can rearrange the soil particles at the surface and form a surface crust, effectively sealing the larger pores (e.g. Römkens *et al.* 1990; Smith *et al.* 1999). A vegetation or litter layer, on the other hand, will protect the surface and also create root channels that may act as pathways for infiltrating water. Bare surfaces of dispersive soil materials are particularly prone to crusting and such crusts, once formed, will persist between storms unless broken up by vegetation growth, freeze–thaw action, soil faunal activity, cultivation or erosion. Studies of crusted soils have shown that in some cases, infiltration rates after ponding might increase over time more than would be expected as a result of the depth of ponding alone (e.g. Fox *et al.* 1998). This was thought to be due to the breakdown or erosion of the crust.

In cold environments, the vegetation may also be important in controlling the degree to which a soil gets frozen before and during the build-up of a snowpack by controlling both the local energy balance of the soil surface and the drifting of a snow cover. This may have important consequences for the generation of runoff during the snowmelt, even though this may, in some cases, take place months later (e.g. Stadler *et al.* 1996). It is worth noting that a frozen topsoil is not necessarily impermeable. There will usually be some reduction in potential infiltration rates due to freezing, but seasonal freeze–thaw processes can also lead to the break-up of surface crusts so as to increase infiltration capacities (e.g. Schumm 1956). The effects of freezing will depend on the moisture content of the soil and the length of the cold period. Even where widespread freezing takes place, infiltration capacities may be highly variable.

It has long been speculated that during widespread surface ponding, air entrapment and pressure build-up within the soil could have a significant effect on infiltration rates. This has been shown to be the case in the laboratory (e.g. Wang *et al.* 1998) and, in a smaller number of studies, in the field (e.g. Dixon and Linden 1972). It has also been suggested that air pressure effects can cause a response in local water tables (e.g. Linden and Dixon 1973) and that the lifting force due to the escape of air at the surface might initiate the motion of surface soil particles. The containment of air will be increased by the presence of a surface crust of fine material but significant air pressure effects would appear to require ponding over extensive areas of a relatively smooth surface. In the field, surface irregularities (such as vegetation mounds) and the presence of macropores might be expected to reduce the build-up of entrapped air by allowing local pathways for the escape of air to the surface.

In the absence of a surface crust, the underlying soil structure and particularly the macroporosity of the soil, will be an important control on infiltration rates. Since discharge of a laminar flow in a cylindrical channel varies with the fourth power of the radius, larger pores and cracks may be important in controlling infiltration rates (Beven and Germann 1981). However, soil cracks and some other *macropores*, such as earthworm channels and ant burrows, may only extend to limited depths so that their effect on infiltration may be limited by storage capacity and infiltration into the surrounding matrix as well as potential maximum flow rates. An outlier in the data on flow rates in worm holes of Ehlers (1975), for example, was caused by a worm still occupying its hole! Some root channels, earthworm and ant burrows can, of course, extend to depths of metres below the surface. The Jarrah trees of Western Australia are again a particularly remarkable example.

Overland flow may also occur as a *saturation excess* mechanism. Areas of saturated soil tend to occur first where the antecedent *soil moisture deficit* was smallest. This will be in valley bottom areas, particularly headwater hollows where there is convergence of flow and a gradual decline in slope towards the stream. Saturation may also occur on areas of thin soils where storage capacity is limited or in low permeability and low slope areas that will tend to stay wet during recession periods. The area of saturated soil will tend to expand with increased wetting during a storm, and reduce again after rainfall stops at a rate controlled by the supply of water from upslope. This is the *dynamic contributing area* concept. Any surface runoff on such a saturated area may not all be due to rainfall but may also be due to a return flow of subsurface water, and there is some evidence that, even under such conditions of a saturated surface, rainfall may still locally infiltrate into the soil (see the tracing experiments of Henderson *et al.* 1996). In this way, surface runoff may be maintained during the period after rainfall has stopped. When overland flow, by whatever mechanism, is generated, some surface depression storage may need to be satisfied before there is a consistent downslope flow. Even then, surface flow will tend to follow discrete pathways and rills rather than occurring as a sheet flow over the whole surface.

A similar concept may be invoked in areas where responses are controlled by subsurface flows. When saturation starts to build up at the base of the soil over a relatively impermeable bedrock, it will start to flow downslope. The connectivity of saturation in the subsurface will, however, initially be important. It may be necessary to satisfy some initial bedrock depression storage before there is a consistent flow downslope. The dominant flow pathways may be localized, at least initially, related to variations in the form of the bedrock surface (e.g. McDonnell *et al.* 1996). Some catchments, with high infiltration capacities and reasonably deep soils, may have responses dominated by *subsurface stormflow*. It is worth remembering that a 1m depth of soil, with an average porosity of 0.4 has a storage capacity of 400 mm of water. Thus, if the infiltration capacity of the soil is not exceeded, a large 100 mm rainstorm could, in principle, be totally absorbed by that 1 m soil layer (ignoring the effects of any downslope flows), even if the antecedent storage deficit is only a quarter of the porosity.

It is a common (and very convenient) assumption that the bedrock underlying small upland catchments is impermeable. This is not always the case, even in rocks that have little or no primary permeability in the bulk matrix. The presence of secondary permeability in the form of joints and fractures can provide important flow pathways and storage that may be effective in maintaining stream baseflows over longer periods of time. It is very difficult to learn much about the nature of such pathways; any characteristics are often inferred from the nature of the geochemistry of baseflows since the bedrock can provide a different geochemical environment, and long residence times can allow weathering reactions to provide higher concentrations of some chemical constituents (see for example the study of Neal *et al.* (1997) in the Plynlimon research catchments in Wales).

There is an interesting possibility that connected fracture systems that are full of water could act as pipe systems, transmitting the effects of recharge very rapidly. Remember that if water is added to one end of a pipe full of water, there will be an almost instantaneous displacement of water out of the other end, whatever the length of the pipe and even if the velocities of flow in the pipe are relatively slow. The reason is that the transmission of the pressure effect of adding the water is very much faster than the actual flow velocity of the water. Such displacement effects are an explanation of rapid subsurface responses to storm rainfalls (see next section).

The perceptual model briefly outlined above represents a wide spectrum of possible hydrological responses that may occur in different environments or even in different parts of the same catchment at different times. Traditionally, it has been usual to differentiate between different conceptualizations of catchment response based on the dominance of one set of processes over another, for example, the *Hortonian model* in which runoff is generated by an infiltration excess mechanism all over the hillslopes (Figure 1.4(a)). This model is named after Robert E. Horton (1875–1945), the famous American hydrologist (he may be the only modern hydrologist to have a waterfall named after him), who worked as both hydrological scientist and consultant. I am not sure that he would have totally approved of such widespread use of the infiltration excess concept. Although he frequently used the infiltration excess concept as a way of calculating the volume of runoff production from a particular rainfall (e.g. Horton 1933), he also had a hydrological laboratory in his wooded back garden in Voorheesville, New York State (Horton 1936) where he would surely not have observed infiltration excess overland flow very often. Horton was an excellent scientist who published papers on a wide variety of hydrological and meteorological phenomena. His perceptual model surely involved a much wider range of processes than the model that now bears his name (see, for example, the process descriptions in Horton 1942).

In the same period as Horton, however, Charles R. Hursh was working in the Coweeta watersheds in Georgia in the United States. These Southern Appalachian catchments are forested with soils that are deeply weathered and have generally high infiltration capacities. Surface runoff is restricted mainly to the channels, so here the storm runoff production must be controlled by subsurface responses (Figure 1.4(d)). Hursh published a number of articles dealing with subsurface responses to rainfall (e.g. Hursh and Brater 1941). A later director of the Coweeta laboratory, John Hewlett, was also influential in getting the importance of subsurface stormflow more widely recognized in the 1960s (Hewlett and Hibbert 1967; Hewlett 1974).

Independently in the 1960s, studies within the Tennessee Valley Authority (which at that time served as one of the major hydrological agencies in the eastern United States) were revealing that it was very difficult to predict runoff production in many catchments under the assumption that infiltration excess surface runoff was produced everywhere on the hillslopes. The information on infiltration capacities of the soils and rainfall rates could not support such a model. Betson (1964) suggested that it would be more usual that only part of a catchment would produce runoff in any particular storm and that since infiltration capacities tend to decrease with increasing soil moisture and the downslope flow of water on hillslopes tends to result in wetter soils at the base of hillslopes, then the area of surface runoff would tend to start close to the channel and expand upslope. This *partial area model* (Figure 1.4(b)) allowed for a generalization of the Horton conceptualization. It is now realized that the variation in overland flow velocities and the heterogeneities of soil characteristics and infiltration rates are important in controlling partial area responses. If runoff generated on one part of a slope flows onto an area of higher infiltration capacity further downslope it will infiltrate (the *run-on* process). If the high intensity rainfall producing the overland flow is of short duration, then it is also possible that the water will infiltrate before it reaches

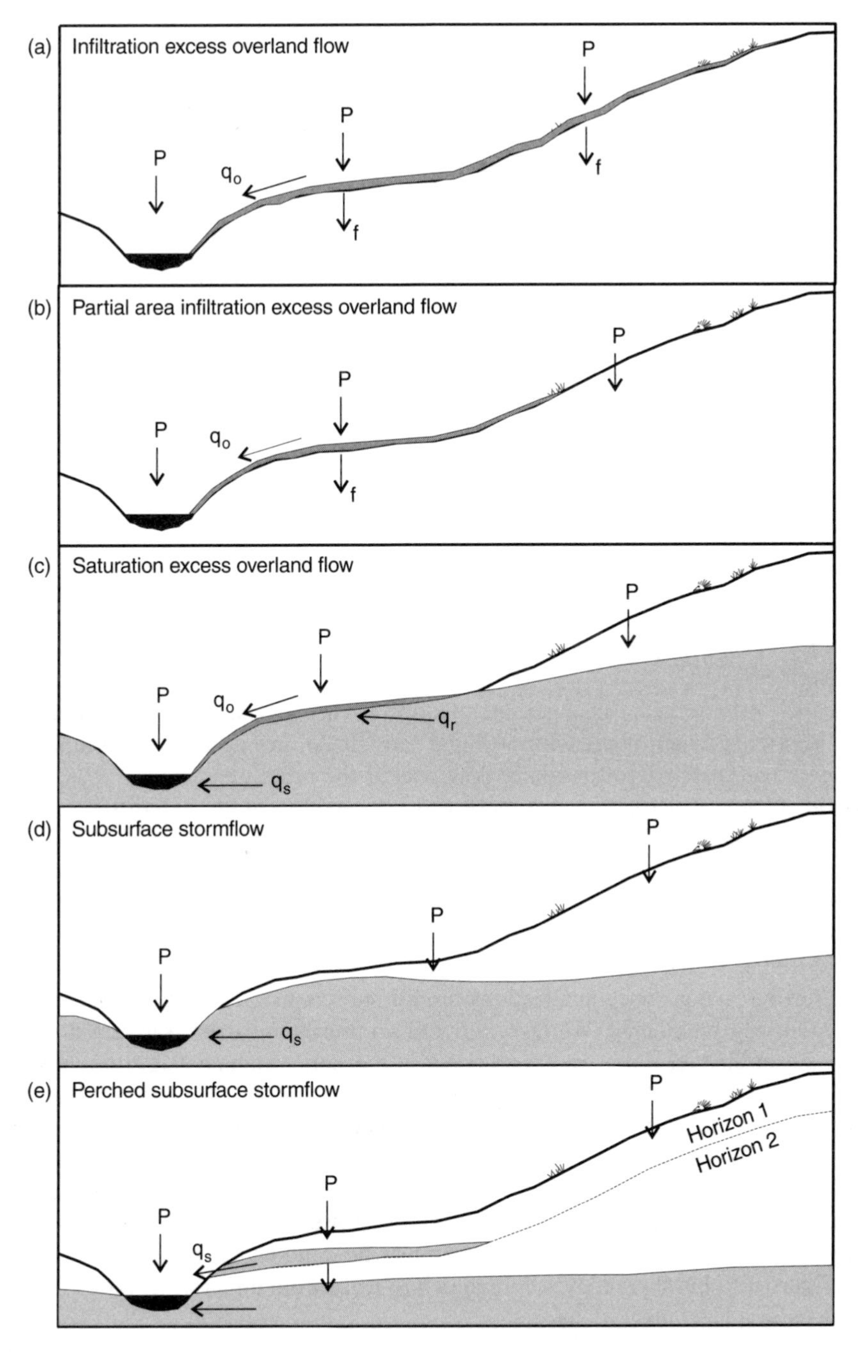

Figure 1.4 *A classification of process mechanisms in the response of hillslopes to rainfalls. (a) Infiltration excess overland flow (Horton 1933). (b) Partial area infiltration excess overland flow (Betson 1964). (c) Saturation excess overland flow (Cappus 1960; Dunne 1970). (d) Subsurface stormflow (Hursh 1936; Hewlett 1961). (e) Perched saturation and throughflow (Weyman 1970)*

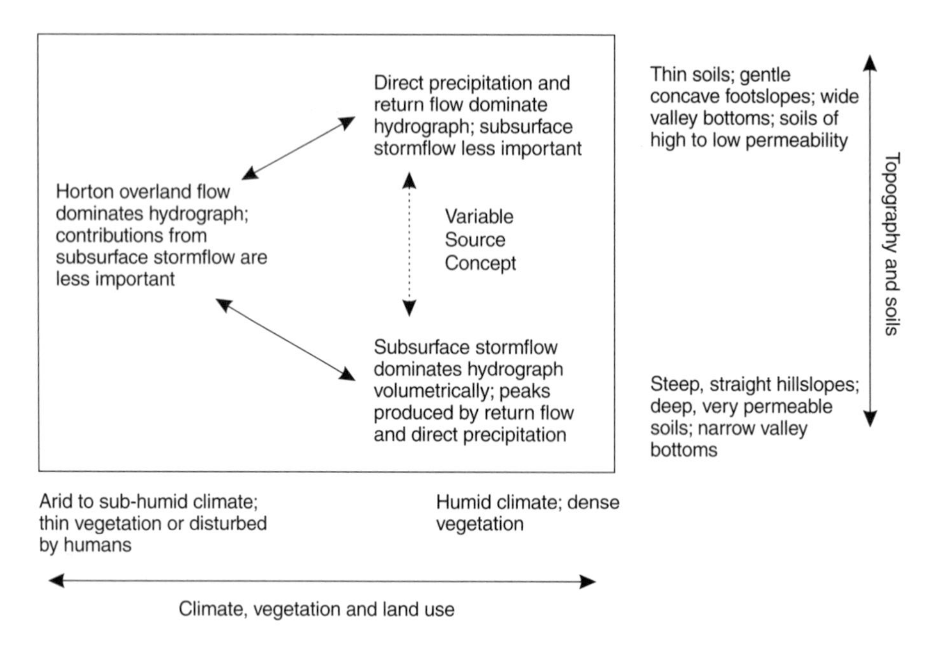

Figure 1.5 *Dominant processes of hillslope response to rainfall (after Dunne 1978)*

the nearest rill or stream channel. Bergkamp (1998), for example, estimated that for some plot-scale experiments with artificial rainfalls at an intensity of 70 $\mathrm{mm\,h^{-1}}$, the average travel distance for overland flow was of the order of 1 m!

Another form of partial area response was revealed by studies in a different environment by Dunne and Black (1970) working in Vermont. They observed surface runoff production, but on soils with high surface infiltration capacities. The surface runoff resulted from a saturation excess mechanism (Figure 1.4(c)), a type of response that had been previously suggested by Cappus (1960) (but published in French and only recently rediscovered).

These four major conceptualizations are all subsets of the more general perceptual model outlined previously. We now know that infiltration excess, saturation excess or purely subsurface responses might all occur in the same catchment at different times or different places due to different antecedent conditions or soil characteristics or rainfall intensities. In addition, an infiltration excess mechanism might take place within the soil at a permeability break, perhaps associated with a horizon boundary. This might lead to the generation of a perched water table and even to saturation at the surface of a soil that may be unsaturated at depth (e.g. Figure 1.4(e); Weyman 1970). Attempts have been made to suggest which mechanisms might be dominant in different environments (e.g. Figure 1.5) but there may still be much to learn from direct observations of runoff processes in a catchment of interest.

1.5 Flow Processes and Geochemical Characteristics

One of the mostly influential factors in revising hydrological thinking in the last 20 years has been the use of geochemical characteristics to provide additional information on flow processes. Some characteristics, in particular the use of artificial tracers, can provide direct information on flow velocities; others, such as the various environmental tracers, require a greater degree of inference. Even the results of artificial tracers may be difficult to interpret since most tracers are not ideal for following water movement over the wide range of time-scales involved and it is difficult to apply artificial tracers at the catchment scale. Thus, any experiment will tend to sample only some of the possible flow pathways.

The environmental isotopes of oxygen and hydrogen are often used in catchment-scale studies (see the review of Sklash 1990). They have the advantage that they are part of the water molecule and will therefore follow the flow pathways of water in the catchment directly. There remain some difficulties of interpretation of the results due to spatial and temporal variations in the concentrations of the isotopes in the rainfall inputs, the effects of vegetation on the input concentrations, and the spatial variability of concentrations of water stored in different soil horizons and parts of the catchment. However, at least in ideal conditions when there is a strong difference between the concentrations observed in rainfalls and the concentrations of water stored in the catchment before an event, the measured concentrations can be used in a simple two-component mixing model to differentiate between the contribution to the hydrograph for an event of the rainfall and the contribution of the water stored in the catchment prior to the event.

Some of the first hydrograph separations of this type were published by Sklash and Farvolden (1979) and revealed that the contribution of stored water (often called the 'pre-event' or 'old' water component) was surprisingly high (e.g. Figure 1.6). This result has been confirmed by many other studies for a wide range of different catchments, although the number of reports is dominated by results from humid temperate catchments and small to moderate rainfall events. The technique can be extended using other environmental tracers to three-component mixing to differentiate the rainfall contribution from 'soil water' and 'deep groundwater' components, where these components can be differentiated geochemically (e.g. Bazemore *et al.* 1994). Again, pre-event water is often found to be a major component, even in some cases for very rapidly responding processes such as pipe flows in wet soils (Sklash *et al.* 1996).

This pre-event water is displaced from storage by the effects of the incoming rainfall. This must therefore necessarily involve subsurface flow processes. The fact that the rising limb of the hydrograph is often dominated by the pre-event water component reveals that this displacement can take place rapidly, despite the fact that subsurface flow velocities are generally assumed to be much slower than surface flow velocities. This perception is, in fact, one of the reasons for the continuing use of the Hortonian conceptualization of runoff production, even now. If subsurface velocities are so slow, how can subsurface flow make a major contribution to the hydrograph?

The answer lies in the physics of the flow processes, and, in particular, in the saturated zone. It can be shown that there is a difference between the flow velocity of water and the velocity with which a disturbance to the saturated zone gets propagated as a pressure wave, which is called the *wave speed* or *celerity*. The type of disturbance of interest here is the addition of recharge due to rainfall during an event. The theory suggests that an infinitely small disturbance at the water table will be propagated infinitely quickly. Larger disturbances will have a much smaller wave velocity, the

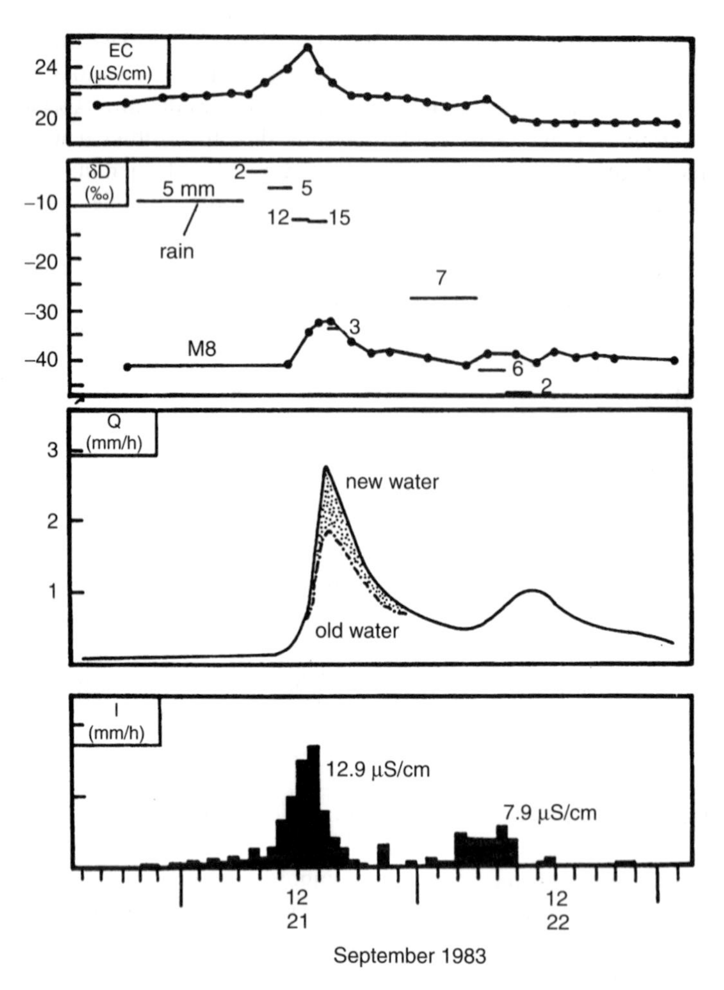

Figure 1.6 *Hydrograph separation based on the concentration of environmental isotopes (after Sklash 1990). Reprinted with permission of John Wiley & Sons Limited*

magnitude of which is a function of the inverse of the *effective storage capacity* in the soil (the difference between the current soil moisture in the soil immediately above the water table and saturation). In a wet soil or close to a water table, the effective storage capacity may be very small so that the wave velocity may be very much faster than the actual flow velocity of the water (see Section 5.4.3). The increase in discharge to the stream during an event will then depend more on the response of the hydraulic potentials in the system, which will be controlled by the local wave velocities, than the actual flow velocities of the water. Thus if discharge starts to increase before the recharging water has had time to flow towards the channel, it will be water stored in the profile close to the stream that flows into the channel first. This water will be predominantly pre-event water, displaced by the effects of the rainfall.

Similar effects may take place in unsaturated soil, but here the picture or perception is made more complicated by the relative mobility of water stored in different parts of the pore space and by the effects of preferential flows within the structural voids of the soil. The important message to take from this section is that in many catchments, particularly in humid environments, an important part of the hydrograph may be made up of 'old' water and may not be rainfall flowing directly to the stream. Certainly, it should not be assumed that fast runoff is always the result of overland flow or surface runoff on the hillslopes of a catchment.

1.6 Runoff Production and Runoff Routing

The evidence discussed in the previous two sections has been primarily concerned with the processes of runoff generation, both surface and subsurface. Runoff generation controls how much water gets into the stream and flows towards the catchment outlet within the time-frame of the storm and period immediately following. There is also, however, a further component to consider, which is the routing of the runoff from the source areas to the outlet. The boundary between runoff generation and runoff routing is not a very precise one. It would be more precise if it was possible to measure or predict the timing of inflows into the stream network itself accurately. Then the routing would only depend on the flow processes within the stream, which can be reasonably well predicted on the basis of hydraulic principles (although in arid areas it may also be necessary to take account of the infiltration of some of the water into the stream bed). Unfortunately, it is generally not possible to predict the volume and timing of the inflows precisely, so the routing problem becomes one of the velocities of surface and subsurface flows on the hillslope as well as in the stream channel. It may be very difficult then to separate out the effects of the different possible flow pathways that different waters take on the timing of the hydrograph at the stream outlet.

However, every hydrological model requires two essential components: one to determine how much of a rainfall becomes part of the storm hydrograph (the *runoff production* component); the other to take account of the distribution of that runoff in time, to form the shape of the storm hydrograph (the *runoff routing* component). These two components may appear in many different guises and different degrees of complexity in different models but they are always there in any rainfall–runoff model, together with the difficulty of clearly separating one component from the other.

In general it is accepted that the runoff production problem is the more difficult of the two. Practical experience suggests that the complexities and nonlinearities of modelling the flow generation processes are much greater than for the routing processes, and that relatively simple models for the routing may suffice (see discussion in Section 2.2).

1.7 The Problem of Choosing a Conceptual Model

The majority of hydrologists will be model users rather than model developers. Having said that, there has been no shortage of hydrologists, particularly those undertaking research for a doctorate, who have set themselves the task of developing a model. This is understandable; even now, the obvious approximation inherent in today's models suggests that it should be possible to do better! However, given the range of models consequently available in the literature or, increasingly, as software packages, the problem of model choice is not so different for the model user as for a researcher wanting to develop a new and improved version. The question is how to decide what is satisfactory and what are the limitations of the models available. We will take a preliminary look at this question in this section but will return to the problem in Chapter 10.

Let us first outline what the 'generic' choices are in terms of a basic classification of model types. There are many different ways of classifying hydrological models (see e.g. Clarke 1973; O'Connell 1991; Wheater *et al.* 1993; Singh 1995). We will concentrate only on a very basic classification here. The first choice is whether to use a *lumped* or *distributed* modelling approach. Lumped models treat the catchment as a single unit, with state variables that represent averages over the catchment area, such as average storage in the saturated zone. Distributed models make predictions that are distributed in space, with state variables that represent local averages of storage, flow depths or hydraulic potential, by discretizing the catchment into a large number of elements or grid squares and solving the equations for the state variables associated with every element grid square. Parameter values must also be specified for every element in a distributed model. There is a general correspondence between lumped models and the 'explicit soil moisture accounting' (ESMA) models of O'Connell (1991) (see Section 2.4), and between distributed models and 'physically based' or process-based models. Even this correspondence is not exact, however, since some distributed models use ESMA components to represent different subcatchments or parts of the landscape as *hydrological response units* (see Section 6.2), while even the most distributed models currently available must use average variables and parameters at grid or element scales greater than the scale of variation of the processes. They are consequently, in a sense, lumped conceptual models at the element scale (see Beven 1989). There is also a range of models that do not make calculations for every point in the catchment but for a distribution function of characteristics. TOPMODEL, discussed in Chapter 6, is a model of this type, but has the feature that the predictions can be mapped back into space for comparison with any observations of the hydrological response of the catchment. It could therefore be called, perhaps, a semi-distributed model.

A second consideration is whether to use a *deterministic* or *stochastic* model. Deterministic models permit only one outcome from a simulation with one set of inputs and parameter values. Stochastic models allow for some randomness or uncertainty in the possible outcomes due to uncertainty in input variables, boundary conditions or model parameters. The vast majority of models used in rainfall–runoff modelling are used in a deterministic way, although again the distinction is not clear-cut since there are examples of models which add a stochastic error model to the deterministic predictions of the hydrological model and there are models that use a probability distribution function of state variables but make predictions in a deterministic way. A working rule is that if the model output variables are associated with some variance or other measure of predictive dispersion the model can be considered stochastic; if the output values are single valued at any time step the model can be considered deterministic, regardless of the nature of the underlying calculations.

There is one other modelling strategy based on fuzzy methods that looks highly promising for the future. The number of fuzzy models is currently few (see Bárdossy *et al.* 1995) but the range of application would appear to be large. In particular, fuzzy models would appear to offer the potential for a more direct translation from the complexity of the perceptual model into a procedural model. Applications to date, however, have used an intermediate conceptual model to formulate the fuzzy rules and have defuzzified the results so as to run as essentially deterministic solutions.

So, these are the broad generic classes of rainfall–runoff model. lumped or distributed; deterministic or stochastic. Within each class there is a range of possible model structures. How then to go about choosing a particular model structure for a particular application? The following procedure is suggested based, in essence, on considerations of the *function* of possible modelling structures:

1. Prepare a list of the models under consideration. This list may have two parts: those models that are readily available, and those that might be considered for a project if the investment of time (and money!) appeared to be worthwhile.
2. Prepare a list of the variables predicted by each model and those required. Decide whether the model under consideration will produce the outputs needed to meet the aims of a particular project. If you are interested in the rise in the water table in valley bottoms due to deforestation, for example, a model predicting the lumped response of the catchment may not fulfil the needs of the project. If, however, you are only interested in predicting the discharge response of a catchment for real-time flood forecasting, then it may not be necessary to choose a distributed modelling strategy.
3. Prepare a list of the assumptions made by the model (see the guides in the chapters that follow). Are the assumptions likely to be limiting in terms of what you know about the response of the catchment you are interested in? Unfortunately the answer is likely to be yes for all models, so this assessment will generally be a relative one, or at best a screen to reject those models that are obviously based on incorrect representations of the catchment processes (e.g. any reasonable hydrologist should not try to use a model based on Hortonian overland flow to simulate the Coweeta catchments mentioned in Section 1.4).
4. Make a list of the inputs required by the model, for specification of the flow domain, for the specification of the boundary and initial conditions and for the specification of the parameter values. Decide whether all the information required can be provided within the time and cost constraints of a project.
5. Determine whether you have any models left on your list. If not, review the three previous steps, relaxing the criteria used. If predictions are really required for an application, one model at least will need to be retained at this stage!

1.8 Model Calibration and Validation Issues

Once one or more models have been chosen for consideration in a project, it is necessary to address the problem of parameter calibration. It is unfortunate that it is not, in general, possible to estimate the parameters of models by either measurement or prior estimation. Studies that have attempted to do so have generally found that, even using intensive series of measurements of parameter values, the results have not been entirely satisfactory (e.g. Beven *et al.* 1984; Refsgaard and Knudsen 1996; Loague and Kyriakidis 1997). Prior estimation of feasible ranges of parameters also often results in ranges of predictions that are wide and may still not encompass the measured responses all of the time (e.g. Parkin *et al.* 1996).

There are two major reasons for these difficulties in calibration. The first is that the scale of the measurement techniques available is generally much less than the scale at which parameter values are required. For example, there may be a hydraulic conductivity parameter in a particular model structure. Techniques for measuring hydraulic conductivities of the soil generally integrate over areas of less than 1 m^2. However, even the most finely distributed models require values that effectively represent the response of an element with an area of 100 m^2 or, in many cases, a much larger area. For saturated flow, there have been some theoretical developments that suggest how such effective values might change with scale, given some underlying knowledge of the fine-scale structure of the conductivity values. In general, however, carrying out the experimental measurements required to use such a theory at the hillslope or catchment scale would be very time-consuming and expensive, and would result in a large number of holes in the hillslopes! Thus it may be necessary to accept that the small-scale values that it is possible to measure and the effective values required at the model element scale are different quantities (a technical word is that they are *incommensurate*) – even though the hydrologist has traditionally given them the same name. The effective parameter values for a particular model structure will then still need to be calibrated in some way.

Most calibration studies in the past have involved some form of optimization of the parameter values by comparing the results of repeated simulations with whatever observations of the catchment response are available. The parameter values are adjusted between each run of the model, either manually by the modeller or by some computerized optimization algorithm until some 'best fit' parameter set has been found. There have been many studies of different optimization algorithms and measures of goodness of fit or *objective functions* in hydrological modelling (see Chapter 7). The essence of the problem is to find the highest peak in the *response surface* in the *parameter space* defined by one or more objective functions. An example of such a response surface is shown in Figure 1.7. The two basal axes are two different parameter values, varied between specified maximum and minimum values. The vertical axis is the value of an objective function, based on the sum of squared differences between observed and predicted discharges, that has the value 1 for a perfect fit. It is easy to see from this example why optimization algorithms are sometimes called 'hill climbing' algorithms, since the highest point on the surface will represent the optimum values of the two parameters. Such a response surface is easy to visualize in two-parameter space. It is much more difficult to visualize the response surface in an N-dimensional parameter hyperspace. Such surfaces can often be very complex and much of the research on optimization algorithms has been concerned with finding algorithms that are robust with respect to the complexity of the surface in an N-dimensional space and will find the *global optimum* set of parameter values.

However, for most hydrological modelling problems, the optimization problem is ill-posed in that if the optimization is based on the comparison of observed and simulated discharges alone, there may not be enough information in the data to support the robust optimization of the parameter values. Experience suggests that even a simple model with only four or five parameter values to be estimated may require at least 15 to

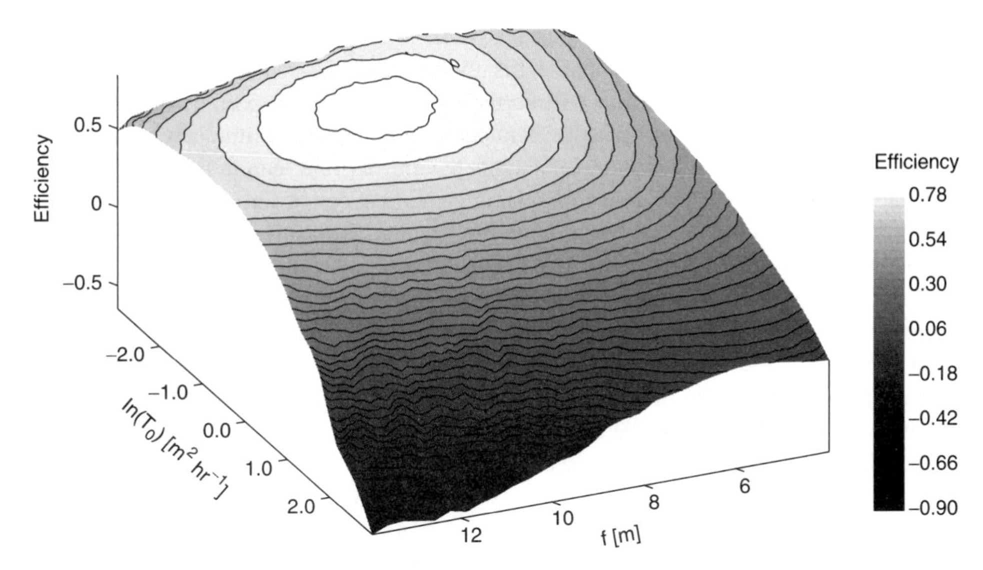

Figure 1.7 *Response surface for two TOPMODEL parameters (see Box 6.2) in an application to modelling the stream discharge of the small Slapton Wood catchment in Devon, UK. The objective function is the Nash–Sutcliffe efficiency which has a value of 1 for a perfect fit of the observed discharges*

20 hydrographs for a reasonably robust calibration, and if there is strong seasonal variability in the storm responses a longer period still (see, e.g. Kirkby 1975; Gupta and Sorooshian 1985; Hornberger *et al.* 1985; Yapo *et al.* 1996). For more complex parameter sets, many more data and different types of data may be required for a robust optimization unless it might be possible to fix many of the parameters beforehand by independent measurement. This has proven to be very difficult to achieve in practice.

These are not the only problems with finding an optimum parameter set. Optimization generally assumes that the observations with which the simulations are compared are error-free and that the model is a true representation of that data. We know, however, at least for hydrological models, that both the model structure and the observations are not error-free. Thus, the optimum parameter set found for a particular model structure may be sensitive both to small changes in the observations, or to the period of observations considered in the calibration, and possibly to changes in the model structure such as a change in the element discretization for a distributed model. (e.g. Refsgaard 1997; Saulnier *et al.* 1997(a)).

There are a number of important implications that follow from these considerations:

- The parameter values determined by calibration are effectively valid only inside the model structure used in the calibration. It may not be appropriate to use those values in different models (even though the parameter may have the same name) or in different catchments.
- The concept of an optimum parameter set may be ill-founded in hydrological modelling. While one optimum parameter set can often be found there will usually be many other parameter sets that are very nearly as good, perhaps from very

different parts of the parameter space. It is most unlikely that, given a number of parameter sets that give reasonable fits to the data, the ranking of those sets in terms of the objective function will be the same for different periods of calibration data. Thus to decide that one set of parameter values is the optimum is a somewhat arbitrary choice. Some examples of such behaviour will be seen in the *dotty plots* of Chapter 7 where the possibility of rejecting the concept of an optimum parameter set in favour of a methodology based on the *equifinality* of different model structures and parameter sets will be considered.

- If the concept of an optimum parameter set must be superseded by the idea that many possible parameter sets (and perhaps models) may provide acceptable simulations of the response of a particular catchment, then it follows that validation of those models may be equally difficult. In fact, rejection of some of the acceptable models given additional data may be a much more practical methodology than suggesting that models might be validated (see Section 7.8).

The idea of equifinality is an important one in what follows, particularly from Chapter 7 onwards. It suggests that, given the limitations of both our model structures and observed data, there may be many representations of a catchment that may be equally valid in terms of their ability to produce acceptable simulations of the available data. In essence then, different model structures and parameter sets used within a model structure are competing to be considered acceptable as simulators. Some may be rejected in the process of evaluation of different model structures suggested in Section 1.7, but even if only one model is retained then the evaluation of the performance of different parameter sets against the observed data will usually result in many parameter sets that produce acceptable simulations.

The results with different parameter sets will not, of course, be identical either in simulation or in the predictions required by the modelling project. An optimum parameter set will give only a single prediction. Multiple acceptable parameter sets will give a range of predictions. This may actually be an advantage since it allows the possibility of assessing the uncertainty in predictions, as conditioned on the calibration data, and then using that uncertainty as part of the decision-making process arising from a modelling project. A methodology for doing this is outlined in Chapter 7.

The rest of this book builds upon this general outline of the modelling process by considering specific examples of conceptual models and their application within the context of the type of evaluation procedures, for both model structures and parameter sets, outlined above. We take as our starting point that, a priori, all the available modelling strategies and all feasible parameter sets within those modelling strategies are potential models of a catchment for a particular project. The aims of that project, the budget available for the project, and the data available for calibrating the different models will all limit that potential range of simulators. The important point is that choices between models and between parameter sets must be made in a logical and scientifically defensible way. It is suggested, however, that at the end of this process, there will not be a single model of the catchment but a number of acceptable models (even if only different parameter sets within one chosen model structure) to provide predictions.

There are clearly implications for other studies that depend on models of rainfall–runoff processes. Predictions of catchment hydrogeochemistry, sediment production and transport, the dispersion of contaminants, hydroecology, and, in general, integrated catchment decision support systems depend crucially on good predictions of water flow processes. To keep my task manageable, I have not chosen to address the vast literature that deals with these additional topics, but each additional component that is added to a modelling system will add additional choices in terms of the conceptual representation of the processes and the values of the parameters required. In that all these components will depend on the prediction of water flows, they will be subject to the types of uncertainties in predictive capability that have been outlined in this chapter and will be discussed in more detail later. This is not only a research issue. In the UK, uncertainties in model predictions have already played a major role in decisions made at public inquiries into proposed developments. The aim of this volume is to help provide a proper basis for rainfall–runoff modelling across this range of predictive contexts.

1.9 Key Points from Chapter One

- There are important stages of approximation in the modelling process in moving from the *perceptual model* of the response of a particular catchment, to the choice of a *conceptual model* to represent that catchment, and the resulting *procedural model* that will run on a computer and provide quantitative predictions.
- One particular perceptual model has been outlined as a basis for comparing the descriptions of different models that are given in the following chapters.
- Studies of the geochemical characteristics of runoff, and particularly the use of environmental tracers, have resulted in an increased appreciation of the importance of subsurface storm runoff in many catchments.
- A basic classification of modelling strategies has been outlined, differentiating between *lumped* and *distributed* models and *deterministic* and *stochastic* models.
- Some preliminary guidelines for the choice of a conceptual model for a particular project have been outlined. This problem will be reconsidered in Chapter 10.
- The problem of the calibration of parameter values has been outlined. The idea of an optimum parameter set has been found to be generally ill-founded in hydrological modelling and can be rejected in favour of the concept of the *equifinality* of different models and parameter sets.
- It is expected that, at the end of the model evaluation process, there will not be a single model of the catchment but a number of acceptable models (even if only different parameter sets within one chosen model structure) to provide predictions.
- Prediction of other processes that are driven by water flows, such as hydrogeochemistry, erosion and sediment transport and ecology, will introduce additional choices about conceptual model structures and parameter values and will be subject to the uncertainty arising in the rainfall–runoff predictions.

2

Evolution of Rainfall–Runoff Models: Survival of the Fittest?

Everything of importance has been thought of before by someone who did not invent it.

Alfred North Whitehead, 1920

2.1 The Starting Point: The Rational Method

It is worth remembering that rainfall–runoff modelling has a long history and that the first hydrologists attempting to predict the flows that could be expected from a rainfall event were also thoughtful people who had insight into hydrological processes, even if their methods were limited by the data and computational techniques available to them. We can go back nearly 150 years to the first widely used rainfall–runoff model, that of the Irish engineer Thomas James Mulvaney (1822–1892) and published in 1851. The model was a single simple equation but, even so, manages to illustrate most of the problems that have made life difficult for hydrological modellers ever since. The equation was as follows:

$$Q_p = CA\overline{R} \tag{2.1}$$

The Mulvaney equation does not attempt to predict the whole hydrograph but only the hydrograph peak Q_p. This is often all an engineering hydrologist might need to design a bridge or culvert capable of carrying the estimated peak discharge. The input variables are the catchment area, A, a maximum catchment average rainfall intensity, $\overline{R}$, and an empirical coefficient or parameter, C. Thus, this model reflects the way in which discharges are expected to increase with area and rainfall intensity in a rational way. It has become known as the *rational method*. In fact, variations on equation (2.1) were published by a variety of authors based on different empirical data sets (for a summary, see Dooge 1957), and are still in use today (e.g. Hromadka and Whitley 1994).

The scaling parameter C will reflect the fact that not all the rainfall becomes discharge, but here the method is not quite so rational since it makes no attempt to separate the different effects of runoff production and runoff routing that will control the relationship between the volume of rainfall falling on the catchment in a storm, effectively $A\overline{R}$, and the discharge at the hydrograph peak. In addition, the coefficient C is required to take account of the nonlinear relationship between antecedent conditions and the profile of storm rainfall and the resulting runoff production. Thus C is not a constant parameter, but will vary from storm to storm on the same catchment, and from catchment to catchment for similar storms. The easiest way to get a value for C is to back-calculate it from observations of rainfall and peak discharge (the very simplest form of model calibration). Predicting the correct value for a different set of conditions, perhaps more extreme than those that have occurred before, or for a catchment that has no observations is a much more difficult task.

Similar difficulties persist to the present day, even in the most sophisticated computer models. It is still difficult to take proper account of the nonlinearities of the runoff production process, particularly in situations where data are very limited. It is still easiest to obtain effective parameter values by back-calculation or calibration where observations are available; it remains much more difficult to predict the effective values for a more extreme storm or ungauged catchment. There are still problems of separating out the effects of runoff production and routing in model parametrizations (and in fact this should be expected because of the real physical interactions in the catchment).

However, it is not impossible to make predictions, even with such simple models. Even in the pre-computer era, the rational method evolved into the graphical estimation technique (see Linsley *et al.* (1949) or Chow (1964) for full details). This was an attempt to summarize a wide range of analyses carried out for catchments in the United States into a set of graphs or *nomograms* that could be used to predict peak discharges under different rainfall and antecedent conditions (Figure 2.1). This approach has been used as a design tool for many years and has been put into mathematical form by, for example, Plate *et al.* (1988).

2.2 Practical Prediction: Runoff Coefficients and Time Transformations

In Chapter 1 and the previous section, the problem of separating the effects of runoff production and runoff routing was raised. This differentiation of two sets of processes was the essence of the first attempts to model hydrographs, starting back in the 1920s. It must be remembered that all the calculations at that time had to be done by hand, without even the benefit of electronic calculators. At this time, the word 'computer' meant a human being who did calculations. The calculating aids available were limited to log tables. Thus the calculations had to be simple.

In a paper published in 1921, Ross was perhaps the first to attempt to use a distributed hydrological model. His idea was to split the catchment up into zones on the basis of travel time to the catchment outlet. Zone 1 would be the area for which runoff could reach the outlet within one time step (e.g. one hour). Zone 2 would be the area with a travel time of two time steps, and so on (see Figure 2.2). Ross argued that if the production of runoff could be calculated for each area then it was a relatively simple matter to route that runoff to the catchment outlet to obtain a prediction of the hydrograph. Different antecedent conditions and different rainfall rates would result in

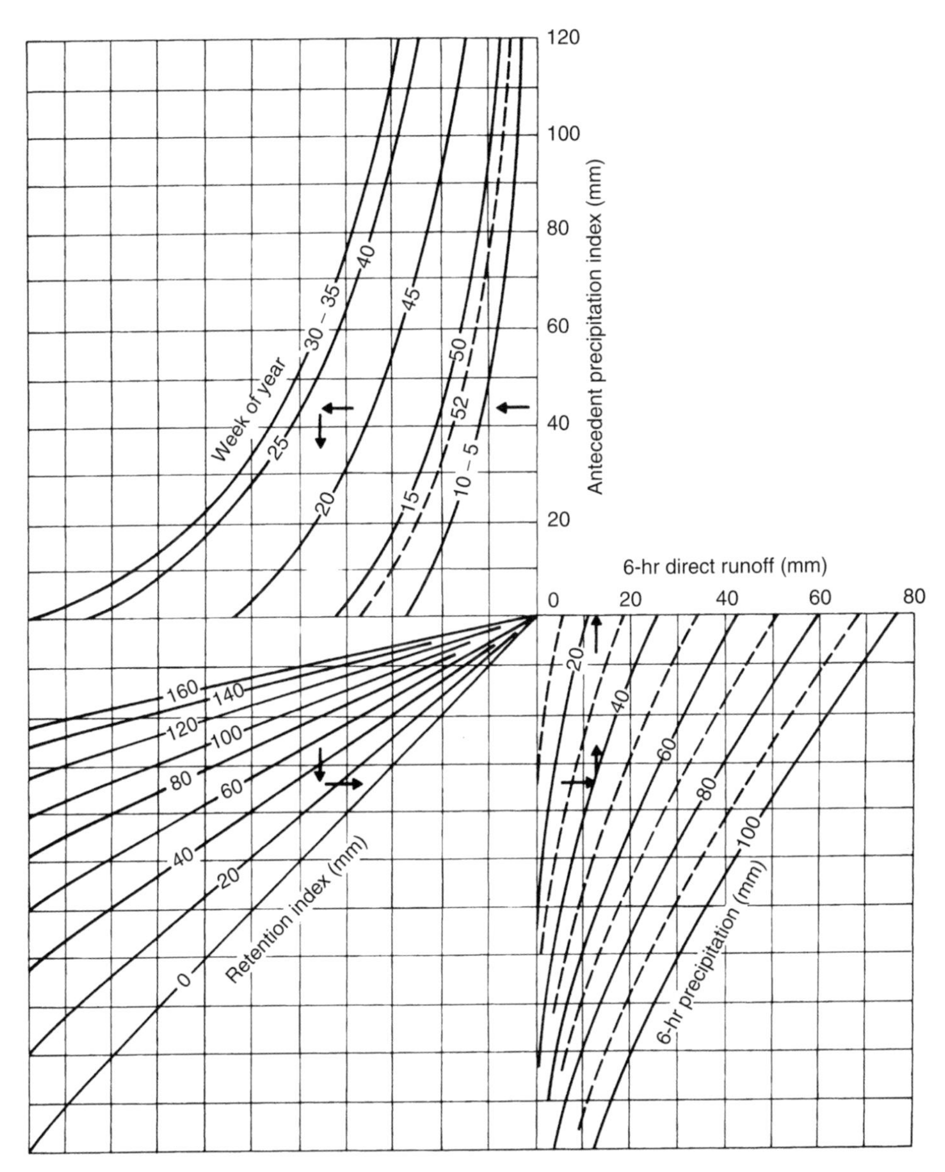

Figure 2.1 *Graphical technique for the estimation of incremental storm runoff given an index of antecedent precipitation, week of the year, a soil water retention index and precipitation in the previous 6 hours. Arrows represent the sequence of use of the graphs (after Linsley et al. 1949)*

different amounts of runoff production and, after the routing, different hydrographs. The resulting time–area diagram represents the delays for runoff from each portion of the catchment. A similar concept was used in the USA by Zoch (1934), Turner and Burdoin (1941) and Clark (1945), and in the UK by Richards (1944) in one of the first books on rainfall–runoff modelling and flood estimation to be published. These

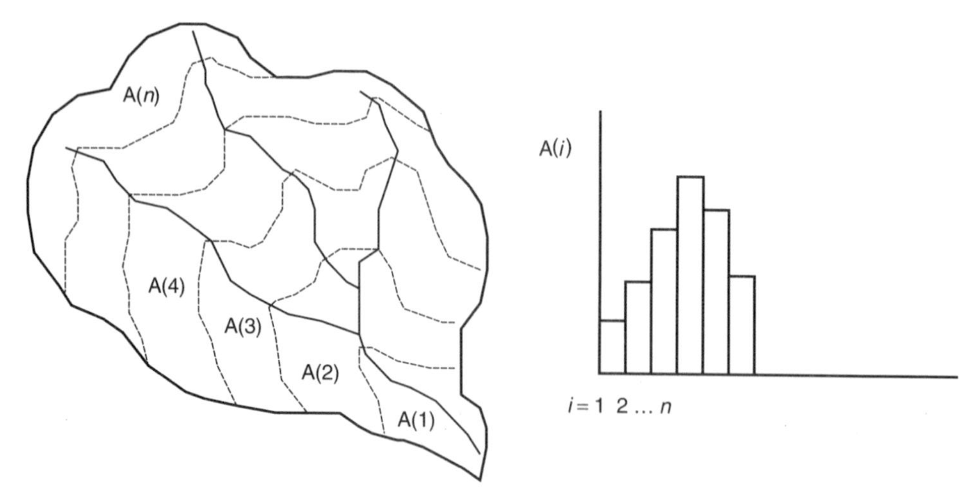

Figure 2.2 *Creating a time–area histogram by dividing a catchment into* n *areas at different travel times from the outlet,* i = *1, 2, . . .* n

ideas still underlie some of the distributed models being used today, and Kull and Feldman (1998), for example, have demonstrated how the Clark method can be used with distributed rainfall inputs derived from the NEXRAD radar system.

Note that these early studies made an assumption of *linearity* in routing the runoff. Linearity means that the routing times for the different zones are always the same, regardless of the amount of runoff being routed: the routing process is then mathematically a linear operation (see Box 2.1). This was an approximation. It has been known for some centuries that flow velocities change in a nonlinear way with flow rate or flow depth. The assumption of linearity, however, makes the computations very much easier.

It also works surprisingly well, as will be shown in Chapter 4. Any inaccuracies due to the linearity assumption for routing the runoff are generally less than the inaccuracies associated with deciding how much of the rainfall to route, i.e. the problem of estimating the *effective rainfalls* or *runoff coefficient* for an event. Effective rainfall is that part of a rainfall event that is equal in volume to the runoff generated by the event. The runoff coefficient is that proportion of the total rainfall in an event that becomes runoff. The way in which runoff production is predicted is generally nonlinear, with a runoff coefficient that depends on both antecedent conditions and rainfall.

The major problem with the Ross time–area concept was much more the difficulty of deciding which areas of the catchment would contribute to the different zones, since there was little information on velocities of flow for all the possible surface and subsurface flow pathways. This problem was avoided by Sherman (1932), who used the idea that the various time delays for runoff produced on the catchment to reach the outlet could be represented as a time distribution without any direct link to the areas involved. Because the routing procedure was linear, this distribution could be normalized to represent the response to a unit of runoff production, or effective rainfall, generated over the catchment in one time step. He called this function the unitgraph. We now know it as the *unit hydrograph*, and it has become one of the most commonly used hydrograph modelling techniques in hydrology, being simple to understand and easy to apply (especially with the benefit of modern computers). The unit hydrograph represents a discrete *transfer function* for effective rainfall to reach the basin outlet, lumped to the scale of the catchment.

The unit hydrograph remains a linear routing technique such that the *principle of superposition* can be applied. Thus, two units of effective rainfall in one time step will produce twice as much predicted runoff in the hydrograph at the catchment outlet as one unit, with the same time distribution (Box 2.1). The calculated outflows from effective rainfalls in successive time steps can be distributed in time by appropriately delayed unit hydrographs and added up to calculate the total hydrograph at the outlet. It is also generally assumed that the form of the unit hydrograph does not change over time.

There remains the more difficult problem of how to determine the amount of effective rainfall to route. This is definitely a nonlinear problem that involves a variety of hydrological processes and the heterogeneity of rainfall intensities, soil characteristics and antecedent conditions in the same way as the coefficient C of the rational formula of the last section. Thinking about the problem of estimating effective rainfalls was the start of thinking about modelling the rainfall–runoff process on the basis of an understanding of hydrological processes. It is not yet, however, a solved problem and there remain a number of competing models for estimating effective rainfalls based on different assumptions about the nature of the processes involved.

A major step in tackling this problem was made when just a year after Sherman introduced his unitgraph, Robert Horton published his paper on the generation of runoff when the infiltration capacity of the soil is exceeded (Horton 1933). Horton's work was based on experiment, and he used an empirical function to describe the decrease in infiltration capacity over time that he found in his experiments (e.g. A in Figure 2.3), even though simplified solutions of the Darcy flow equation for flow through soils had been available at least since the paper of Green and Ampt (1911). Since then, many other infiltration equations have been proposed, mostly based on various simplifications of the nonlinear Darcy flow problem (see, for example, the review of Parlange and Haverkamp (1989) and Box 5.2).

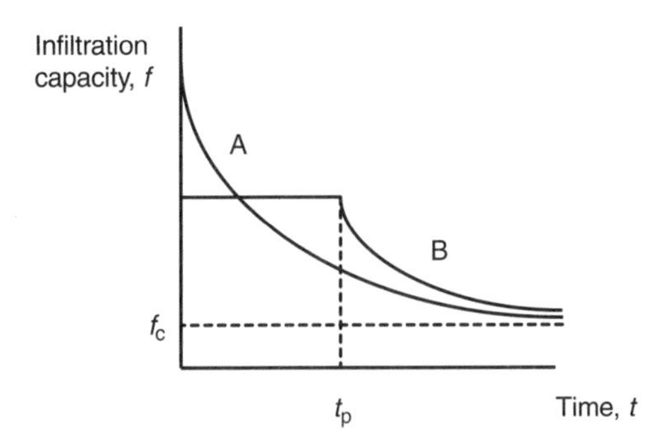

Figure 2.3 *Decline of infiltration capacity with time since the start of rainfall. A, Rainfall intensity higher than initial infiltration capacity of the soil. B, Rainfall intensity lower than initial infiltration capacity of the soil so that the infiltration rate is equal to the rainfall rate until time to ponding* t_p. f_c *is the final infiltration capacity of the soil*

All of these equations provide estimates of the local limiting infiltration capacity of the soil over time. When rainfall during a storm exceeds the infiltration capacity, then water will start to pond at the surface and, perhaps after the storage in local depressions has been filled, may start to run downslope as overland flow. Comparison of rainfall rates with infiltration capacities therefore provides a means of estimating the effective rainfall for a storm (e.g. B in Figure 2.3) – if runoff is actually being generated by an infiltration excess mechanism. However, as we saw in Chapter 1, this is not always the case, and even when surface runoff does occur, infiltration rates may show a high degree of heterogeneity in space. There is little doubt that this type of approach to estimate effective rainfall has often been misapplied, and probably continues to be misapplied (or at least misinterpreted) 60 years after the original formulation of the concepts.

The reasons for this are functional. The infiltration excess model of effective rainfall and the unit hydrograph provide together the necessary functional components of a hydrological model, i.e. a way of estimating how much of the rainfall becomes runoff and a means of distributing that effective rainfall through time to predict the shape of the hydrograph. It is not therefore necessary to apply this method under the assumption that it is actually surface runoff in excess of the infiltration capacity of the soil that is being routed by the unit hydrograph (as in Figure 2.4(a)). The simplest models of effective rainfall, assuming either that there is a constant 'loss' rate (the Φ-index method) (Figure 2.4(b)) or that a constant proportion of the rainfall is effective rainfall (Figure 2.4(c)), are also still widely used but are less obviously surface runoff models; rather, they are very simple ways of deriving an approximate runoff coefficient. This type of estimation of effective rainfall serves the functional requirement of having a loss function that is nonlinear with respect to total rainfall, regardless of whether the runoff generation process is actually due to an infiltration excess mechanism. Other ways of calculating effective rainfalls, with similar functionality, are also commonly used. Both methods involve only one parameter, but will lead to different patterns of effective rainfalls in time for the same event.

A further empirical method for estimating effective rainfalls is the USDA Soil Conservation Service (SCS) Curve Number approach (McCuen 1982). This is also often interpreted as an infiltration equation (e.g. Yu 1998; Mishra and Singh 1999), but in fact has its origins in the analysis of runoff volumes from small catchments by Mockus (1949) and which may not only have included infiltration excess overland flow as the runoff generating mechanism. The critical assumption of the SCS method is that the ratio of the actual runoff to the potential runoff (rainfall less some initial abstraction) is equal to the ratio of the actual retention to the potential retention. There is no physical justification for this assumption. Mockus himself suggested only that it produced rainfall–runoff curves of the type found on natural watersheds. It is therefore a purely empirical function for estimating a runoff coefficient and any process interpretation equating the retention to infiltration and the runoff to surface runoff has been made since the original work. This illustrates how deeply the Hortonian concepts of runoff generation have permeated the development of rainfall–runoff modelling in the past. The SCS method remains widely used within a number of current distributed models and will be covered in more detail in Chapter 6 including an alternative process intepretation (see Box 6.1).

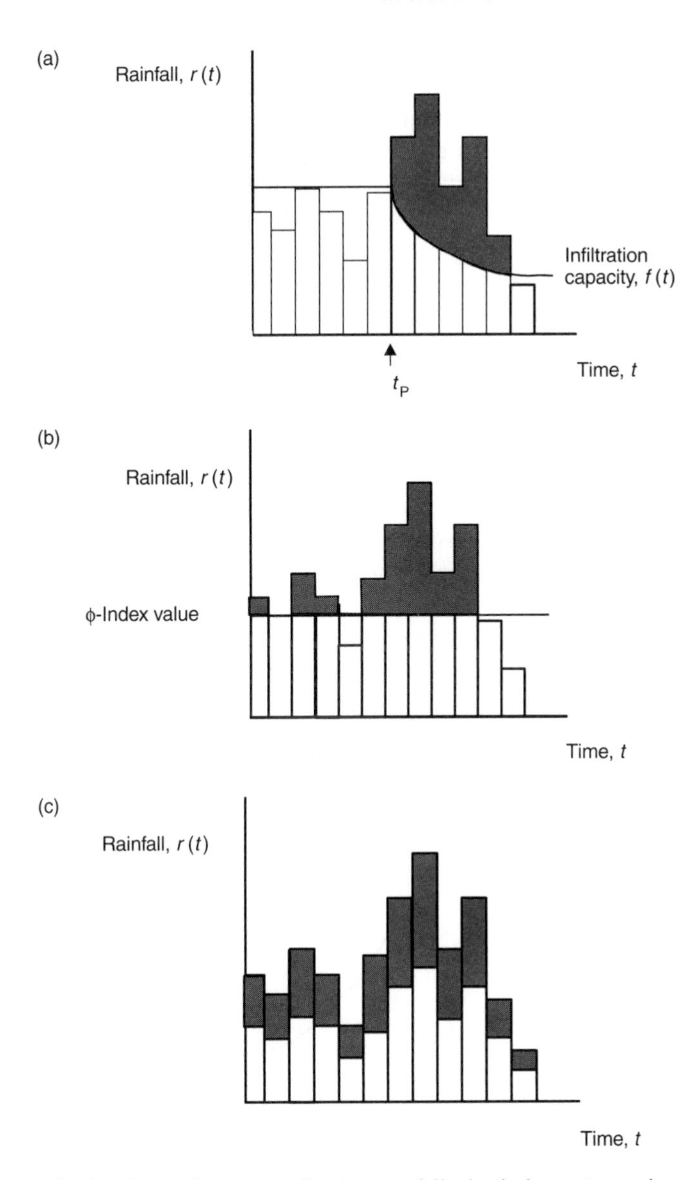

Figure 2.4 *Methods of calculating an effective rainfall (shaded area in each case): (a) When rainfall intensity is higher than the infiltration capacity of the soil, taking account of the time to ponding if necessary; (b) When rainfall intensity is higher than some constant 'loss rate' (the ϕ-index method); (c) When effective rainfall is a constant proportion of the rainfall intensity at each time step*

The calculation of effective rainfalls is a major problem in the use of the unit hydrograph technique, especially since it is inherently linked to decisions about hydrograph separation to determine the amount of storm runoff for an event (see below). However, since use of the unit hydrograph is a linear operation, given a sequence of rainfalls and a separated storm hydrograph, once a unit hydrograph is available for a catchment it can be used in an inverse way to estimate a pattern of effective rainfall. Indeed, by using an iterative process starting with some initial estimate of the form of the unit hydrograph, both effective rainfall sequences and the unit hydrograph can be calibrated without making any assumptions about the nature of the runoff generation processes (see Box 4.2). Unfortunately, this does not appear to make it any easier to interpret the effective rainfalls derived in this way so that it is easier to predict the effective rainfalls for other storms.

There is one further problem in applying the unit hydrograph technique. In any storm hydrograph, some of the stream discharge would have occurred even without the rainfall. This is usually called the *baseflow* component of the stream, and if there has been a dry period since the previous storm event, baseflow is usually assumed to be derived from subsurface flow. Early on it was found that the effective rainfall is more linearly related to stream discharge if the total hydrograph is separated into a baseflow component and a *storm runoff* component (e.g. Figure 2.5(a)). Hydrograph separation then became an important component of the application of the unit hydrograph model: the problem is that there are no satisfactory techniques for doing hydrograph separation. There are, indeed, some very strange methods of hydrograph separation reported in the literature (see review in Beven 1991b). About the only physically justifiable technique for hydrograph separation is to try to estimate the flow that might have occurred if the storm had not happened. However, such a procedure tends to lead to storm runoff hydrographs with very long tails and can get quite complicated in the case of several storms in quick succession (Figure 2.5(b)) so is not commonly used (but see Reed *et al.* 1975). In fact, the best method of dealing with hydrograph separation is to avoid it all together, as will be discussed in the next section.

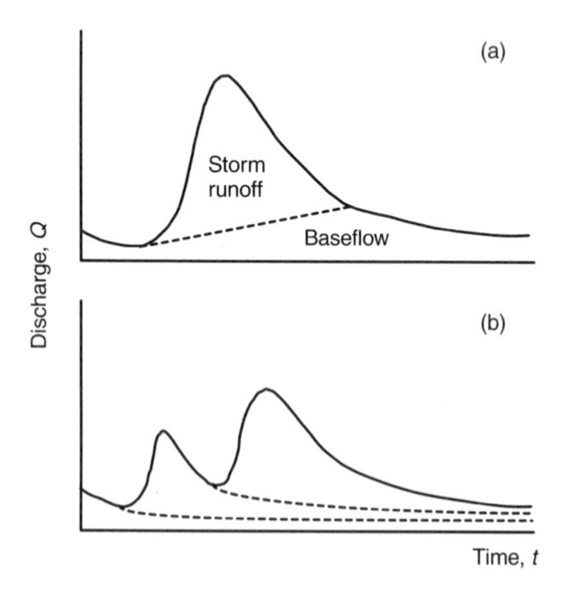

Figure 2.5 *Hydrograph separation into 'storm runoff' and 'baseflow' components. (a) Straight line separation (after Hewlett 1974). (b) Separation by recession curve extension (after Reed et al. 1975)*

In many texts, the storm runoff component is called the 'surface runoff' component. This is appropriate in that the total stream hydrograph is measured as surface runoff in the channel but the name also subtly perpetuates the generally incorrect conceptual link with the idea that runoff is generated by an infiltration excess mechanism. This nomenclature should be discouraged, as should the idea that the effective rainfall is the same water that forms the discharge hydrograph. Tracer information generally suggests that it is not (as discussed in Section 1.5).

Despite all these limitations, the unit hydrograph model works for predicting discharges. As noted above, it has the basic functional components needed. There are modern variants of the technique that work extremely well both in short-term flood forecasting and in prediction over longer periods (see Chapter 4). There are also now variants linked into geographical information systems (GIS) that have returned to concepts similar to the original Ross time–area diagram formulation. The approach could be considered as the 'model to beat' for discharge prediction and we will return to it a number of times in what follows.

2.3 Variations on the Unit Hydrograph

As experience in the application of the unit hydrograph approach was gained, a number of difficulties with the approach came to be appreciated, both in calibration to a particular catchment and in prediction. Two major problems arise in calibration. One has already been mentioned, that of hydrograph separation. However, once a technique has been chosen, the volume of storm runoff arising from the calibration storms can be calculated. Because of the linearity assumption, comparison of this runoff volume with the storm rainfall volume means that the runoff coefficient for each calibration storm can then be calculated exactly. Thus, in calibration at least, the choice of a method for separating the effective rainfall from the total rainfall is not quite so critical, provided that the volumes match. This is one reason why the very simple Φ-index method continues to be used today.

Once effective rainfall and storm runoff time series are available, a second problem in calibration arises from numerical difficulties in calculating the unit hydrograph. If the unit hydrograph is treated as a histogram (Figure 2.6(a)), each ordinate of the histogram is an unknown to be determined, effectively a parameter of the unit hydrograph. However, the histogram ordinates are strongly correlated, especially on the recession limb, and together with the errors inherent in the time series of effective rainfalls and storm runoff, this makes for a mathematically ill-posed problem. Attempts at a direct solution tend to lead to oscillations in the unit hydrograph ordinates, sometimes extreme oscillations, that are physically unacceptable as a representation of the catchment routing.

A number of ways of avoiding such oscillations have been tried, including imposing various constraints on the shape of the hydrograph (e.g. Natale and Todini 1977), or superimposing data from many storms and determining an average unit hydrograph by a least-squares procedure (e.g. O'Donnell 1966). This latter approach is still used in the DPFT-ERUHDIT model of Duband *et al.* (1993).

Another approach is to reduce the number of parameters to be determined. This can be achieved by specifying a particular mathematical form for the unit hydrograph.

The simplest possible shape, with only two parameters, is the triangle (if the base time and time to peak are specified, then the mass balance constraint of having a hydrograph equal to the unit volume means that the peak height of the triangle can also be calculated). The triangle was chosen as a simple model in procedures for predicting the response of ungauged catchments in the UK *Flood Studies Report* (NERC 1975; see also Shaw 1994). It has been retained in the revised procedures in the new UK *Flood Estimation Handbook* (IH, 1999).

This is not the only two-parameter model that can be used, however. One of the most well-known and widely used models is the so-called Nash cascade, which can be visualized as a sequence of N linear stores in series, each with a mean residence time of K time units (Nash 1959). The resulting mathematical form for the unit hydrograph $h(t)$ is equivalent to the gamma distribution:

$$h(t) = \left(\frac{t}{K}\right)^{N-1} \frac{\exp(-t/K)}{K\Gamma(N)} \tag{2.2}$$

where $\Gamma(N)$ is the gamma function ($\Gamma(N) = (N-1)!$ for integer values of N). For different values of N and K, the gamma distribution has quite a flexible range of forms (Figure 2.6(b)). Mathematically N does not have to be an integer number of stores but can also take on fractional values to give a wider range of shapes in fitting the observed data. Dooge (1959) provided a summary of a number of other simple linear models that could be used, including those with time delays.

The advantage of these functions is that, in general, much more stable estimates of the parameters are obtained, while retaining a flexibility in shape for representing a variety of different catchment hydrographs. Attempts have been made to relate the resulting parameters to different variables representing catchment characteristics, but it must be remembered that the parameter values will be dependent on the procedures used in deriving the parameters, and particularly on the hydrograph and rainfall separation techniques used.

In the last few years, however, there have been some successful attempts to avoid the problems inherent in these separation techniques and derive a model that relates the total rainfall to the total discharge, not only for single storms but also for continuous simulation. These models stem from developments in general linear systems analysis pioneered by Box and Jenkins (1970). The general linear model that allows explicit hydrograph separation to be avoided is discussed in detail in Chapter 4. A physical interpretation of the range of models is as one or more linear storage elements arranged in series or parallel (as, for example, the series of equal stores in the Nash cascade above). Given time series of inputs and outputs that are related in a reasonably linear way there are now robust algorithms available for the estimation of the parameters.

The critical phrase here is 'related in a reasonably linear way', since, as we have already noted, total rainfall is not related linearly to total discharge. In the past, there have been a number of attempts to use nonlinear transfer functions based on Volterra series (e.g. Amorocho and Brandstetter 1971; Diskin and Boneh 1973) but still requiring a prior estimation of the effective rainfalls. More recent approaches have attempted to relate total rainfalls directly to total discharge. Such models clearly need to have some form of nonlinear transformation of the rainfall inputs but it has proven possible to retain the linearity assumption in the routing component while still maintaining an

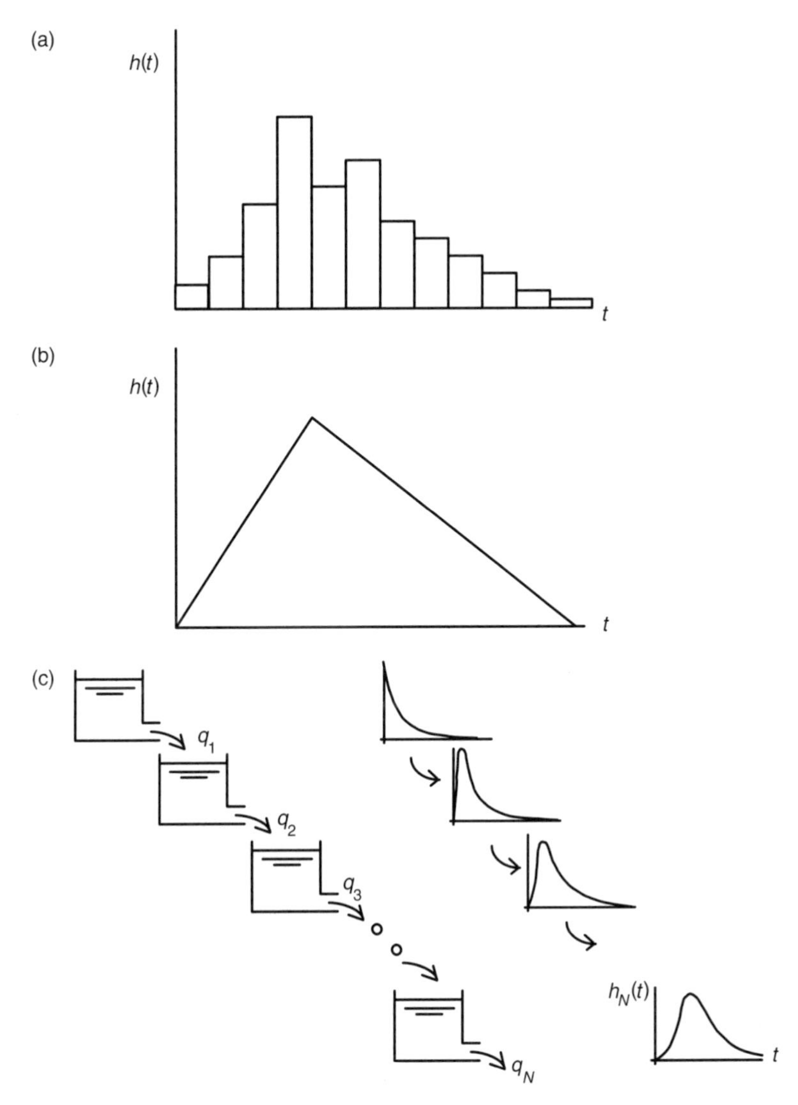

Figure 2.6 *The unit hydrograph as (a) a histogram, (b) a triangle, and (c) a Nash cascade of* N *linear stores in series*

adequate representation of the long time delays associated with the baseflow component. The result is often a parallel model structure, with part of the rainfall being routed through a store with a short mean residence time to model the storm response, and another part through a store with a long mean residence time to model the baseflow. Examples are the IHACRES model of Jakeman *et al.* (1990) and the bilinear power model of Young and Beven (1994), which are very similar in their modelling of the routing component but differ in their approach to modelling the catchment nonlinearity. More detailed examination of this type of general *transfer function* model will be given in Section 4.3.

These approaches are based on letting an analysis of the data suggest what the form of the model should be. Another line of recent development of unit hydrograph theory has been the attempt to relate the unit hydrograph more directly to the physical structure of the catchment and in particular of the channel network of the catchment with the aim of developing models that will provide accurate simulations of ungauged catchments. Two lines of approach may be distinguished (see Section 4.7): one that is based on analysis of the actual structure of the network (using the *network width function*) and one that uses more generalized geomorphological parameters to represent the network (the *geomorphological unit hydrograph* (GUH) approach). Both are concerned primarily with the routing problem, and not with the estimation of the effective rainfall.

The availability of modern GIS databases has also allowed a return to the original Ross concept of the time–area diagram representation of the unit hydrograph. Overlays of different spatial databases of soil, vegetation and topography data within a GIS results in a classification of parcels of the landscape with different functional responses. Amerman (1965) called these parcels *unit source areas* but they are now more commonly known as *hydrological response units* (HRUs, e.g. Figure 2.7) or *hydrotopes*. The topography of the catchment can also be used to define flow directions and distances to the outlet for each hydrological response unit, which can provide the basis for a routing algorithm (either linear or nonlinear). A representation of the response for each HRU will allow a calculation of the effective rainfall to be routed to the outlet to form the predicted hydrograph.

This type of distributed model is not often presented within the context of unit hydrograph models but the similarities with the Ross time–area diagram concept are clear,

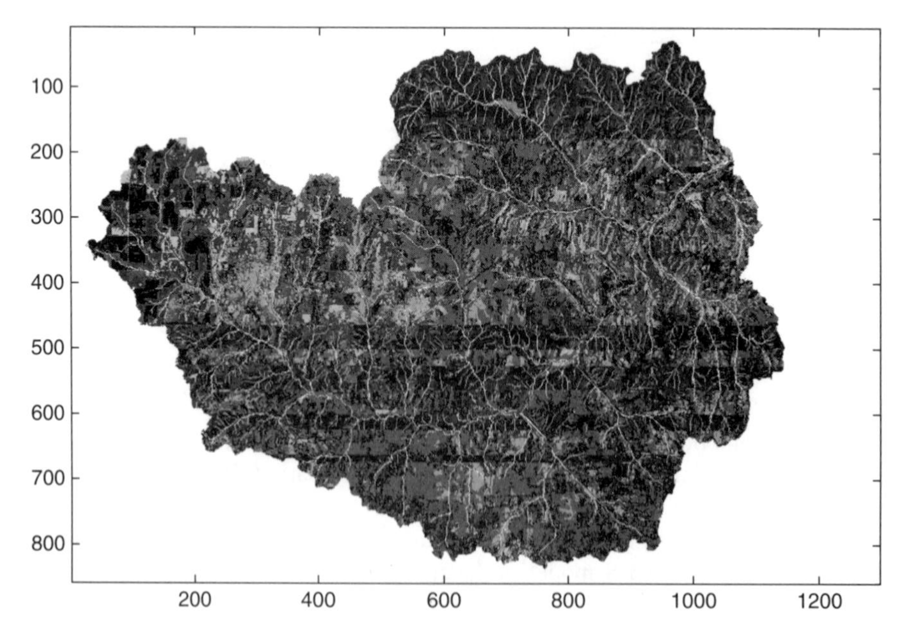

Figure 2.7 *A map of hydrological response units in the Little Washita catchment, Oklahoma, USA, formed by overlaying maps of soils and vegetation classifications within a raster geographical information system with pixels of 30 m*

particularly if a linear algorithm is used to route runoff generated on each HRU to the outlet. The technology used has changed dramatically, of course, with the availability of GIS databases and modern computer graphics for pre- and post-simulation data processing. Figure 2.7 is clearly much more impressive than Figure 2.2 but, underneath, the approaches are conceptually very similar. It also has to be remembered that the definition of 'response units' by GIS overlays does not solve the problem of determining how much of the rainfall becomes runoff, since the classification of the landscape by its soil and vegetation characteristics does not give the hydrological parameters needed to describe the processes operating at the response unit scale directly. Many such models use simple conceptual components to describe each HRU, similar to the type of explicit soil moisture accounting models that have been used very widely at the catchment scale since the very earliest days of rainfall–runoff modelling on digital computers. Such an approach can also be used to predict spatial variations in both evapotranspiration and snowmelt (e.g. Gurtz *et al.* 1999).

2.4 Early Digital Computer Models: The Stanford Watershed Model and its Descendants

The computational constraints on rainfall–runoff modelling persisted until the 1960s when digital computers first started to become more widely available. Even so, those computers that were available were expensive, very slow by today's standards, and had very limited memory available. Even the biggest and most expensive were much less powerful than a simple portable PC of today. The types of program that could be run were still limited in size and complexity. During this period there was, however, a very rapid expansion in the number of hydrological models available. For the most part they were of similar form: a collection of storage elements representing the different processes thought to be important in controlling the catchment response, with mathematical functions to describe the fluxes between the storages. One of the first and most successful of these models was the Stanford Watershed model developed by Norman Crawford and Ray Linsley at Stanford University, which later evolved into the Hydrocomp Simulation Program (HSP) which was widely used in hydrological consulting. The model survives, with the addition of water quality components, in the form of the US EPA's Hydrological Simulation Program – Fortran (HSPF; Donigian *et al.* 1995). Models of this type, called *explicit soil moisture accounting* (ESMA) models by O'Connell (1991), varied in the number of storage elements used, the functions controlling the exchanges, and consequently in the number and type of parameters required. The Stanford Watershed model had up to 35 parameters, although it was suggested that many of these could be fixed on the basis of the physical characteristics of the catchment and only a much smaller number needed to be calibrated.

In the early years of digital computing there was a tendency for every hydrologist with access to a computer (there were no personal computers then) to build his own variant of this type of model. It was not, after all, a difficult programming exercise. It was one of the ways that I learned how to program computers as an undergraduate by writing a model to try to *hindcast* runoff generation on Exmoor during the Lynmouth flood event. This model, in 1971, was written in the Algol programming language, stored on punched cards, and run on the Bristol University Elliot 503 computer with

16 kbytes of memory, with all output on a paper lineprinter. This is an indication of how rapidly the resources available to the modeller have changed in the last three decades.

Most of these models had a sufficient number of parameters and flexibility to be able to produce a reasonable fit to the rainfall–runoff data after some calibration. Indeed it was all too easy to add more and more components (and more associated parameters) for different processes. The potential for a confusing plethora of models was soon recognized, and Dawdy and O'Donnell (1965) tried to define a relatively simple 'generic' model structure, with just a few parameters (Figure 2.8). This did not, however, stop a continued expansion in the number of models of this type published in the hydrological literature (for a review, see Fleming 1975). A number of examples

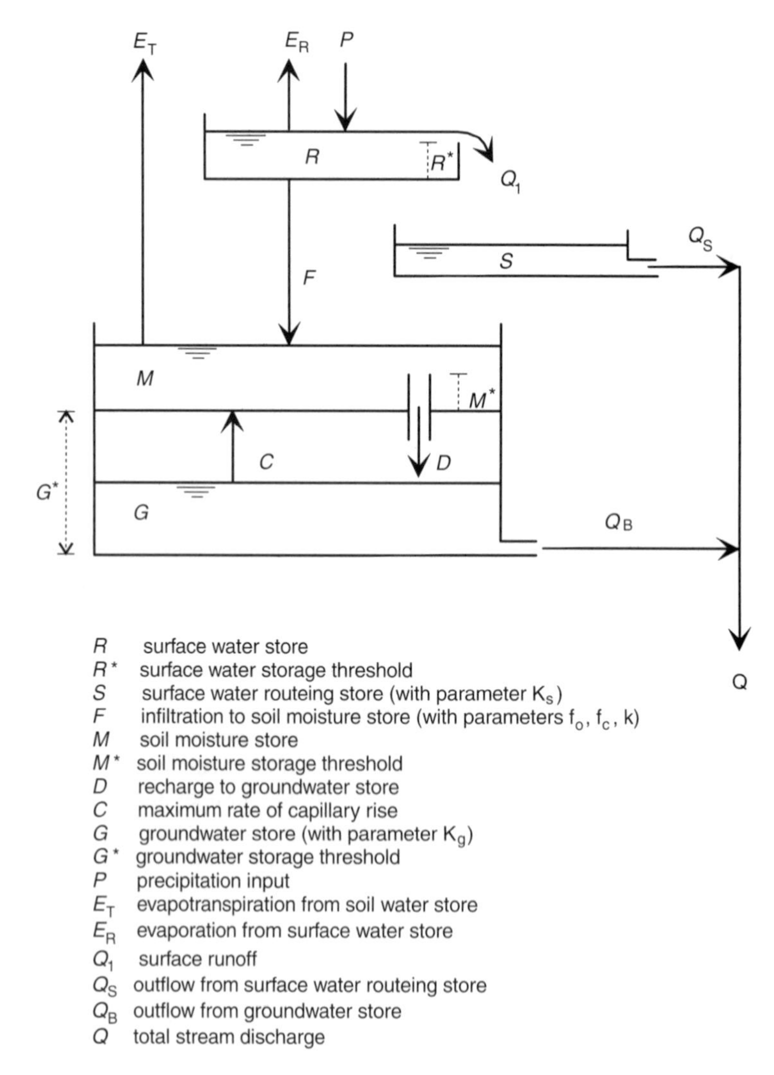

Figure 2.8 *Schematic diagram of the Dawdy and O'Donnell (1965) conceptual or explicit soil moisture accounting (ESMA) rainfall–runoff model*

that are still in current use may be found in the chapters of Singh (1995), including the HSPF, SSARR and Sacramento models from the USA, the HBV model from Sweden, the Tank model from Japan, the UBC model from Canada, and the RORB model from Australia. A comparison of different models reveals the subjectivity involved in defining a particular model structure, albeit that there is often similarity in some components. An example in current use, variously known as the Xinanjiang model or Arno model or variable infiltration capacity (VIC) model, is described in Box 2.2. This model is of interest in that, although it can be classed as an ESMA-type model, the surface runoff generation component can also be interpreted in terms of a distribution function of catchment characteristics (see Section 2.6). It has also been implemented as a macroscale hydrological model or *land surface parametrization* in some global climate models (see Chapter 9).

Functionally, there are also similarities between the modern IHACRES package, mentioned in the last section, and ESMA-type models, since in both cases the runoff generation and runoff routing components are based on storage elements. The difference lies in the modern approach of trying to find the simplest model structure supported by the data (see discussion in Jakeman and Hornberger 1993) and in not necessarily fixing the model structure beforehand. Instead, an analysis of the data should be allowed to suggest what the appropriate structure should be, as in the data-based mechanistic approach of Young and Beven (1994) (see Chapter 4).

Providing some data are available to calibrate parameter values, the results from even simple ESMA models can be quite acceptable, both in modelling discharges (Figure 2.9) and in soil moisture deficit modelling (Figure 2.10). The performance demonstrated in Figure 2.10 is particularly impressive if it is remembered that 1976 was one of the driest summers on record in the UK. A number of comparisons of ESMA models have been published, though with the limitations of both model structures and input and output data it has not generally been possible to conclude that one model consistently performs better than another after the model parameters have been calibrated (see for example the studies of Franchini and Pacciani 1991; Chiew *et al.* 1993; and Editjatno *et al.* 1999).

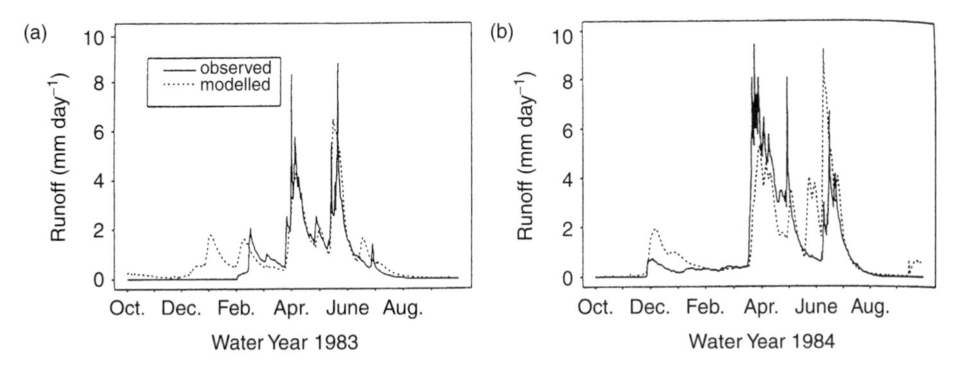

Figure 2.9 *Observed and predicted discharges for the Kings Creek, Kansas (11.7 km²) using the VIC-2L model of Box 2.2. Note the difficulty of simulating the wetting-up period after the dry summer (after Liang et al. 1996). Reprinted from Global Planetary Charge 13: 195–206, copyright (1996), with permission from Elsevier Science*

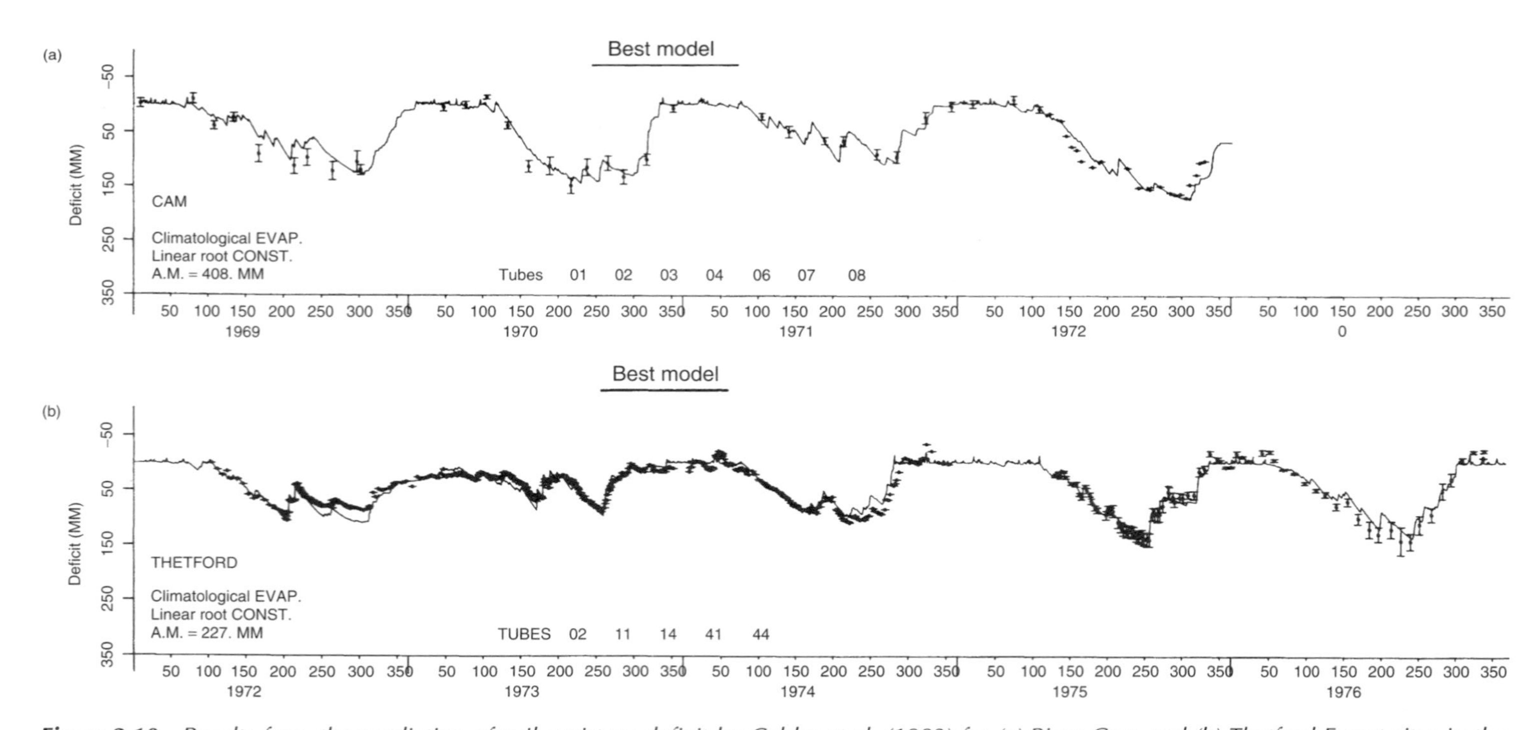

__Figure 2.10__ Results from the prediction of soil moisture deficit by Calder et al. (1983) for (a) River Cam and (b) Thetford Forest sites in the UK. Observed soil moisture deficits are obtained by integrating over profiles of soil moisture measured by neutron probe. Input potential evapotranspiration was a simple daily climatological mean time series. Reprinted from Journal of Hydrology 60, 329–355, copyright (1993), with permission from Elsevier Science

ESMA models are also being used for the prediction of the impacts of climate change in different countries (e.g. Bultot *et al.* 1988, 1992; Arnell 1996). In this context, their use is more problematic because the accuracy of the predictions will rely heavily on the availability of data for calibration. Thus, if the parameter values of a model are successfully calibrated under current conditions, and then the model is used with a different range of climatic inputs, representing perhaps one possible scenario for conditions later in this century, there is no guarantee that the current accuracy will be maintained, especially if the changed conditions are more extreme. No data are available, however, for calibration under the changed conditions. Thus, the impact predictions should be expected to be more uncertain than the current day simulations and any impact predictions should be associated with an estimation of uncertainty. This is not yet commonly done (although see Section 9.4).

ESMA-type models are also still used for representing the GIS-derived hydrological response units. Models of this type include the SLURP model of Kite and Kouwen (1992); the USGS Precipitation–Runoff Modelling System (PMRS) of Leavesley and Stannard (1995); and the USDA-ARS SWAT model of Arnold *et al.* (1998). We run into a few model classification problems here: such models aim to represent hydrological processes in a distributed manner, but using functional components in the style of ESMA models at the scale of the GIS-derived HRU, rather than attempting full process descriptions. It is a modelling technology that has been driven by the availability of GIS and remote sensing data rather than any real advances in the understanding of how to represent hydrological processes. They can perhaps be considered most appropriately with the class of distribution function models of Section 2.6, and we will return to them briefly in Chapter 6.

2.5 Distributed Process Description Based Models

It will be realized from the discussion in the previous section that there is considerable subjectivity associated with the definition of an ESMA-type model, even if the hydrologist is attempting to reflect her or his perceptual model of how catchments work as far as possible; hence the wide range of ESMA models available. Another early (in the digital computer era) response to this subjectivity was the attempt to produce models based directly on equations describing all the surface and subsurface flow processes in the catchment. A blueprint for such a model was described in the article of Freeze and Harlan (1969).

In their seminal paper, Freeze and Harlan wrote down the equations for different surface and subsurface flow processes and showed how they could be linked by means of common boundary conditions into a single modelling framework. Their analysis is still the basis of the most advanced distributed rainfall–runoff modelling systems today. The equations are all nonlinear partial differential equations (i.e. differential equations that involve more than one space or time dimension; see Box 2.3). For the type of flow domains and boundary conditions of interest in rainfall–runoff modelling, such equations can normally only be solved by approximate numerical methods which replace the differential terms in the equations by an approximate discretization in time and space (see Chapter 5). However, the descriptive equations that they used for each process required, in all cases, certain simplifying assumptions. Thus, for subsurface flow, it is assumed that both saturated and unsaturated flows can be described by Darcy's law (that flow velocity is proportional to hydraulic conductivity and a gradient of total potential; see Box 5.1), while for surface flows it is assumed that the flow could be treated as a one-dimensional, cross-section averaged flow either downslope over the surface or along a reach of the channel network in a catchment (leading to the St Venant equations; see Box 5.6).

Freeze and Harlan (1969) discuss the input requirements and boundary conditions for these equations in some detail. Meteorological data are required to define rainfall inputs and evapotranspiration losses. Their **model definition input** includes the necessary assumptions on the extent of the flow domain, the assumptions about prescribed potential or prescribed flow boundaries (especially zero flow at impermeable or divide boundaries) and also the way in which the flow domain is divided to create space and time increments for solving the process equations. **Flow parameter inputs** are then required for each element of the solution grid, allowing consideration of heterogeneity of catchment characteristics to be taken into account.

This type of distributed model allows the prediction of local hydrological responses for points within the catchment. The first applications of this type of model were made for hypothetical catchments and hillslopes by Freeze (1972). The calculations required the largest computers available at the time (Al Freeze was then working at the IBM Thomas J. Watson Research Centre at Yorktown Heights), and even then only a limited flow domain and coarse mesh of points could be solved. The first application to a field site of this type of model was published by Stephenson and Freeze (1974), who attempted to model a single hillslope at the Reynolds Creek catchment in Idaho (see Section 5.3). The results were not particularly successful but, as they pointed out, it was a complex hillslope underlain by fractured basalts, with complex flow pathways and limited knowledge of the inputs and initial conditions for the simulation; in addition, computing constraints limited the number of simulations they could actually try. Arising from these difficulties, they were also among the first to discuss the difficulties of validation of hydrological models.

Distributed models of this type have the possibility of defining parameter values for every element in the solution mesh. Even with continuing limitations imposed by computational constraints, there may be many thousands of such elements. In addition, the process equations require many different parameters to be specified for each element. With so many parameter values, parameter calibration by comparison with the observed responses in a catchment becomes very difficult. Which parameter values should be changed to try and improve the simulation? The choice is not always obvious because of the interactions between the effects of different parameters that follow from the physical basis of the model.

In principle, of course, parameter adjustment of this type should not be necessary. If the process equations used are valid, it should follow that the parameters should be strongly related to the physical characteristics of the surface, soil and rock. Techniques are also available for measuring such parameters, although, as noted in Chapter 1, there are problems of scale in such measurements. Most measurement techniques can only be used to derive values at scales much smaller than the element mesh used in the approximate solution. The model requires effective values at the scale of the elements. If the soil was homogeneous this would not matter too much, but unfortunately soils and surface vegetation tend to be very heterogeneous at the measurement scale such that establishing a link between measurements and element values is difficult, even theoretically. Indeed, for the case of coupled surface and subsurface flow processes it has been suggested that the concept of effective values of element-scale parameters may not be valid (Binley *et al.* 1989). This remains an important topic in distributed modelling and requires further research.

Despite these difficulties, there has been a strong surge in the use of distributed modelling over the last decade. This has been partly because the increase in computer power, programming tools and digital databases has made the development and use of such models so much easier, and partly because there is a natural tendency for a model development team to try to build in as much understanding from their perceptual model of the important processes as possible. Thus, there is an obvious attraction of distributed process modelling. There are also very good scientific reasons underlying the effort. One is the need for distributed predictions of flow pathways as a basis for other types of modelling, such as the transport of sediments or contaminants. It may not be possible to make such predictions without a distributed model of some sort.

Another reason is the use of models for impact assessment. Changes in land use, such as deforestation or urbanization, often affect only part of a catchment area. With a distributed model it is possible to examine the effects of such piecemeal changes in their correct spatial context. There is also an argument that because of the physical basis of the model we might be able to make a better assessment of the effects of changing characteristics of a catchment because it will be easier to adjust parameter values that have physical meaning. The difficulties of specifying effective values of parameter values at the element scale, however, rather undermines this argument.

Recent examples of distributed process based models include the SHE model (Système Hydrologique Europèen), originally a joint project between the Institute of Hydrology in the UK, the Danish Hydraulics Institute and SOGREAH in France, but now being developed separately (see Abbott *et al.* 1986a; Bathurst *et al.* 1995; Refsgaard and Storm 1995). The UK Institute of Hydrology also developed the IHDM model (Institute of Hydrology Distributed Model; Calver and Wood 1995). In Australia there is the THALES model (Grayson *et al.* 1995) and the CSIRO TOPOG-dynamic model (Vertessy *et al.* 1993; Zhang *et al.* 1999); and there are a number of others. They differ primarily in the way they discretize a catchment and solve the process equations (sometimes with simplifications), but all are essentially based on the original Freeze and Harlan blueprint from 1969 as a description of the flow processes (see Chapter 5 for more details).

It has to be said that neither the process descriptions, nor the questions demanding further research, have changed very much in the thirty years since the Freeze and Harlan blueprint was published. Certainly the tremendous advances in computer power, numerical methods for solving partial differential equations, and advances in programming techniques, have allowed more robust solutions with finer spatial and temporal resolutions to be implemented in applications to larger catchments, but computational constraints on the full implementation of three-dimensional solutions still remain.

It must be remembered that the blueprint remains a major simplification of the perceptual model discussed in Chapter 1. Some processes, such as preferential flow in soil macropores, are omitted altogether, for good reason: there is no adequate descriptive equation to integrate the effects of macropore flow at the element scale (although see Bronstert and Plate (1997) and Faeh *et al.* (1997) for recent attempts to do so). Similarly the idea of describing surface runoff as a sheet flow of uniform depth and velocity across the slope, as used in most process models, is clearly a gross simplification of reality. Only recently have attempts been made to take account of variations in depth on flow velocities and infiltration rates (e.g. Dunne *et al.* 1991; Tayfur and Kavvas 1998). Thus these models are based on flow physics in name, but it is a very approximate physics and is likely to remain so until measurement and visualization techniques for studying flow processes in the field improve to allow better descriptions, particularly of subsurface flow.

2.6 Simplified Distributed Models Based on Distribution Functions

The fully distributed models discussed above are complex and both computationally and parametrically demanding. The mathematical descriptions of processes on which they are based are, however, still simplifications of reality. A separate strand in rainfall–runoff modelling can be distinguished that attempts to maintain a distributed description of catchment responses but which does so in a much simpler way, without the detailed process representations of SHE and the other fully distributed models. This type of model generally uses a form of distribution function to represent the spatial variability of runoff generation. The distribution may be based on a purely statistical description, as in the probability distributed model (PDM) of Moore and Clarke (1981); on a simple functional form, as in the Xinanjiang/ARNO/VIC model of Box 2.2; on GIS-derived hydrological response units as described in Section 2.4 above; or on some simplified physical reasoning leading to a distribution of an index of hydrological similarity as in TOPMODEL (the topography-based model of Beven and Kirkby (1979); see Section 6.4). It is worth noting that a uniform distribution function for infiltration capacity was included as one of the components of the original Stanford Watershed model and that many of these models have ESMA-type components.

In all these models the distribution function component is an attempt to make allowance for the fact that not all of the catchment can be expected to respond in an exactly similar way. Volumes of runoff generation, for example, should be expected to vary with position in the catchment. This is in keeping with a generally held perceptual model of hydrological responses. In representing these different functional responses as a distribution function, it is being recognized that this type of variability is important but that it might be very difficult to specify exactly where in the catchment runoff generation is happening in any particular event. The aim is to get the bulk responses at the catchment scale modelled correctly. TOPMODEL does allow for the distribution function calculations to be mapped back into the catchment, but it is not expected that the predictions will be any more than approximately correct in space.

One of the advantages of this distribution function approach is that some important nonlinearities of the runoff generation process can be reflected in the distribution function but without introducing the large number of parameter values needed for fully distributed models. This will generally make model calibration much easier where some observed data are available for comparison with model predictions.

2.7 Recent Developments: What is the Current State of the Art?

Computers continue to get more powerful. As a result there is absolutely no doubt that distributed hydrological models will get more detailed, more complex, and will become closely coupled to geographical information systems for the input of data and display of results (see, for example, Refsgaard and Storm 1995). This, in one sense, is the current state of the art. There is a question, however, as to whether this type of development will lead to better hydrological predictions. The answer to this question is not at all clear. More complexity mean more parameters, more parameters mean more calibration problems, more calibration problems will often mean more uncertainty in the predictions, particularly outside the range of the calibration data.

This still leaves open the possibility that other, parametrically simpler, models may have much to offer. If the interest is in discharge prediction only, it would appear that, where calibration data are available, simple lumped parameter models such as IHACRES can provide just as good simulations as complex physically based models. For distributed predictions, no study has yet demonstrated that a fully distributed model can do better in predicting the distributed responses in a catchment than a much simpler distribution function model such as TOPMODEL, where the assumptions of the simpler model are reasonably valid (Franchini and Pacciani 1991; see discussion in Chapter 7).

In essence, the application of all types of models is limited by the available data on how hydrological systems work. Models are data constrained because of the strong limitations of current measurement techniques. However, even if improved measurement techniques lead to a better understanding of complex flow processes, it appears that it will be necessary in the foreseeable future to distinguish between models developed for understanding, which describe those processes in detail at small scales, and models developed for prediction at catchment scales. The former will certainly depend on detailed (even if statistical) descriptions of the geometry of the flow domain. The latter will not be able to demand such inputs because they will not be measurable practically or economically at the larger scales at which predictions are required. Models for prediction will necessarily reflect the types of data that are readily available.

In the next chapter, we will look at the data available for modelling in more detail. Chapter 4 will then take a look at modern lumped catchment-scale models, Chapter 5 discusses fully distributed physically based models, and Chapter 6 looks at distribution function models.

2.8 Key Points from Chapter 2

- Any hydrological model must include functional components that account for the relationship between total rainfall and runoff generation in an event, and the routing of the generated runoff to the catchment outlet.
- The volume of rainfall equivalent to the generated runoff is called the effective rainfall. It has a strongly nonlinear dependence on the antecedent state of the catchment. Linear routing methods, such as the unit hydrograph, often work quite well.
- Following the work of Robert Horton, early applications of the unit hydrograph technique assumed that all storm runoff was generated by an infiltration excess mechanism. This is not generally true but the methods continue to be applied

successfully, despite difficulties of hydrograph and rainfall separation, because they have the functionality needed for discharge prediction at the catchment scale.

- With calibration of parameter values, even simple storage element (ESMA) models can produce good predictions of streamflow hydrographs and soil moisture deficits.
- Modern transfer function models aim to overcome some of the problems of defining the storage elements of an ESMA model correctly by letting the data available determine an appropriate structure and level of complexity while avoiding the problem of hydrograph separation (see Chapter 4).
- The earliest distributed models were based on the time–area concept of Ross. Recent work based on the definition of distributed hydrological response units by overlays of different data types in a GIS system is based on essentially similar concepts.
- Fully process based distributed models allow the prediction of local hydrological responses within a catchment but have many parameter values that must be specified for every grid element. This makes parameter calibration difficult but direct measurement or estimation of effective parameter values at the grid scale is also difficult due to heterogeneity of catchment characteristics and the limitations of available measurement techniques (see Chapter 5).
- Simpler models based on distributions of responses within a catchment may still have much to offer for prediction at the catchment scale and some, such as TOPMODEL, have the potential to map those responses back into the catchment to allow additional evaluation of the simulations (see Chapter 6).

Box 2.1 Linearity, Nonlinearity and Nonstationarity

Unit hydrograph and linear transfer function models in general are based on assumptions of linearity and stationarity in time. In this context this may be simply understood in terms of the relationship between inputs and outputs. A response that is stationary in time means that a unit of input will always produce the same output response (remember that for the unit hydrograph, the inputs are *effective* rainfalls not total rainfalls). A response that is linear means that if two units of input fall in the same time step, we would expect twice the output response. If two units of input fall in successive time steps the two associated output responses, suitably delayed, can be simply added together to produce a total output. More complex input sequences can be dealt with as the simple addition of unit responses. This is called the *principle of superposition*. It follows directly from the assumption of a linear model.

In a nonlinear model, the principle of superposition breaks down since it cannot be assumed that a unit of input will always produce the same output. In rainfall–runoff modelling, nonlinear responses are primarily due to two causes. The most important is the effect of antecedent conditions. Thus the relationship between total rainfall and runoff is generally considered to be nonlinear because the wetter the catchment prior to a unit input of rainfall, the greater the volume of runoff that will be generated.

A secondary cause of nonlinearity is due to the change of flow velocity with discharge. In general, for both surface and subsurface flow processes, average flow velocities increase with flow in a nonlinear way. Faster flow velocities mean that the runoff will get to a measurement point more quickly so that the time distribution of runoff (e.g. the shape of the unit hydrograph) will change as runoff increases. This was shown for a small catchment in the classic study of Minshall (1960). Minshall showed a dependence of the shape of the unit hydrograph derived for a small catchment on the volume of effective rainfall. Larger storm sizes resulted in a faster time-to-peak and higher peak discharge in the unit hydrograph (Figure B2.1.1). In larger catchments the effects of the routing nonlinearity are not always easy to distinguish in data analysis. The nonlinearity is not always a greater than linear increase with increasing inputs. For example, when a river overtops its banks in a flood, the slow-moving water on the flood plain may lead to a decrease in the average velocity of the discharge. In semiarid areas transmission losses due to infiltration into a dry channel bed may lead to responses that become more nonlinear with increasing catchment area (Goodrich *et al.* 1997).

It is sometimes also difficult to distinguish between simple nonlinear and *nonstationary* responses, and indeed the difference may be only one of interpretation. A nonstationary response is one for which the relationship between inputs and outputs is changing over time. One obvious reason for this would be long-term changes in the characteristics of a catchment due to changes in land use such as urbanization or the installation of field drainage. However, the effects of antecedent conditions might also be considered as a nonstationary effect, especially if the nature of the processes involved in runoff generation is changing. Such a relationship will be called nonstationary if it cannot be represented as a simple nonlinear function of the inputs or other available variables.

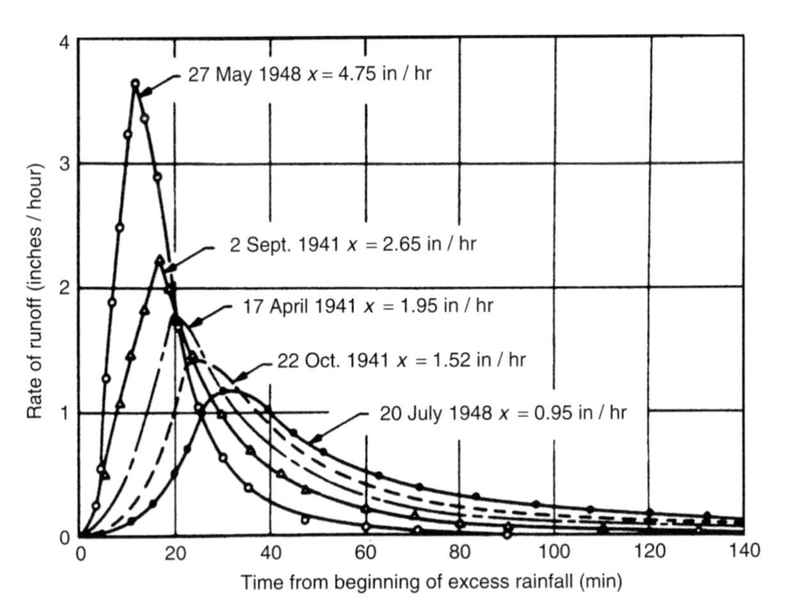

Figure B2.1.1 *Nonlinearity of catchment responses revealed as a changing unit hydrograph for storms with different volumes of rainfall inputs (after Minshall 1960). Reproduced by permission of the American Society of Civil Engineers*

Box 2.2 The Xinanjiang/Arno/VIC Model

A description of the class of models variously named the Xinanjiang (Zhao and Liu 1995), Arno (Todini 1996) or variable infiltration capacity (VIC) model (Wood *et al.* 1992; X. Liang *et al.* 1994; Lohmann *et al.* 1998a) is included here as one example of an explicit soil moisture accounting (ESMA) or 'conceptual' rainfall–runoff model. ESMA models are typically constructed from connected storage elements, with parametric functions controlling the exchanges between elements, losses to evapotranspiration and discharges to the stream. In general, all the parameters are effective catchment-scale parameters and are calibrated on the basis of a comparison of observed and predicted discharges, adjusting the values of the parameters until a best fit is obtained (see Chapter 7 for a discussion of this type of model calibration). This class of models has been chosen for presentation in greater detail here over all other ESMA models because of one interesting feature: in the VIC type models there is a function that attempts to allow for the heterogeneity of fast runoff production in the catchment. Hence the 'variable infiltration capacity' name, although it should be noted that there is no necessary inference that the fast runoff is produced by an infiltration excess mechanism (but see Zhao and Liu (1995) for a process interpretation in this way).

The original idea for this class of models originated in China in the 1970s, where it has been widely applied (see Zhao *et al.* 1980; Zhao 1992). The idea was later adapted for use in a flood forecasting system for the Arno River in Italy by Todini (1996). Its simplicity has also seen it used in large-scale hydrological modelling and as the land surface component in atmospheric circulation and global climate models (GCMs) (see Dumenil and Todini 1992; X. Liang *et al.* 1994; Lohmann *et al.* 1998a,b). The intention of applying the model in this macroscale context was to try and improve the prediction of runoff production and routing in large-scale land surface parametrizations in GCMs. It had two advantages in this respect: first, in at least attempting to take account of some of the heterogeneity in soil moisture storage and runoff generation at the large scale (which had been lacking in earlier land surface parametrizations); and secondly, by having more realistic routing of runoff, more rigorous comparisons with observed discharges from large basins could be made. It is also important in this context that these are achieved in a computationally efficient way, since the complexity of individual components of GCMs is still constrained by the limitations of currently available supercomputers. Applications to the Mississippi basin (Liston *et al.* 1994), Arkansas–Red River basin (Abdullah *et al.* 1996; Abdullah and Lettenmaier 1997) and Weser River (Lohmann *et al.* 1998b) have been reported, and VIC-2L was included in the Project for the Intercomparison of Landsurface Parametrization Schemes (PILPS) (see for example the comparsion of runoff predictions reported in Lohmann *et al.* 1998c).

The version of the model described here is the VIC-2L (two-layer) structure of X. Liang *et al.* (1994; see also Lohmann *et al.* 1998a). The form of the fast runoff production function is perhaps seen most clearly in Figure B2.2.1. The curved function represents the distribution of local total storage capacities in the basin. As rainfall is added, more and more of the storage capacities are filled and, once filled, any excess rainfall on that part of the catchment is assumed to become fast runoff. Between rainstorms, it is assumed that all the storages gradually drain, thereby setting up the antecedent conditions prior to the next storm. It is interesting to note that one of the earliest ESMA models, the Stanford Watershed model of Crawford and Linsley (1966), had a similar storage capacity function for the prediction of fast runoff, but assumed that the distribution was always distributed uniformly between some minimum and maximum capacities across the basin. The variable infiltration capacity form allows for a non-uniform distribution according to the power function:

$$i = i_m[1 - (1 - A_i)^{\frac{1}{b}}] \tag{2.3}$$

where i is the infiltration storage capacity, i_m is a maximum infiltration storage capacity for the area, A_i is the fraction of the area with infiltration capacity less than i, and b is a shape parameter controlling the form of the distribution. For $b = 1$ the infiltration capacity will be uniformly distributed, as assumed in the Stanford Watershed model. This distribution function has two parameters, i_m and b.

This equation is applied to the upper soil layer. For any level of storage in the upper soil layer, an equivalent threshold for saturation, i_o, and the equivalent area of the catchment that will be saturated, A_s, can be calculated (see Figure B2.2.1). The distribution of local storage deficits for the area that is not yet saturated can be then defined as:

$$d_i = i - i_o = i_m[1 - (1 - A_i)^{\frac{1}{b}}] - i_o \qquad i > i_o \tag{2.4}$$

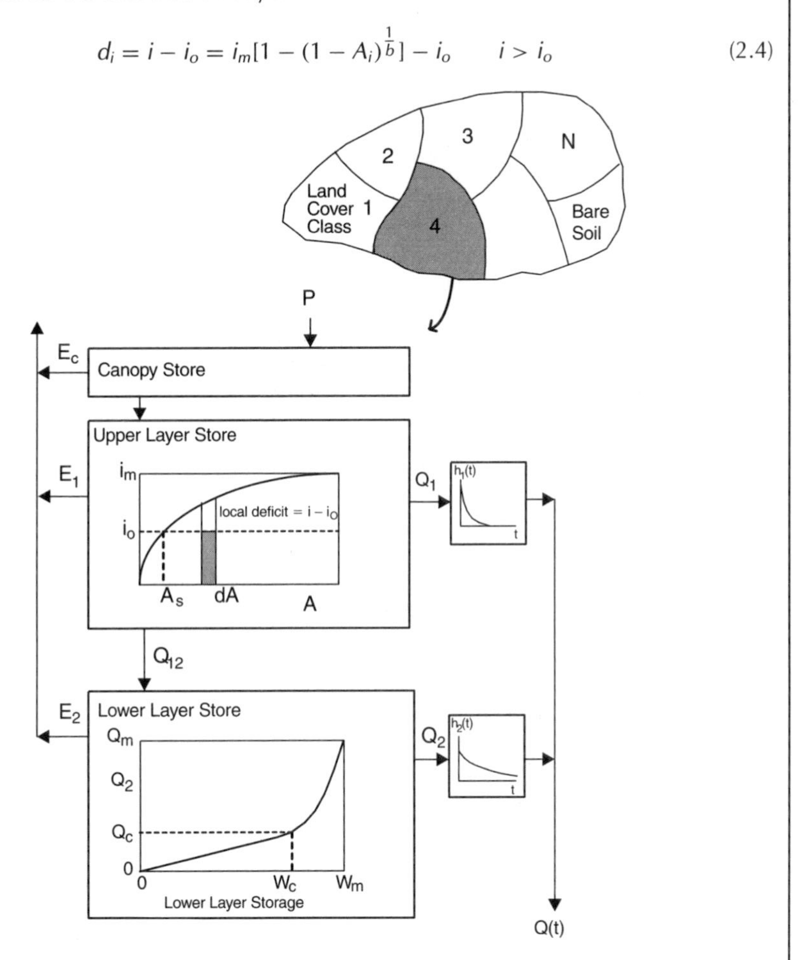

Figure B2.2.1 *Schematic diagram of the VIC-2L model (after X. Liang et al. 1994)*

During a rainstorm, average catchment rainfall in excess of any canopy interception losses is added to the upper soil layer storage at each time step. The saturated area is calculated as that part of the catchment for which upper layer storage exceeds i for that time step. Any rainfall in excess of saturation is assumed to reach the stream as fast runoff. Evapotranspiration is assumed to take place at the potential rate for the area that is saturated and at a reduced rate depending on storage deficit for the remaining part of the area. At the end of the time step, the upper layer storage is depleted by drainage to a lower soil water storage, assuming that drainage is due to gravity alone and that the unsaturated hydraulic conductivity in the upper layer can be described as a function of storage in that layer by the Brooks–Corey relation (see Box 5.4). This requires three further parameters to be specified: a hydraulic conductivity, K_s; a residual moisture content, θ_r; and a pore size distribution index, B.

The lower storage is recharged by this drainage from the upper layer and loses water according to a baseflow function that is linear for low storage values but becomes nonlinear at higher storage values so as 'to represent situations where substantial subsurface stormflow occurs'. The function has four parameter values: the maximum water content in the lower store, W_m; the maximum flow rate from the lower store, Q_m; and the storage and flow rate at the upper limit of the linear part of the function, W_c and Q_c.

Three types of evapotranspiration are included in the model: evaporation from a wet vegetation canopy, transpiration from the vegetation and evaporation from bare soil. The interception capacity of the canopy is taken as a linear function of leaf area index (LAI) with a parameter K_L. The evapotranspiration component uses a Penman–Monteith formulation (see Box 3.2) which also requires specification of aerodynamic and architectural resistance parameters, r_a and r_o, and a canopy resistance. Canopy resistance is allowed to vary as a simple function of soil moisture, with five parameters: the minimum value of canopy resistance when water is not limiting, r_{oC}; the moisture content at which transpiration starts to decline as a result of water stress in each layer, W_1^c and W_2^c; and the wilting point at which transpiration has declined to zero for each layer, W_1^w and W_2^w. The proportion of roots active in transpiration can also be divided between the two layers, adding one extra parameter. Thus, this far the model has 17 parameters for a single land cover element. It may be argued that many of these parameters may be physically meaningful and can be estimated on the basis of soil and vegetation characteristics but one of the limitations of using this type of model at large scales is knowing how local variability should be reflected in effective values of the parameters (see Section 1.8).

An additional snow storage and melt component can be added to the model at the expense of additional parameters. X. Liang *et al.* (1994) also show how the model can be applied separately to each of a number of different land cover classes. Land cover classes will vary in their leaf area index, aerodynamic resistance, architectural resistance, minimum canopy resistance, soil layer storage capacities, and relative fraction of roots in the two soil water stores. Increasing the number of land cover classes therefore means that more parameter values must be specified.

The model is completed by routing of the fast runoff production and the baseflow to the catchment outlet. This is carried out using two linear storage elements with different mean residence times, that for the baseflow being longer than that for the fast runoff. These parameters are currently difficult to estimate from physical measurement or reasoning and will generally be calibrated by comparison of observed and predicted discharges. An example of the predictive capability of the VIC-2L model is shown in Figure 2.9 in the main text.

As well as the link back in time to the Stanford Watershed model noted above, there are some interesting similarities between features of this model and others to be discussed in Chapters 4, 5 and 6. The distribution of storage capacities used in the Xinanjiang/Arno/VIC-2L models is also a feature of the probability distributed model (PDM) described in Section 6.2. The form of the distribution of equation (2.3) can be interpreted as one particular form of distribution function. There are also similarities in this respect with TOPMODEL (see Section 6.4) where the distribution of storage deficits is made a function of catchment topography through a simple approximate theory for flow on the hillslopes. The model could also be thought of as a conceptual nonlinear rainfall filter that calculates an effective rainfall for the type of parallel transfer function routing models discussed in Sections 4.4 and 4.5.

Summary of model assumptions

- A1. Infiltration storage capacity is distributed in space according to a power law distribution.
- A2. Vertical recharge to the lower store is calculated as gravity drainage depending on storage in the upper layer store using a Brooks–Corey hydraulic conductivity function.
- A3. Drainage from the lower layer store is calculated from a function that has both linear and nonlinear segments.
- A4. Calculated actual evapotranspiration is based on the Penman–Monteith equation (see Box 3.1) and takes account of wet and dry canopies, storage in upper and lower soils, and proportion of active roots in each store.
- A5. Runoff generation from the upper store and baseflow drainage from the lower store are routed through parallel linear stores with different time constants.
- A6. Different vegetation types may be modelled separately, regardless of their spatial distribution in the catchment (and at the expense of introducing more parameters).

Box 2.3 Control Volumes and Differential Equations

An easy way of understanding the development of the type of differential equations that are used in the physically based descriptions of surface and subsurface flow processes is through a control volume approach. Here, we will start by considering the mass balance equation for a one-dimensional channel flow, with x representing distance in the downstream direction, and t representing time. We will define a control volume of length Δx and look at the changes that take place over a small time step Δt (see Figure B2.3.1). Over the time step, the change in storage in the control volume will depend on the net balance of the input from upstream Q_i, the lateral recharge into the channel per unit length of channel q (which might be negative if there is infiltration into the channel bed), and the output discharge from the control volume Q_o. Thus the mass balance for the control volume may be written:

$$
\begin{aligned}
\Delta S \Delta x &= \Delta t Q_i - \Delta t Q_o + \Delta t \Delta x q \\
&= -\Delta t (Q_o - Q_i) + \Delta t \Delta x q \\
&= -\Delta t \Delta Q + \Delta t \Delta x q
\end{aligned}
$$

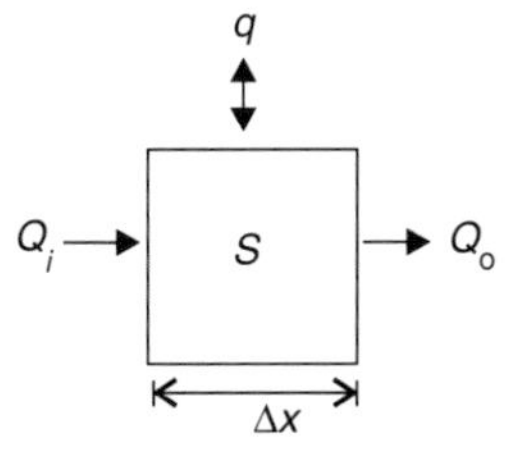

Figure B2.3.1 *A control volume with local storage S, inflows Q_i, local source or sink, q, output Q_o and length scale Δx in the direction of flow*

where ΔS is the incremental change of storage per unit length of channel over the time step Δt, and ΔQ is the incremental change in discharge across the control volume in the downstream direction (i.e. if the inflow is greater than the outflow, ΔQ will be negative and $-\Delta Q$ will be positive). Dividing through by $\Delta t \Delta x$ gives

$$\frac{\Delta S}{\Delta t} = -\frac{\Delta Q}{\Delta x} + q$$

or, if we make the increments very small, this may be written in differential equation form as:

$$\frac{\partial S}{\partial t} = -\frac{\partial Q}{\partial x} + q$$

This type of equation is called a partial differential equation because it involves differentials with respect to more than one variable (here x and t).

The control volume approach may be used to derive all the differential equations used in process descriptions in hydrology (such as the channel flow momentum balance equation of Box 5.6). If we extend this mass balance to a control volume in three spatial dimensions, x, y and z, using local velocity v rather than discharge as the solution variable, we would have

$$\frac{\partial S}{\partial t} = -\frac{\partial v}{\partial x} - \frac{\partial v}{\partial y} - \frac{\partial v}{\partial z} + s$$

where s is a local source or sink flux. This can also be written as

$$\frac{\partial S}{\partial t} = -\nabla v + s$$

where ∇ is called the differential operator.

All these forms of the mass balance equation involve more than one unknown and cannot be solved without more information. This is usually to assume that the local inflow term q is known, and that understanding of the flow process can be used to relate Q or v to S, either directly or through some intermediate variable. Boundary conditions for the flow domain must also be specified. Many such process descriptions result in nonlinear partial differential equations that are not easily solved analytically but must be solved by approximate numerical methods (see Box 5.3).

3

Data for Rainfall–Runoff Modelling

It may seem strange to end a review of modelling with an observation that future progress is very strongly linked to the acquisition of new data and to new experimental work but that, in our opinion, is the state of the science.

George Hornberger and Beth Boyer, 1995

Ultimately, the success of a hydrological model depends critically on the data available to set it up and drive it. In the first two chapters of this book there were several references to the fact that hydrology is limited as a science by data availability and measurement techniques. In some areas of data collection relevant to rainfall–runoff modelling, techniques have improved in recent years. We now have a much better idea of spatial rainfall variations due to the development of rainfall radar; improvements in transducers and data loggers have led to more reliable and more continuous measurements of discharges, water tables and soil moisture; there are techniques available for the direct estimation of evapotranspiration rates; and remote sensing techniques have led to a range of spatial data sets being available for use in modelling. This book is not about data collection for hydrological purposes and will not cover measurement techniques in detail, but the rest of this chapter will consider the issues associated with the main types of data that are available to the rainfall–runoff modelling process.

3.1 Rainfall Data

Rainfall–runoff modelling still depends heavily on the records from point raingauges, both recording raingauges giving estimates of rainfall intensities at time steps of one hour or better, and daily raingauges. In large catchments, models using a daily time step may be perfectly adequate for application purposes; the spatial variation in inputs is then generally more important than the temporal variation. In small catchments, a daily time step may be longer than the storm response time of the catchment and finer time resolution may be required for adequate modelling of the dynamics of the response and the hydrograph peak. Recording raingauges then become more important but they are more expensive to operate and much fewer in number. Thus it may still be necessary to use daily gauges to get an estimate of the total volume of rainfall over a catchment, using the nearest recording raingauge to give an approximate idea of the distribution of rainfall in time, i.e. the *storm profile*.

Even on small catchments, daily rainfalls may be adequate for obtaining acceptable predictions of runoff volumes (rather than hydrograph peaks), especially when volumes over longer times, such as monthly steps, are required. This is implied by the successful simulation of *soil moisture deficits* (Figure 2.10), since discharge will complete the water balance for these sites. Hydrograph prediction, however, will be particularly difficult when a storm spans two daily measurements, since the fixed daily measurement period (often 9 a.m. to 9 p.m.) is hydrologically arbitrary.

Raingauge-measured volumes may be subject to error. In particular they will depend on the design of the raingauge in relation to wind conditions at the site and rainfall intensities. The best design is thought to be a raingauge with the orifice set at ground level and surrounded by an anti-splash grid, but this is not always practical, particularly in environments with frequent snow. A variety of designs of wind shield have been used in different countries to try to mitigate this wind effect. The wind effect can be large; estimates of reductions of up to 20 percent have been reported at windy sites for gauges only 30 cm above the ground compared to ground-level gauges (e.g. Rodda and Smith 1986). High rainfall intensities can also cause problems for some types of recording raingauge, such as the tipping bucket; if the tips start to occur too rapidly the buckets start to bounce, so that high intensities may require a specific calibration.

Rainfall volumes and intensities can vary rapidly in space and time, particularly in convective rainfall events (e.g. Figure 3.1). Thus, as well as an interpolation of rainfall volumes in time to produce the storm profile, it may also be necessary to interpolate in space since raingauge measurements represent only point measurements. A number of techniques are available for such spatial integration, including simple averaging, Thiessen polygons, inverse distance weighting and a variety of others that are covered in most introductory hydrology texts (e.g. Shaw 1994). None of these techniques can be more than an approximation to the actual volume of rainfall over the catchment, and accuracy of a particular technique is likely to change from storm to storm.

The development of radar rainfall measurement has led to a much greater appreciation of the temporal and spatial variation of rainfall intensities than was previously available from raingauge measurements alone. Much of Europe and large areas of the United States are now routinely monitored by ground-based radar rainfalls. The radar has a revolving antenna that sends regular electromagnetic pulses at a low angle into the atmosphere. A detector measures the strength (and in some cases frequency attenuation) of the return signal. The principle is that the return signal to the radar is strongly dependent on the intensity of the precipitation in the path of the radar beam at different distances from the measurement site. A calibration function then allows the intensity of rainfall at each distance to be estimated; the estimated intensities are then normally interpolated onto a square grid, commonly with a resolution of 2 or 4 km for operational radars.

This would appear to be a very important development in the data available for rainfall–runoff modelling, and indeed it is, but there are some important limitations

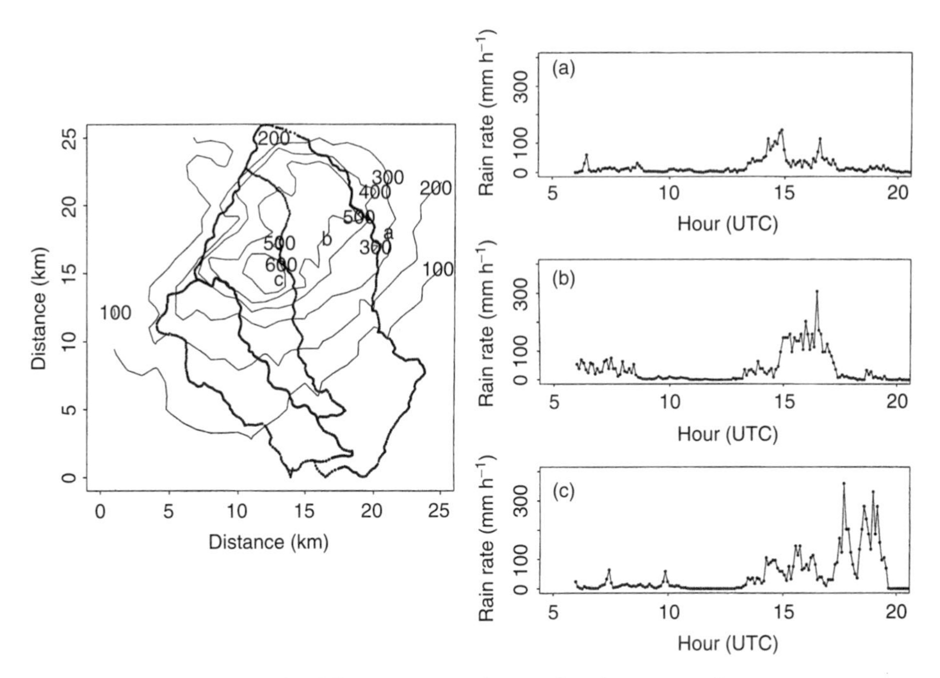

Figure 3.1 *Variations in rainfall in space and time for the storm of 27 June, 1995, over the Rapidan catchment, Virginia (after Smith et al. 1996). Reproduced from Water Resources Research 32(10): 3099–3113, 1996, copyright by the American Geophysical Union*

that must be recognized. The first is that the radar does not measure the rainfall at ground level but at some distance above the ground (often hundreds of metres and increasing away from the radar station). There is thus potential for changes in the patterns of intensities at ground level, particularly where winds are strong and where there is a strong orographic effect. Secondly, the radar returns from rainfalls far from the source may be significantly attenuated by rainfalls nearer to the source. Thus the pattern of radar returns must be corrected for this attenuation effect. Thirdly, the calibration of the radar depends not only on rainfall intensity but also on the type of precipitation, particularly the drop size distribution and whether the rain is all liquid water or a mixture of water and ice (which produce very different return signals). Thus it may be necessary to 'correct' the basic calibration of the radar in different ways. This is most often done by continuously adjusting the estimates of intensity produced by the radar using on-line data from recording raingauges at the ground surface. In some sense then the radar becomes an expensive (but effective) spatial interpolation technique. It is effective because it can give an indication of cells of high rainfall intensity that might be completely missed by the network of ground-level gauges. This may be particularly important in high-magnitude localized rainfall events such as that shown in Figure 3.1.

The estimation of rainfall is very important in rainfall–runoff modelling, since no model, however well-founded in physical theory or empirically justified by past performance, will be able to produce accurate hydrograph predictions if the inputs to the model do not adequately characterize the rainfall inputs (the well-established GIGO principle of 'garbage in, garbage out' applies). A good example is reported in Hornberger *et al.* (1985). In their study, a calibrated rainfall–runoff model for the 5 km^2 White Oak Run catchment in Virginia which had previously performed well in reproducing observed discharges, failed completely to predict a later 'validation' storm. In fact, the volume of rainfall recorded by the raingauges was far less than the volume of discharge recorded in the stream (there was a line of raingauges at different elevations on one side of the catchment, but the intense storm was centred on the other side). It is difficult in such circumstances for any model to predict the response accurately. Inadequate estimation of the rainfall inputs to a catchment must therefore increase the uncertainty of runoff predictions.

There have been a number of models reported in the literature that have included a rainfall multiplier as a parameter to be calibrated as one way of trying to allow for the fact that the raingauge data available might not be a good characterization of the rainfall inputs to a catchment area. It is not clear that this is generally a good strategy in that it may be only some events for which the rainfall inputs are not well estimated. For extreme events, as in the White Oak Run example above, it may be quite obvious that there is a problem. For more moderate events it may be suspected that some events are not well estimated, but it may not be clear which events might be problematic. Thus, a constant rainfall multiplier would not be an appropriate way of adjusting the catchment inputs. It would be better to implement some quality controls (see next section) and, if necessary, exclude some periods of data from the modelling exercise as unreliable.

However, there may be cases when an adjustment might be justified. It is a fairly common situation, for example, in mountainous terrain, that one or more raingauges might be available in the valley bottoms, but none in the higher elevations where it is expected that precipitation inputs might be greater. The average catchment rainfall input might then be consistently higher than that recorded by the valley bottom gauges. Some adjustment would then be necessary to achieve a reasonable water balance. Even in such a case, however, calibration of a rainfall multiplier might not be the best solution, since there would be a distinct possibility that the calibration process would result in interaction with other parameters being calibrated at the same time. Thus, it might be better to make a prior adjustment on the basis of physical reasoning rather than allowing the multiplier to vary during calibration. This would also allow for the possibility of making different adjustments in different periods, although it will be rare that there would be sufficient information on which to base such a variable adjustment. However, calibration of any other parameter values would still be conditional on the adjusted inputs. The general case, as noted in Section 1.8, is that any calibrated parameter values must be conditional on the sequence of inputs used, even if no such adjustments are made.

The estimation of precipitation inputs in the form of snow raises a whole range of additional problems. The hydrologist is interested in the water equivalent of the snow, which depends on both its depth and profile of density, both of which change over time as the snowpack structure evolves and ripens. Snow water equivalent may be measured directly at a point or on transects known as *snow courses* by field measurements of snow depths and density profiles but this can be arduous and expensive to maintain at frequent intervals. The best continuous measurement method available is to measure the weight of snow above a point using a pressure measurement device such as the snow pillow.

An increase in the pressure will indicate a new fall of snow; a decrease will indicate loss by *sublimation* or, more importantly, melting. The continuous measurement of pressure can then give a good indication of the rates of melt which are required for hydrograph modelling.

Unfortunately such installations are expensive and remain relatively rare. Like rain-gauges, they only give an indication of conditions at a single point and snowpacks are renowned for their variability in terms of both water equivalent and rates of melt, particularly in mountainous terrain and where vegetation extends above the pack. Factors such as the redistribution of snow by wind; the effects of topography and vegetation on snow collection, temperature and insolation conditions; freeze–thaw cycles; and changing pack albedos over time, all affect this variability and make modelling snowmelt very difficult indeed (see for example the study of Bathurst and Cooley (1996) in the case study of Section 5.3). This is one area of hydrology where remote sensing has proven especially useful (see Section 3.7 below).

3.2 Discharge Data

The availability of discharge data is important for the model calibration process. Discharge data are, however, generally available at only a small number of sites in any region. It is also an integrated measure in that the measured hydrograph will reflect all the complexity of flow processes occurring in the catchment. It is usually difficult to infer the nature of those processes directly from the measured hydrograph, with the exception of some general characteristics such as mean times of response in particular events. Rainfall–runoff modelling for sites where there are no discharge data is a very much more difficult problem. This ungauged catchment problem is one of the real challenges for hydrological modellers in the twenty-first century.

There are many different ways of measuring discharges (e.g. Herschy 1995). Except for very small flows, it is difficult to make a direct measurement. The level of water in a channel is, however, relatively easy to measure and most methods for estimating discharges require a conversion of a water-level measurement to flow. If this is done as the water flows through a well-maintained weir or flume structure this conversion can be accurate to better than 5 percent. If there is no such structure, or if a structure is overtopped in a high flow, then the accuracy may be very much worse than this. In the worst case of extreme floods, the water-level measuring device may itself be washed away and then the only resort is to try to estimate the maximum flow using the *slope–area method*, in which the cross-sectional area of the flow and the slope of the water surface are estimated from the trash lines indicating the maximum extent of the flow, and a uniform flow roughness equation used to determine an average velocity. Since at the crest of a flood, the flow may be non-uniform, highly turbulent, and with a high sediment load in a dynamically changing cross-section, it may be difficult to estimate an effective roughness coefficient and cross-sectional area, and hence the average velocity and discharge. Errors in discharge estimates will then be much higher.

These potential errors tend to get forgotten when the discharge data are made available as a computer file for use in rainfall–runoff modelling. There is always a tendency for the modeller to take the values as perfect estimates of the discharge. To some extent this is justified: the data are the only indication of the true discharges and the best data available for calibrating the model parameters. However, if a model, any model, is calibrated using data that are in error, then the effective parameter values will be affected and the predictions for other periods, which depend on the calibrated parameter values, will be affected. This will be an additional source of uncertainty in the modelling that we will return to in Chapter 7.

For now, it is worth stressing that prior to applying any model the rainfall–runoff data should be checked for consistency. Some errors, of course, may not be obvious, but the following types of simple checks can be made:

- Calculate the total volumes of rainfall and runoff for different periods in the record, choosing periods separated by similar low flows where possible so that the calculated volumes are not greatly affected by recession discharges. Is the runoff coefficient (the ratio of runoff to rainfall volume) consistent with expected seasonal changes? Lower values would be expected in the summer, higher values in the winter.
- Are the runoff coefficients consistent for increasing storm volumes (allowing for seasonal variations)? For example, are any runoff coefficients greater than 100 percent? This would indicate that one or other of the measurements is in error since mass balance makes it difficult for a catchment to produce more runoff outputs than rainfall inputs.
- If more than one discharge gauge or raingauge are available, check for consistency between the gauges (normalizing for differences in area for discharges). Compare runoff coefficients or use *double mass curves* to check for changes in slope in the accumulated volumes at different gauges.
- Check for any obvious signs that infilling of missing data has taken place. A common example is where measured rainfall intensity is apparently constant for a period of 24 hours, suggesting that a volume from a daily raingauge has been used to fill in a period where the recording raingauge was not working. Hydrographs with long flat tops are also often a sign that there has been a problem with the measurements.

These types of simple checks are easy to make and at least allow some periods of data with apparently unusual behaviour to be checked more carefully or eliminated from the analysis. There is a danger, of course, of rejecting periods of data on the basis that a chosen model cannot be made to give a good simulation of that period. Unless there is some other reason for rejection, this should not be considered good practice, since it is normally the case that the modeller learns more about the limitations of a model from situations where it cannot give good simulations than where it does. The reader should remain aware, however, that most of the hydrological modelling literature tends (quite naturally) to report the best simulations with any given model rather than the worst!

3.3 Meteorological Data and the Estimation of Interception and Evapotranspiration

3.3.1 Estimating Potential Evapotranspiration

In many environments, evapotranspiration makes up a larger proportion of the catchment water balance than stream discharge. Thus, for longer periods of rainfall–runoff simulation, it will generally be necessary to estimate actual evapotranspiration losses from a catchment in order to have an adequate representation of the antecedent state of the catchment prior to each rainfall event.

We must distinguish here between estimates of *potential evapotranspiration* and *actual evapotranspiration*. Potential evapotranspiration is the loss expected over a surface with no limitation of water. It is a function of the *atmospheric demand*, i.e. the rate at which the resulting water vapour can be moved away from the surface. The atmospheric demand depends primarily on the energy available from net radiation to convert liquid water to vapour, the humidity gradient in the lower atmosphere, the wind speed, and the roughness of the surface. Wet rough surfaces, such as forests, will have higher potential evapotranspiration rates than smooth surfaces, such as a lake, given similar radiation, humidity and wind conditions. In general, actual evapotranspiration rates will be at the potential rate until the water supply from the soil becomes limiting.

Methods for estimating potential evapotranspiration range from using a simple annual sine curve to the more physics-based Penman–Monteith equation detailed in Box 3.1. A simple seasonal sine curve for daily potential evapotranspiration, regardless of the variations in the weather, would appear to be far too simplistic to be a useful model of potential evapotranspiration. The study of Calder *et al.* (1983), however, showed that such a curve could give equally good results as more complex formulations requiring more data in modelling soil moisture deficits for several sites in the UK. They treated the mean daily potential evapotranspiration, the only parameter required by the model, as a parameter to be calibrated and found that similar values could be used for all their study sites. A seasonal sine curve can be defined generally in the form:

$$E_p = \overline{E}_p \left[1 + \sin\left(\frac{360i}{365} - 90\right)\right] \tag{3.1}$$

where $\overline{E}_p$ is the climatological mean daily potential evapotranspiration rate in mm day^{-1}, and i is day of the year. Thus, the only parameter needed to apply the sine curve method is the mean annual daily potential evapotranspiration rate. If the diurnal variation in evapotranspiration is also of interest, then the daily values can be redistributed in time using a separate sinusoidal variation over the potential sunshine hours each day.

This method has the advantage that it requires no meteorological variables to be available, but clearly can therefore take no account of the effects of changing temperatures, humidity and cloudiness from day to day and hour to hour on the potential evapotranspiration estimates. The effective mean daily evapotranspiration rate, $\overline{E}_p$, will then become a parameter to be estimated, although the Calder *et al.* (1983) study suggested that, at least in a humid temperate environment, this might be a relatively conservative quantity in space.

A number of empirical approaches for estimating potential evapotranspiration have been suggested based on different levels of data availability. For example, if only mean daily temperature data are available, the empirical equation of Hamon (1961) can be used to estimate daily potential evapotranspiration rates. There are many others based on the Dalton evapotranspiration law that require data on the humidity deficit (i.e. both wet and dry bulb temperatures (see Bras 1990; Calder 1990; Singh and Yu 1997). All empirical methods will be conditional on the range of conditions used in their calibration and care should be taken not to use them outside that range. An intercomparison of a variety of methods for estimating potential evapotranspiration rates is provided by Federer *et al.* (1996).

The best simple physics-based approach available is the Penman–Monteith equation (Monteith 1965; Box 3.1). This equation attempts, in a simple way, to take into account the energy balance of the surface and the way in which turbulence in the lower atmosphere controls the movement of vapour away from the surface, but is more demanding in terms of both data requirements and parameter values. It requires meteorological data on net radiation, air (dry bulb) temperature, humidity (or wet bulb temperature) and wind speed. It also requires an estimate of two resistance coefficients: the *aerodynamic resistance*, r_a, which is an expression of the roughness of the canopy surface, and a *canopy (or surface) resistance*, r_c, which is an effective parameter for the surface as a whole expressing how easily water vapour moves from the stomata of leaves or the pores of a bare soil surface into the air. For a wet canopy or surface, this surface resistance will be zero. For a dry canopy, but without a limitation on supply of water, a typical value of r_c might be 50 m s^{-1}. This implies that there should be different potential rates of evapotranspiration depending on whether the canopy is wet or dry (see next section).

These are representative of the methods available for the calculation of potential evapotranspiration rates. However, as we discussed in Chapter 1, a limitation on the supply of water available for evapotranspiration during long dry periods will mean that the actual evapotranspiration rate may be much smaller than the potential rate. In the Penman–Monteith equation this would be reflected in an increasing canopy resistance as the soil dries. Most rainfall–runoff models contain components, more or less sophisticated, that attempt to simulate this reduction in actual evapotranspiration as the soil dries and water becomes limiting. Models intended primarily for the prediction of discharges have tended to use relatively simple components based on relating soil moisture storage to a ratio of actual to potential evapotranspiration; but it has been suggested that a better approach is to predict actual evapotranspiration rates directly through the use of the effective canopy resistance (e.g Wallace 1995). Recent developments in so-called SVAT models (soil–vegetation–atmosphere transfer models), which aim to predict the fluxes of latent and sensible heat to the atmosphere as a boundary condition for atmospheric circulation models, have resulted in highly complex model structures with multiple soil layers and multiple vegetation layers. In effect, these models aim to predict the way in which the Penman–Monteith canopy resistance changes with water availability and other factors such as solar radiation, leaf temperature, carbon dioxide concentration, vapour pressure deficit, and position in the canopy. Hence the complexity of the models that try to take account of all these effects. Some recent examples are the Biosphere–Atmosphere Transfer Scheme (BATS, e.g. Dickinson and Henderson-Sellers 1988; Gao *et al.* 1996), SECHIBA (Ducoudre *et al.* 1993), ISBA (Interactions between Soil, Biosphere and Atmosphere; Manzi and Planton 1994), the SiB2 model (Simple Biosphere model, version 2; of Sellers *et al.* 1996) and others.

Such models have a very large number of parameters for each of the soil and vegetation layers that may be very difficult to estimate a priori (and may, in fact, change over time). Interestingly, the runoff generation components of such models tend to be rather simple (e.g. Lohmann *et al.* 1998c). This is a good example of how what is considered important in a model depends, initially at least, on the perceptions and background of the modeller.

It is worth noting that the availability of meteorological data may be a problem in applying some of the more demanding methods, including the Penman–Monteith equation. The need for net radiation, temperature, humidity and wind-speed data necessitates either an automatic weather station to be installed within the catchment of interest, or at very least a high-quality meteorological station to be nearby. When this is not the case, some of the simpler methods, even the simple sine curve approach, may still have value.

At many principal meteorological stations an evaporation pan may also be available, with measurements of the depth of water lost from a reservoir open to the atmosphere, usually on a daily basis. There are several different sizes of pans in use, even within the USA. Such measurements can give an index of the rate of potential evapotranspiration at a site but the measured rate does depend on the way in which the pan is exposed, and the nature of the surroundings at the site. In general, such pans will evaporate more water than would be lost from the surrounding surface, even for non-limiting water conditions. Thus, in general, pan evaporation estimates must be multiplied by an empirical pan coefficient to improve the estimate for potential evapotranspiration for a particular type of surface. For a US 'Class A' pan, for example, the coefficient is of the order of 0.7. Pan coefficients are tabulated in many hydrological texts (e.g. Bras 1990), while the UN FAO have proposed a widely used set of coefficients to adjust measured pan evaporation to water use by different crops (for a recent review, see Pereira *et al.* 1999).

3.3.2 Evaporation of Water Intercepted by the Vegetation Canopy

The very low canopy resistance for wet canopy conditions noted above is why, given a source of energy, the loss of intercepted water from a wet canopy tends to be at higher rates than transpiration from a dry canopy. The effect will be particularly marked for rough forest canopies in windy conditions. This can be taken account of within the Penman–Monteith formulation of Box 3.1 by allowing high rates of evaporation at $r_c = 0$, from a conceptual interception storage until that storage is dry, after which transpiration rates are predicted using dry canopy r_c values. The evaporation of intercepted water from leaf surfaces in rough canopies can be very efficient and a significant component of the total water balance in some environments (e.g. Calder 1990). It has also been suggested that there could even be significant losses during some storm events where the rain falls through unsaturated air such that there is still a humidity deficit above the canopy. A number of models of the interception process have been proposed, these being of varying degrees of complexity (e.g. Rutter *et al.* 1975; Gash 1979; Calder 1986). The most widely used is probably the Rutter model which is described in detail in Box 3.2 together with the Calder stochastic model. In general, without specific measurements of throughfall and stemflow below a vegetation canopy, it will not be possible to identify the parameters of an interception model independently such that estimating the parameters of the model will then depend on finding a study of a 'similar' vegetation type reported in the literature, although extrapolation from one site to another should be done with care (see Chapter 7).

3.3.3 Direct Estimation of Actual Evapotranspiration

There are now methods available for the direct measurement of actual evapotranspiration rates over a surface using the *eddy correlation method* (e.g. Shuttleworth *et al.* 1988) but, although a global network of eddy correlation measurement stations is expanding, the main use of such instruments has been during short field campaigns studying land–atmosphere interactions (Figure 3.2). More generally, indirect methods of estimating evapotranspiration are used.

Another promising technique is the use of the laser scintillometer. This uses measurements of the disturbance of a laser beam by rapid variations in the density of the lower atmosphere to estimate the transfers of sensible heat away from the surface. Coupled with a closure of the energy budget for the surface, this can also give an estimate of the latent heat fluxes and evapotranspiration. One nice feature of this technique is that it can integrate the fluxes over the length of the laser beam and therefore provides a

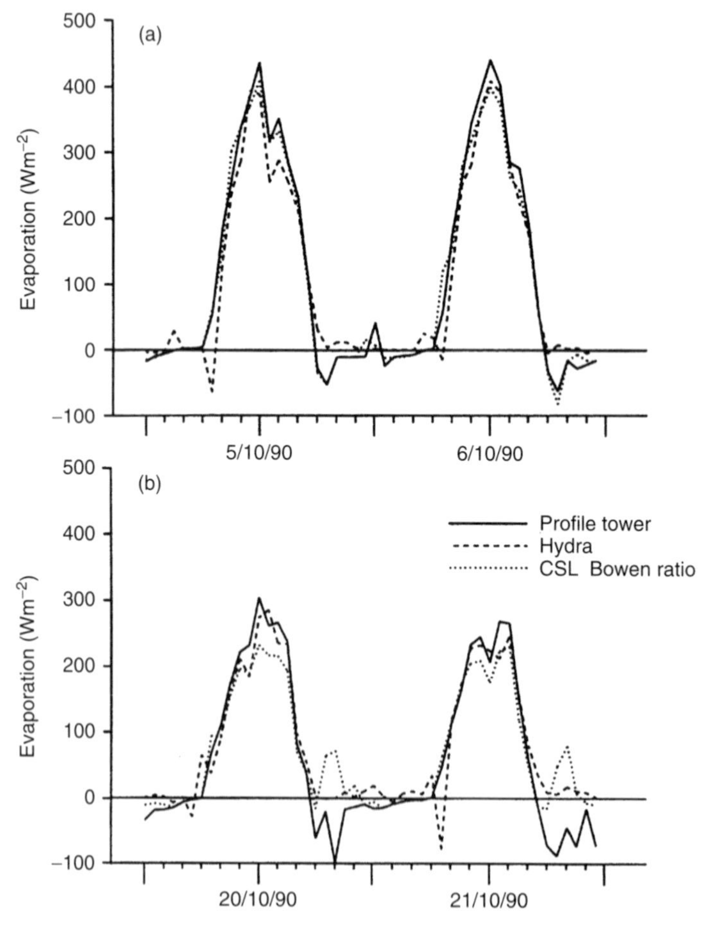

Figure 3.2 Measurements of actual evapotranspiration by profile tower, eddy correlation and Bowen ratio techniques for a ranchland site in Central Amazonia (after Wright et al. 1992)

larger-scale measurement of evapotranspiration than eddy correlation measurements at a site (although both will be affected by the nature of the evapotranspiration over some variable upwind fetch length). The method has been shown to provide good estimates of sensible heat fluxes over both homogeneous (De Bruin *et al.* 1995; McAneny *et al.* 1995) and heterogeneous (Chehbouni *et al.* 1999) surfaces.

The evapotranspiration at a point is necessarily affected by the nature of the surrounding surface. Following original work by Bouchet (1963), Morton (1978) has suggested that pan measurements can be used to derive estimates of the actual evapotranspiration of the surrounding area since, under given energy inputs, the lower the actual evapotranspiration of the surrounding area, the drier the air will be and the greater the resulting pan measurement will be. This *complementarity approach* has been used in catchment models such as SLURP (Kite 1995), and extensively tested by Morton (1983a,b). Kite (1995) reports that methods have been developed to apply the complementarity approach at large scales for different land uses using satellite data. The complementarity approach has been criticized as inaccurate (for example, by LeDrew 1979) but has a certain intrinsic appeal and may yet require re-evaluation as the difficulties of the more physically based but parameter-rich approaches become more widely appreciated.

3.4 Meteorological Data and the Estimation of Snowmelt

In many environments, snowmelt may be the source of the annual maximum discharge in most years and may be a major cause of flooding. Meteorological data will also be required in the modelling of snow accumulation and melt. Again, different types of snow models demand different types of data. The very simplest snow model is the temperature index or *degree-day method*. This model, in its simplest form, is based on the hypothesis that snowmelt is proportional to the difference between air temperature and a threshold melt temperature (Box 3.3). Thus data on air temperature are required as an input, and the threshold temperature is effectively a parameter.

A typical modern variant for snowmelt–runoff modelling, the Swiss SNOW1-ETH4 model, has been proposed by Hottelet *et al.* (1993; see also Ambroise *et al.* 1996). This still only requires temperature as an input but tries to take account of whether precipitation is falling as rain or snow and the heat deficit of the pack that must be satisfied before significant melt will try to occur. This variant increases the number of parameters that must be specified. The degree-day method is applied within the snowmelt–runoff model (SRM) of Rango and Martinec (1995) which has been widely applied in both the United States and Europe (e.g. Mitchell and DeWalle 1998). The SRM makes calculations for different elevation bands within the catchment and takes account of the depletion of the snow-covered area as the melt season proceeds (see Box 3.3).

The degree-day method is obviously a very simple approach, but has the advantage of demanding only temperatures as an input. The method is most accurate when melt is dominated by heat input due to radiation and the pack is ripe at 0 °C and ready to melt. The method is least accurate when melt is dominated by heat advected by an air mass (see Braun and Lang 1986). A combination energy budget approach to modelling the snowpack and snowmelt, similar to that used for evapotranspiration in the

Penman–Monteith equation of Box 3.1, was originally proposed by Anderson (1968). This is much more demanding in both meteorological data (net radiation, temperatures, humidity and wind speed) and parameter values. Models that attempt to model the changing structure of the snowpack as a result of freezing and thawing conditions have also been proposed (e.g. Morris 1991), requiring even more parameters. A number of distributed energy budget models for predicting snowmelt have been developed (e.g. Blöschl *et al.* 1991; Marks and Dozier 1992). However, an intercomparison of snowmelt models carried out by the World Meteorological Organisation (WMO 1986) showed no clear advantage of the more complex models when comparisons were made for a variety of catchments and over a number of years of data.

3.5 Distributing Meteorological Data Within a Catchment

One of the problems in applying all these evapotranspiration and snowmelt models at the catchment scale is taking account of the variability of meteorological conditions within the catchment. Insolation will depend on the angle and aspect of different slopes; wind speeds will depend on wind direction and pressure gradients in relation to the form of the topography; temperatures will depend on elevation; and humidities will depend on evapotranspiration upwind. Distributed rainfall–runoff models will have the potential to take such variations into account, but this requires a further model to distribute the meteorological data measured at one or, at best, a small number of points, to other points in the catchment using digital elevation and other distributed data. This problem will be greatest for hilly and mountainous catchments with a wide elevation range. It is a particular problem for the estimation of snowmelt early in the season, since snowmelt will generally start first at lower elevations on slopes with a southerly aspect and may be significantly delayed at higher elevations. Models for predicting distributed meteorological data within a catchment have been suggested, e.g. in the SAFRAN-CROCUS snowmelt model of Durand *et al.* (1993), in RHESSys (Band *et al.* 1991; Hartman *et al.* 1999), and by Blöschl *et al.* (1991).

Remember that there is also the problem in snowmelt modelling of knowing how much snow is there to melt in the first place, since it is very difficult to obtain information on spatial patterns of snow depth and density to get estimates of snow water equivalent. It is possible to use remote sensing to estimate changing patterns of snow-covered areas which can be used as a constraint on snowmelt models (e.g. Blöschl *et al.* 1991; Rango 1995).

3.6 Other Hydrological Variables

Rainfalls and discharge are the measured hydrological variables most often available to the modeller and certainly are the most useful to the rainfall–runoff modeller. However, in some catchments, other types of hydrological measurements may be available, such as measurements of standing water levels in wells, profiles of soil moisture, or spatial patterns of near-surface soil moisture. Such data will clearly give more information on the hydrological behaviour of a catchment but the amount of information may be limited since, with the exception of a few research catchments, the number of measurement sites is likely to be small. The scale of the measurements is also important in this context. Such internal measurements tend to be small-scale or 'point' measurements, reflecting the hydrological conditions only in the immediate vicinity and to some extent up-gradient. Thus it may be difficult to compare such measurements with the predictions of even the most distributed rainfall–runoff models available. The use of such internal measurements in model calibration and evaluation will be considered again in Sections 5.3 and 6.5.

3.7 Digital Elevation Data

In many developed countries of the world, digital elevation or terrain maps (DEM or DTM) are becoming available at a resolution fine enough to broadly represent the form of hillslopes (50 m in the UK and France; 30 m in the USA; 25 m in Switzerland). DEMs with a fixed grid size are known as *raster* data. Digitized contour maps (*vector* DEMs) may also be available (Figure 3.3(a)). In fact, to date, most raster DEMs have been built by interpolating from digitized contours (Figure 3.3(b)) and as a result may, in places, be subject to significant error, particularly where, in flat topography there are few contours, or where there are short steep slopes. Topography can also be efficiently represented as a *triangular irregular network* (TIN; Figure 3.3(c)). There is potential for developing good-resolution topographic data from photogrammetric analysis of aircraft- or satellite-derived stereo images or directly from aircraft-borne laser altimetry (e.g. Weltz *et al.* 1994). Aircraft platform techniques can now give elevations with a spatial resolution of 2 m × 2 m or better, and a quoted vertical accuracy of 0.1 m. The resulting images, however, give the surface as seen by the sensor, which may include buildings and the canopies of trees.

Water does have a tendency to flow downhill, at least for shallow hydrological systems, so that knowing something about the form of the topography should have some utility in hydrological modelling. Distributed models can clearly use this type of data directly and there are also models, such as TOPMODEL, that are based on a prior analysis of the catchment topography (see Section 6.4 and Box 6.2). Resolution is clearly an issue here. Coarse-resolution DEMs will not be able to provide an adequate description of hillslope flow pathways, while distributed models may not be able to use all the information in a fine-scale DEM because of computational constraints. Variables derived from topographic data, calibrated parameter values, and model predictions within distributed models based on DEMs are known to be sensitive to grid resolution (e.g. Zhang and Montgomery 1994; Bruneau *et al.* 1995; Quinn *et al.* 1995b; Saulnier *et al.* 1997b).

The analysis of a DEM to derive apparent flow pathways has been an interesting topic of research in itself. The methods available depend on whether a raster or vector DEM is available. For raster DEMs, a recent comparison of methods was published by Tarboton (1997). For each grid cell, there are eight possible flow directions. There may be several surrounding grid elements with elevations lower than the cell being considered. The problem is how to distribute the potential flow to these different possible pathways.

(a)

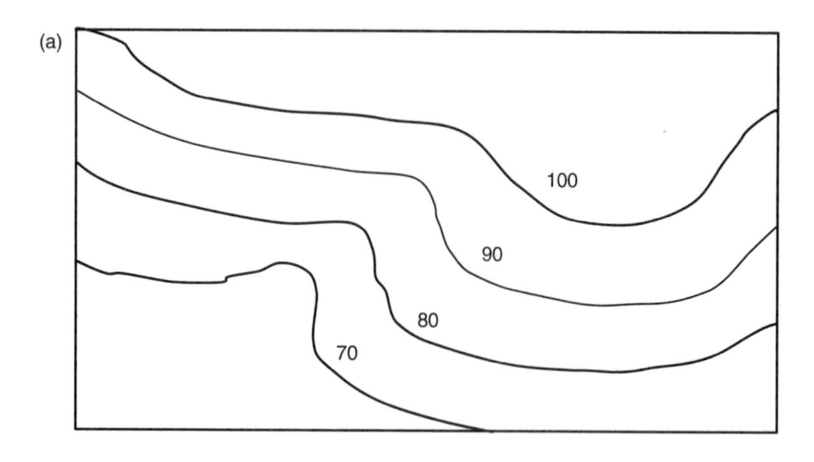

(b)

95	103	103	104	104	105	105	105	104	103	103
86	91	95	95	97	100	102	103	103	103	100
78	80	82	84	84	90	95	100	103	100	95
71	72	72	70	78	86	86	95	96	95	86
69	69	67	68	75	80	82	84	84	82	80
69	66	68	69	70	72	75	76	78	77	76

(c)

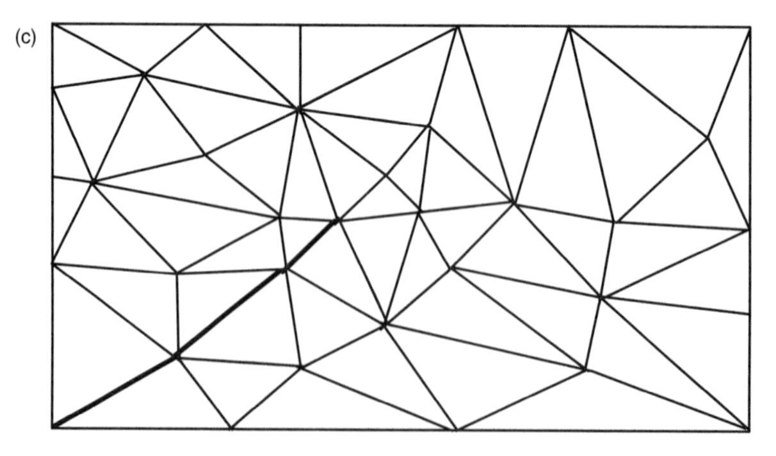

Figure 3.3 *Different forms of digital representation of topography. (a) Vector representation of contour lines. (b) Raster grid of point elevations. (c) Triangular irregular network representation*

Some inaccuracy is clearly inevitable, but the methods that give the best results, at least visually, appear to be the multiple flow direction algorithm of Quinn *et al.* (1995a) (Figure 3.4(b)), and the resultant vector method of Tarboton (1997) (Figure 3.4(c)). An analysis program, DTMAnalysis, based on the multiple direction algorithm, is available to calculate the topographic index distributions required by the TOPMODEL software provided for this book (see Appendix A).

With vector data, the problem is how to derive the lines of greatest slope or *streamlines* for flow on the hillslopes. The idea is that water will follow the same direction of flow as a ball running down the same (smoothed!) surface topography. Since, under this assumption, water should not cross a streamline, it may then be possible to represent the flow between two streamlines, in a *stream tube*, as a one-dimensional flow of varying width in the downslope direction (two dimensions if the vertical is taken into account; Figure 3.5(b)). This is the basis for distributed models such as TOPOG (Vertessy *et al.* 1993) and the Institute of Hydrology distributed model (Calver and Wood 1995). The streamlines should always be at right angles (orthogonal) to the contours. If the contours are available in digital form, calculating the streamlines automatically is a complex problem, but at least one package, TAPES-C, associated with the Australian THALES and TOPOG models, is available (e.g. O'Loughlin 1986; Grayson *et al.* 1995).

TIN DEMs are widely used in GIS systems visualizing a three-dimensional topography on screen, and have been used as the basis for a number of distributed hydrological models, since within each slope facet represented within the TIN, the aspect, slope angle and downslope flow direction are easily calculated (Figure 3.5(c)). The main issue in the construction of a TIN is the best discretization or *tessellation* of the space to represent the topographic form of a catchment most efficiently. The study by Nelson *et al.* (1999) provides a technique for automatically generating a TIN representation of the topography from elevation points or a vector DEM. Once a TIN is defined, algorithms are also available for automatically delineating the river network and catchment area for any point on the network (Palacios-Velez and Cuevas-Renaud 1986; Jones *et al.* 1990).

The idea of analysing the topography of the catchment to give an indication of flow pathways is clearly an attractive one and can result in some attractive computer graphics when model predictions are superimposed back onto a three-dimensional picture of the topography. However, there are some limitations to such analyses that the user must be aware of. Regardless of the algorithms or type of DEM used, all DEM analyses depend

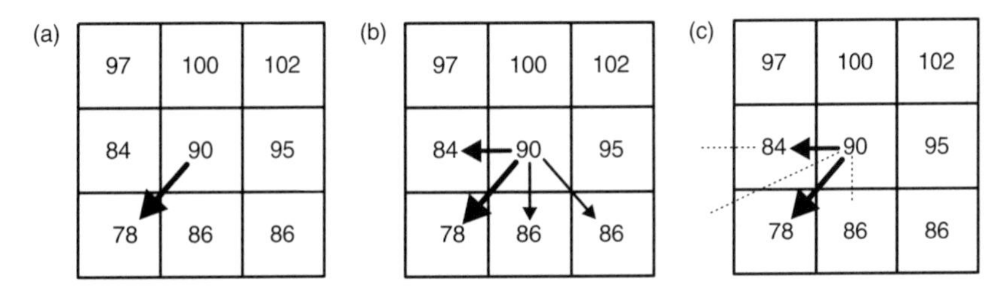

Figure 3.4 *Analysis of flow lines from raster digital elevation data. (a) Single steepest descent flow direction. (b) Multiple direction algorithm of Quinn et al. (1995a). (c) Resultant vector method of Tarboton (1997)*

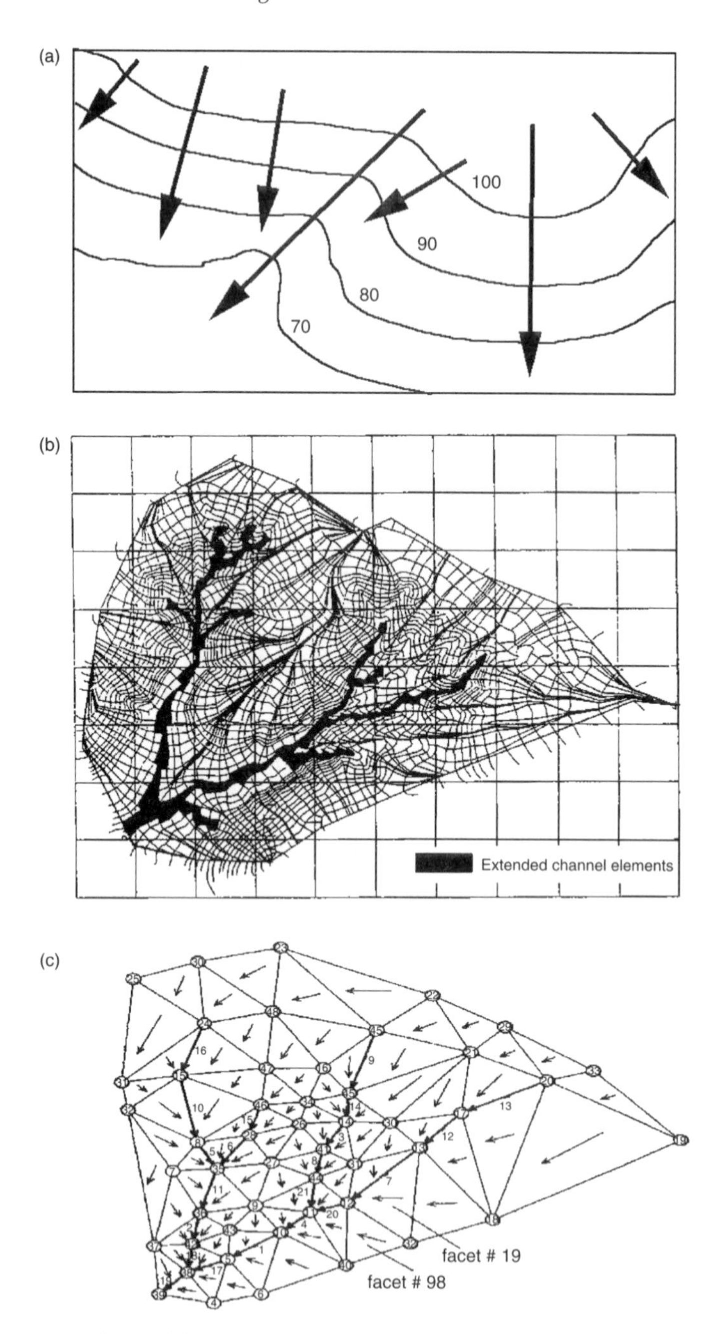

Figure 3.5 *Analysis of flow streamlines from vector digital elevation data. (a) Local analysis orthogonal to vector contour lines. (b) TAPES-C subdivision of streamlines in the Lucky Hills LH-104 catchment, Walnut Gulch, Arizona (after Grayson et al. (1992a). Reprinted from Water Resources Research 28: 2639–2658, 1992a, copyright by the American Geophysical Union (c) TIN definition of flow lines in the Lucky Hills LH-106 catchment (after Palacios-Velez et al. 1998). Reprinted from Journal of Hydrology 211: 266–274, copyright (1998), with permission of Elsevier Science*

crucially on the assumption that the flow pathways will be predominantly controlled by the topography of the catchment. This will only be a good assumption for catchments with relatively shallow soils underlain by impermeable or near-impermeable bedrock. If there are deeper flow pathways they may deviate significantly from those suggested by an analysis of the surface topography. Recent work has also shown that even in shallow systems, the bedrock topography may have a greater control on downslope saturated flow than the surface topography, at least in some catchments (McDonnell *et al.* 1996). Finally, it is worth repeating that, even under ideal hydrological circumstances, an adequate representation of the flow pathways will require a DEM of fine enough resolution to define the shapes of the hillslopes. Analyses suggest that, for raster data, resolutions coarser than 100 m will not suffice.

3.8 Geographical Information and Data Management Systems

Topographic data are just one type of distributed data becoming more readily available to the hydrological modeller in the form of digital geographical information systems (GIS). These are software packages that allow different types of spatial data to be overlain and manipulated. Most GIS do not easily handle data variables that change over time, but some have been specifically designed for time variable data, such as the Institute of Hydrology's Water Information System (WIS). Several have facilities for the analysis of flow directions, although generally limited to single flow direction (steepest descent) algorithms, such as in the ARC-VIEW or GRASS packages.

Other variables that can be easily stored and manipulated within a GIS are maps of vegetation type, soil type and geology. The data can then be used in different ways. The characteristics of each element of a raster grid of arbitrary size could be derived, for example. Or, by overlaying different layers of vector or raster information, irregular *hydrological response units* (HRUs) of different characteristics could be identified (see Section 6.2). A map could then be displayed of all grid elements or HRUs having similar characteristics (see Figure 2.7 for example).

The problem for the hydrological modeller is that the types of information that are generally available in GIS form are only indirectly relevant to the rainfall–runoff processes. Knowing the soil and vegetation type classifications of an HRU is certainly informative, but what parameter values should be used for each class? In principle, such parameter values could be stored directly within a GIS, but only if the values are known. Soil type as mapped by the soil scientist, for example, may not be the same as that required by the hydrologist. In the UK, a hydrological classification of soil types is now available in GIS form (the HOST classification; Boorman *et al.* 1995) but this is based on expectations of soil hydrological behaviour and does not directly give the parameter values required in a hydrological model. In the USA, the USDA STATSGO database defines soil characteristics for the whole of the country, but at a scale of 1 km^2 (USDA SCS 1992).

This is also true of the other GIS data types. In general, another model is required to interpret the GIS data in a form that can be used in a hydrological model. An example is the type of *pedotransfer function* models suggested by Rawls and Brakensiek (1989). Regression analysis is used to provide relationships between soil texture and soil hydraulic parameters. Soil texture is a common characteristic reported in soil surveys, and a soil classification can normally be associated with a texture. A pedotransfer function can then be used to derive values for parameters such as porosity and hydraulic conductivity (see Box 5.5).

However, values derived in this way should be interpreted with care. We have already noted that measured soil hydraulic parameters may be highly variable in space, even within a single soil unit, and that the effective parameter values required for different models may be model structure and scale dependent (Section 1.8). Thus, GIS-derived values of parameters may be associated with a considerable degree of uncertainty which is often ignored. Having said that, a number of studies have reported success in rainfall–runoff modelling based on GIS data (see Section 6.2).

Time series databases can also ease the preparation and use of hydrological data for modelling. A typical example is the ANNIE database system used by the USGS (for a brief description, see Leavesley and Stannard 1995), to link hydrological models to watershed data management (WDM) files that contain rainfall, discharge, meteorological and other data. Time series data, as well as spatial data, are also an integral part of the Institute of Hydrology's WIS package. In WIS, a mouse click on a station site marker on a map will start a dialogue that can be used to display the data available in graphical form or carry out other analysis operations. Romanowicz *et al.* (1993a,b) showed how the rainfall–runoff model TOPMODEL could be integrated into the WIS system.

Within this GIS framework, rainfall–runoff modelling may be just one component of a larger catchment management or decision support system (DSS). Examples are the WATERSHEDSS package developed by the USDA (Chaubey *et al.* 1999) and the UK NERC-ESRC Land-use Programme (NELUP) DSS (Dunn *et al.* 1996). Few GIS packages, however, allow flexible modelling structures to be built directly onto the spatial framework of the GIS. One exception is PC-RASTER which has a built-in programming language that allows model structures, together with input and output data, to be readily incorporated into the GIS. Other general programming languages, such as MATLAB and PV-WAVE, can also be used with large raster spatial databases for modelling (e.g. Clapp *et al.* 1992; Romanowicz 1997).

3.9 Remote Sensing Data

Another source of distributed data for hydrological modelling is remote sensing. Engman and Gurney (1991) and Dubayah *et al.* (1999) provide reviews of the possibilities for the use of remote sensing in hydrology. Remote sensing may be used in the estimation of input data (including topography, rainfall and evapotranspiration rates), state variables (including soil moisture, snow cover and snow water equivalent, and areas of flood inundation) and model parameter values (mostly derived through the classification of soil and vegetation types from remote sensing). In fact, remote sensing derived topography or land cover distribution may contribute to the database stored in a GIS. Many of the same problems apply as for GIS data: remote sensing does not generally give information that is directly hydrologically relevant; a model is required to interpret the digital numbers at each pixel of a remote sensing image into a form that is hydrologically useful (as discussed earlier, for the case of radar-derived estimates of rainfall). It is not often recognized that the interpretation model can be a significant source of uncertainty in the resulting images supplied to the user. Corrections for atmospheric effects for satellite sensors, for example, involve empirical coefficients or parameters that are not known precisely and that may be time variable. These uncertainties are not often quantified and the user has little option but to assume that they are small. However, remote sensing will increasingly be an important source of spatial information and future hydrological models will make increasing use of different types of imaging in both calibration and evaluation of predictions (Schultz 1999).

The main uses of remote sensing in hydrological modelling to date have been in the estimation of precipitation (primarily using ground-based radar), land cover types and vegetation parameters, soil moisture and snow cover (e.g. De Troch *et al.* 1996; Xinmei *et al.* 1995; Schultz 1996). Snow-cover mapping from satellite images is now used operationally in the United States, in conjunction with hydrological modelling and ground surveys, for water resources assessment, flood forecasting and dam regulation (Rango 1995). The utility of the technique remains subject to the limitations of the spatial and temporal resolution of the satellites used, the requirement for generally cloud-free conditions, and the fact that only cover can be easily assessed, rather than snow water equivalent, although attempts have been made to use passive microwave remote sensing to estimate the spatial patterns of water equivalent of snowpacks (Slough and Kite 1992). It is, however, the only method for getting the widespread large-scale coverage that is required for major river basins. Ground-based photographic measurements have also been used in snowmelt modelling (e.g. Blöschl *et al.* 1991).

There are some other remote sensing techniques and interpretation models that may become more useful in the future. Work in this area has become increasingly important with the development of work in *macroscale* hydrology, which has been encouraged by the needs of global atmospheric circulation modelling for hydrological predictions at a large scale. Active and passive microwave techniques for the measurement of soil moisture have been studied for some time. Active sensors transmit a signal to the ground and measure the return signal; passive systems measure only the natural microwave transmission from the surface. The first satellites with active microwave sensors that can be used for soil moisture estimation (ERS1, ERS2 and JERS1) have now been launched. Active microwave systems have also been used from ground- and aircraft-based platforms. The wavelengths normally used will only penetrate at most the first few centimetres of the soil surface. The return signal then depends on the dialectric constant of the surface soil layer. The dialectric constant varies with moisture content, most strongly for soils that are neither too wet nor too dry.

The problem is that for both active and passive microwave sensors the signal also depends on the water content of the vegetation, the roughness of the surface and the state of the atmosphere. In most images from active radar systems, for example, the most obvious features are those associated with the topography and roughness of the surface, such as different vegetation covers. Extracting the soil moisture content signal is then extracting a second-order effect, and often relies on the availability of

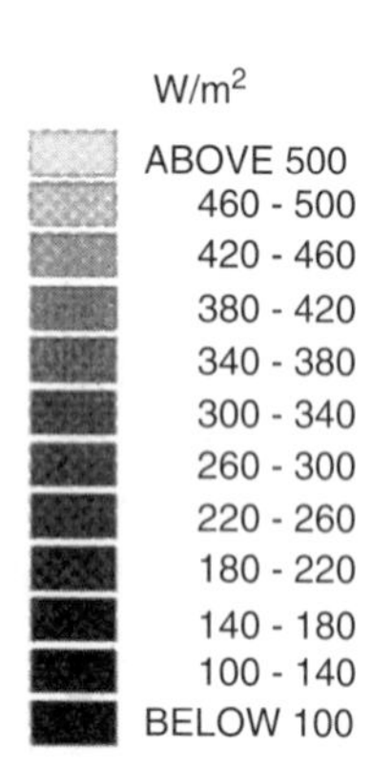

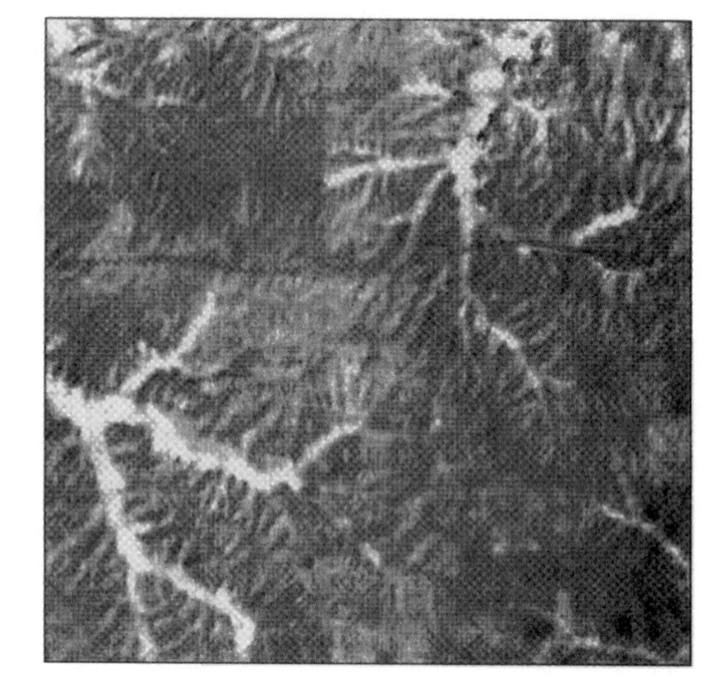

Figure 3.6 *Predicted spatial pattern of actual evapotranspiration based on remote sensing of surface temperatures. Note that these are best estimates of the evapotranspiration rate at the time of the image. The estimates are associated with significant uncertainty (after Franks and Beven 1997)*

ground measurements that can be used to calibrate the interpretation of the radar image (e.g. Lin *et al.* 1994). The approach will work best, therefore, where there is a uniform surface, particularly if there is little vegetation cover. There have, however, been some very interesting results presented from airborne and satellite microwave sensors, using both active (e.g. Verhoest *et al.* 1998) and passive (Schmugge *et al.* 1994; Schmugge 1998) microwave systems. The techniques may well be improved in the future. The wavelength of the signal remains a limitation, in that only surface soil moisture can currently be detected in this way. There is then a problem in relating that surface soil moisture content to the moisture profile and flow processes or to the results of a hydrological model. One of the few studies that has attempted such a comparison at a large scale is that of Wood *et al.* (1993).

There have also been some advances in the use of remote sensing for the estimation of spatial patterns of evapotranspiration. Studies such as those of Holwill and Stewart (1992), Bastiaanssen *et al.* (1994, 1998), Xinmei *et al.* (1995), and Franks and Beven (1997) have used single and multiple images of remotely sensed surface temperatures, together with simple energy balance modelling, to separate sensible and latent heat fluxes to calculate patterns of actual evapotranspiration flux (Figure 3.6). The patterns have, in every case revealed significant and interesting heterogeneity in fluxes, but error calculations suggest that the absolute values of the calculations may be subject to large uncertainty (Franks and Beven 1997).

3.10 Key Points from Chapter 3

- The data available for rainfall–runoff modelling are generally point data and may not be error-free, even though they must often be treated as error-free in applications.
- Data should be checked for consistency before being used in a rainfall–runoff modelling study. Some simple checks can be used to identify periods of unusual behaviour that can be checked more carefully or eliminated from the analysis.
- Methods for directly measuring actual evapotranspiration rates are not yet used routinely. A number of different methods for estimating potential and actual evapotranspiration demanding different levels of data availability are discussed.
- Spatial data are increasingly becoming available through remote sensing such as radar rainfalls and satellite images at different wavelengths, including active and passive microwave sensors used in the estimation of surface soil moistures. In general, such data require some interpretative model to provide hydrologically useful information. This interpretative model may be a source of error in this information.
- Geographical information systems are increasingly being used to store catchment data and interact with distributed hydrological models in setting up model runs and displaying the results. The information stored in a GIS (e.g. soil type, vegetation type) may also require an interpretative model before being useful in hydrological modelling.
- Digital elevation data, in either raster or vector form, can be the basis of distributed modelling of both model inputs and rainfall–runoff processes. The latter may require the derivation of hillslope and channel flow pathways from the digital elevation data. Different methods of analysis and different resolutions of data will give different apparent flow pathways.

Box 3.1 The Penman–Monteith Combination Equation for Estimating Evapotranspiration Rates

The Penman–Monteith equation is based on a combination of a simplified energy balance equation for the surface and equations for the transport of sensible heat and latent heat away from the surface. It is what has been called a *big leaf model,* in that there is an assumption that a complex vegetation canopy can be represented as if it were acting as a single transpiring surface at some effective height above the ground. The energy balance equation, illustrated in Figure B3.1.1, can be written as

$$H = R_n - A - G - S \tag{3.2}$$

where H is the total energy available for evapotranspiration, R_n is net radiation (ranging from -50 Wm^{-2} on a clear night to more than 500 Wm^{-2} on a summer midday), A is heat loss due to advection ($\sim$1 Wm^{-2} for a downwind temperature gradient of 1 °C km^{-1}), G is heat loss into the ground (usually positive during the day and negative at night) and S is the energy flux into physical and biochemical storage in the vegetation (up to 15 Wm^{-2} during the day, and perhaps 3 Wm^{-2} at night).

It is assumed that total available energy can be partitioned into two components: the transport of sensible heat to or from the surface (i.e. energy directly involved in heating or cooling of the air above the surface by conduction and convection); and the transport of

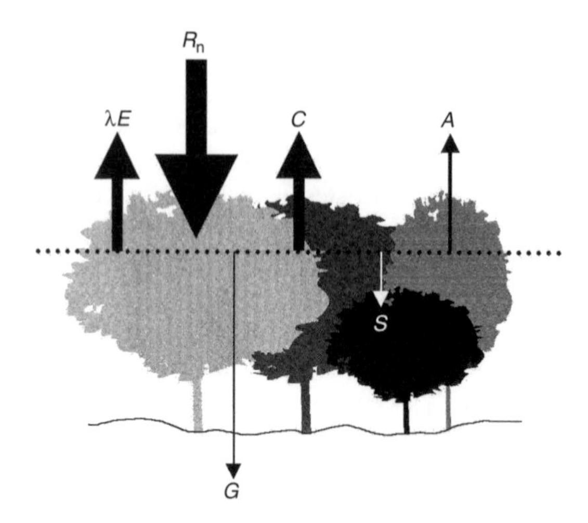

Figure B3.1.1 *Schematic diagram of the components of the surface energy balance.* R_n *is net radiation,* λE *is latent heat flux,* C *is sensible heat flux,* A *is heat flux due to advection,* G *is heat flux due to ground storage,* S *is heat flux due to storage in the vegetation canopy. Dotted line indicates the effective height of a 'big leaf' representation of the surface*

latent heat (i.e. energy used in vaporizing water lost from the surface by evaporation or transpiration). Thus

$$H = C + \lambda E \tag{3.3}$$

where C is the sensible heat flux, λE is the latent heat flux as a product of the latent heat of vaporization λ ($=2.47 \times 10^6$ J kg^{-1}), and E is the evapotranspiration rate (kg m^{-2} s^{-1} $\approx$ mm s^{-1}).

Then

$$\lambda E = H - C = \frac{H}{1+\beta} \tag{3.4}$$

where $\beta = C/\lambda E$ is known as the Bowen ratio. Experimental evidence suggests that the Bowen ratio will often have a fairly constant value for a surface, at least for clear sky conditions without soil water limitations on evapotranspiration (Brutsaert and Sugito 1992; Nichols and Cuenca 1993; Crago and Brutsaert 1996).

The sensible heat flux is a function of the temperature gradient in the air above the vegetation canopy, whereas the latent heat flux is a function of the humidity or vapour pressure gradient above the canopy. Both will also be dependent on factors such as the roughness of the canopy and wind speed (expressed as an aerodynamic resistance to transport). Rough canopies and higher wind speeds (low values of aerodynamic resistance) will result in much more efficient mixing of the air and faster rates of transport. The transport equations are generally assumed to be of the following form:

$$C = \frac{1}{r_{a,H}} \rho_a c_p (T_o - T_z) \tag{3.5}$$

where $r_{a,H}$ is the aerodynamic resistance to transport of heat, ρ_a is the density of the air, c_p is the specific heat capacity of the air, T_o is the temperature of the surface, and T_z is the temperature of the air at some reference height z. The *big leaf* assumption here becomes apparent in the use of the surface temperature, T_o, which must represent some effective value for all the different surfaces of the canopy as a whole.

For the latent heat flux, the equivalent transport equation is

$$\lambda E = \frac{1}{r_{a,V}} \frac{\rho_a c_p}{\gamma} (e_o - e_z) \tag{3.6}$$

where $r_{a,V}$ is the aerodynamic resistance to the transport of vapour, e_o is the vapour pressure at the effective canopy surface, e_z is the vapour pressure at the reference height z, and γ is called the psychrometric constant (=66 Pa K^{-1}). The problem with these equations so far is that the temperature and vapour pressure at the surface are not easily measured. So to make the system of equations solvable, John Monteith came up with the idea of using an additional conceptual expression for the transport of vapour (Monteith 1965) from the interior of the stomata of the leaf surfaces to the free air, as

$$\lambda E = \frac{1}{r_c} \frac{\rho_a c_p}{\gamma} (e_s(T_o) - e_o) \tag{3.7}$$

where r_c is an effective stomatal resistance for the canopy as a whole, generally known as the canopy resistance, and $e_s(T_o)$ is the saturated vapour pressure at the surface temperature T_o. Combining these expressions allows the unknown vapour pressure at the conceptual *big leaf* surface to be eliminated such that

$$\lambda E = \frac{1}{r_{a,V} + r_c} \frac{\rho_a c_p}{\gamma} (e_s(T_o) - e_z) \tag{3.8}$$

There is still the problem of estimating $e_s(T_o)$. This is done by assuming that $e_s(T_o)$ can be approximated by the expression $e_s(T_z) + \Delta_e\{T_o - T_z\}$ where Δ_e represents the slope of the saturation vapour pressure versus temperature curve. The original form of the Penman–Monteith equation uses this linear interpolation of the saturation vapour pressure curve. Milly (1991) has suggested that a higher order approximation will produce more accurate predictions.

In most applications a further approximation is made: that the aerodynamic resistances $r_{a,V}$ and $r_{a,H}$ can both be assumed equal to the equivalent resistance for momentum transport in a well-mixed neutral boundary layer r_a for which a value can be derived from assumptions about the wind speed profile. In particular, for a logarithmic wind speed profile, turbulence theory suggests that

$$r_a = \frac{\ln\{(z - d)/z_o\}^2}{\kappa^2 u_z} \tag{3.9}$$

where d is called the zero plane displacement, z_o is the roughness height, u_z is the wind speed at the measurement height z, and κ is the von Karman constant. This expression assumes a well-mixed boundary layer above the vegetation canopy. Corrections may be required for stable atmospheric conditions.

After these approximations,

$$\lambda E = \frac{1}{r_a + r_c} \frac{\rho_a c_p}{\gamma} (e_s(T_z) - e_z + \Delta_e\{T_o - T_z\}) \tag{3.10}$$

But from the expression for sensible heat flux

$$\{T_o - T_z\} = C r_a/\rho_a c_p = r_a[H - \lambda E]/\rho_a c_p \tag{3.11}$$

so that

$$\lambda E = \frac{1}{r_a + r_c}\left\{\frac{\rho_a c_p}{\gamma}(e_s(T_z) - e_z) + \frac{\Delta_e r_a}{\gamma}[H - \lambda E]\right\} \tag{3.12}$$

Rearranging this equation gives

$$\lambda E\left(1 + \frac{\Delta_e r_a}{\gamma(r_a + r_c)}\right) = \frac{1}{r_a + r_c}\left\{\frac{\Delta_e r_a}{\gamma}R_n + \frac{\rho_a c_p}{\gamma}(e_s(T_z) - e_z)\right\} \tag{3.13}$$

or

$$\lambda E = \frac{\Delta_e H + \rho_a c_p(e_s(T_z) - e_z)/r_a}{\Delta_e + \gamma(1 + r_c/r_a)} \tag{3.14}$$

This is the Penman–Monteith equation. Use of the equation requires measurements of temperature, humidity and wind speed at the reference height z, available energy H, and estimates of the two resistance coefficients r_a and r_c. The approximation $H \approx R_n$ is often made. The equation can be applied with hourly data to provide estimates of the diurnal pattern of evapotranspiration rates.

The resistance coefficients have an important control on predicted evapotranspiration rates, particularly when the resistances are low (e.g. Beven 1979). The variation of predicted evapotranspiration with r_a and r_c for a particular set of meteorological conditions is shown in Figure B3.1.2. Typical values for a dry grass canopy would be $r_a = 50\ \mathrm{s\,m^{-1}}$ and $r_c = 50\ \mathrm{s\,m^{-1}}$, while for a dry tree canopy $r_a = 10\ \mathrm{s\,m^{-1}}$ and $r_c = 50\ \mathrm{s\,m^{-1}}$. The highest actual evapotranspiration rates will be predicted for a rough canopy (low r_a), with intercepted water on the leaf surfaces ($r_c = 0$).

The effects of drying of the soil on evapotranspiration rates can be reflected in an increase in r_c with decreasing soil moisture, although it is known that other factors, such as leaf temperature, carbon dioxide concentration, insolation and even chemical signalling in the plant can also play a role in determining the effective canopy resistance. Calder (1977) suggested an empirical relationship for the change in canopy resistance for transpiration that was a product of a seasonal sinusoid and a function of vapour pressure deficit. Other,

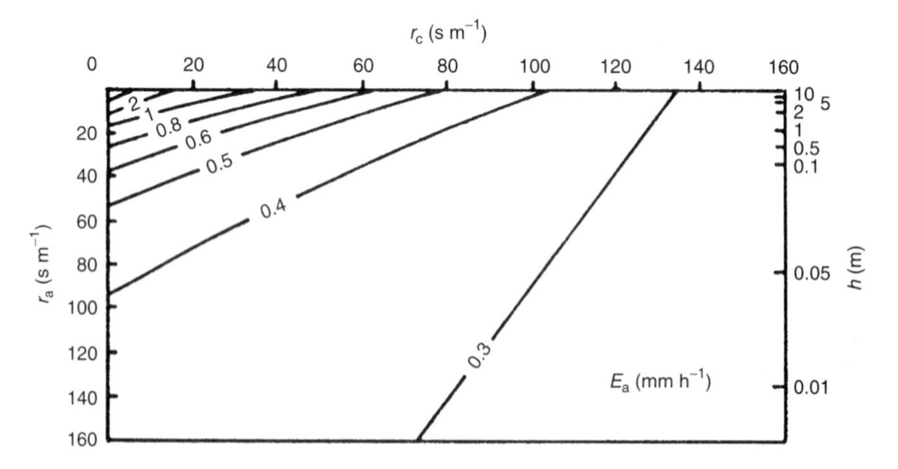

Figure B3.1.2 *Sensitivity of actual evapotranspiration rates estimated using the Penman–Monteith equation for different values of aerodynamic and canopy resistance coefficients (after Beven 1979a). Reprinted from Journal of Hydrology 44, 169–190. Copyright (1979), with permission of Elsevier Science*

more complex relationships have been proposed by Jarvis (1976), Sellers (1985) and Tardieu and Davies (1993). These types of relationships, and their links to carbon dioxide exchanges are included in the current generation of SVAT models, such as SiB2 (Sellers *et al.* 1996). Such models, however, require many more parameter values and there has also been a move, led by John Monteith, to investigate the possibility of using simple models with fewer parameters (Monteith 1995a,b) based on the observation that stomata appear to respond more directly to the rate of transpiration rather than the humidity deficit. Thus, Monteith has suggested a relationship between canopy resistance and transpiration rate of the form:

$$\frac{a}{r_c} = \left(1 - \frac{E}{E_{max}}\right)^{-1} \tag{3.15}$$

where a is an extrapolated maximum canopy conductance when transpiration rate is zero and E_{max} is an extrapolated maximum transpiration rate when canopy conductance $(1/r_c)$ is zero. The parameters a and E_{max} can be obtained from measurements of canopy resistance and transpiration rates at different humidity deficits. Monteith (1995b) suggests that they will vary with temperature, CO_2 concentration and, particularly, soil moisture content.

The assumptions of the Penman–Monteith model for evapotranspiration can be summarized as follows:

- A1. The energy available for evapotranspiration, H, can be estimated from knowledge of net radiation, ground heat flux, advective heat flux and storage of heat in the vegetation canopy. In applications this is often reduced to an estimate based on the net radiation alone, with other terms neglected as small.
- A2. The latent heat and sensible heat fluxes can be estimated from first-order gradient flux relationships with the gradients of vapour pressure and temperature respectively being determined between some measurement height and some conceptual height within the canopy (the 'big leaf' assumption).
- A3. The aerodynamic resistance for the movement of vapour and heat away from the canopy are similar and can be estimated from the aerodynamic resistance for the downwards flux of momentum as determined under the assumptions of a logarithmic velocity profile.
- A4. The controls of stomata on evapotranspiration rates can be represented by some effective canopy resistance for the movement of vapour from stomata to the air within the canopy.
- A5. The saturated vapour pressure of the air within the stomata is assumed to be consistent with the effective canopy temperature and may be predicted by extrapolating from the measurement height using the slope of the known theoretical saturated vapour pressure–temperature relationship.

Box 3.2 Estimating Interception Losses

In Section 3.3, we have made the point that in many environments evapotranspiration is a greater proportion of the catchment water balance than stream discharge. In turn, an important component of the total actual evapotranspiration from a catchment can be due to evaporation from water intercepted on the vegetation canopy, particularly from rough canopies that are frequently wetted. In forest canopies subject to frequent wetting in the windy environment of upland UK, interception may amount to more than 20 percent of the rainfall inputs (Calder 1990). In such circumstances, the vegetation canopy may have an important effect on the amount and pattern of intensity of rainfall reaching the ground.

Regression Models of Throughfall and Stemflow

There have been many experimental studies of interception, many of which have reported results in the form of regression equations for storm throughfall and stemflow as follows:

$$V_{TF} = B_{TF} P_{storm} - C_{TF}$$

$$V_{SF} = B_{SF} P_{storm} - C_{SF}$$

where P_{storm} is the storm rainfall total, the subscripts TF and SF refer to throughfall and stemflow respectively, V_{TF} and V_{SF} are the volumes of throughfall and stemflow in the storm, C_{TF} and C_{SF} are minimum storage capacities for throughfall and stemflow, and B_{TF} and B_{SF} are coefficients. This type of regression relationship can obscure considerable scatter in the measurements for individual storms so that in making predictions the uncertainty in estimated throughfall and stemflow predictions should be assessed. The storage capacities and coefficients vary with vegetation type and seasonal vegetation growth. Not all such studies have differentiated between throughfall and stemflow.

Most of these experiments were based on volumetric collection of throughfall and stemflow with storm-by-storm measurements. Increasingly, however, hydrologists have made measurements on a more continuous basis and this has allowed the development of more dynamic interception models.

The Rutter Model

Perhaps the most widely used model of interception is that proposed by Rutter *et al.* (1971; see also Rutter *et al.* 1975; Calder 1977; Gash and Morton 1978). A schematic diagram of the model is shown in Figure B3.2.1. The model is based on two storage components, one for canopy interception and one for stemflow. Incoming precipitation is partitioned into direct throughfall that reaches the ground without interacting with the canopy, input to the canopy interception and stemflow stores. The partition coefficients p and p_t shown in the Figure B3.2.1 will be vegetation type and seasonally dependent, but are often assumed constant in applications. Drainage from the stemflow store begins when the minimum storage capacity is exceeded. For the canopy store, drainage takes place at the following rate:

$$D_t = D_s \exp[b\{S_{TF} - C_{TF}\}] \quad (3.16)$$

where D_s is the drainage rate when the storage depth S_{TF} just equals the capacity C_{TF} and b is a coefficient. Note that if applied in this form the results will be dependent on the time steps used and that it is better to integrate the equation over the required time step to calculate a storage at the end of that time step, from which the integrated drainage may be obtained from the change in storage. For the stemflow store, all storage in excess

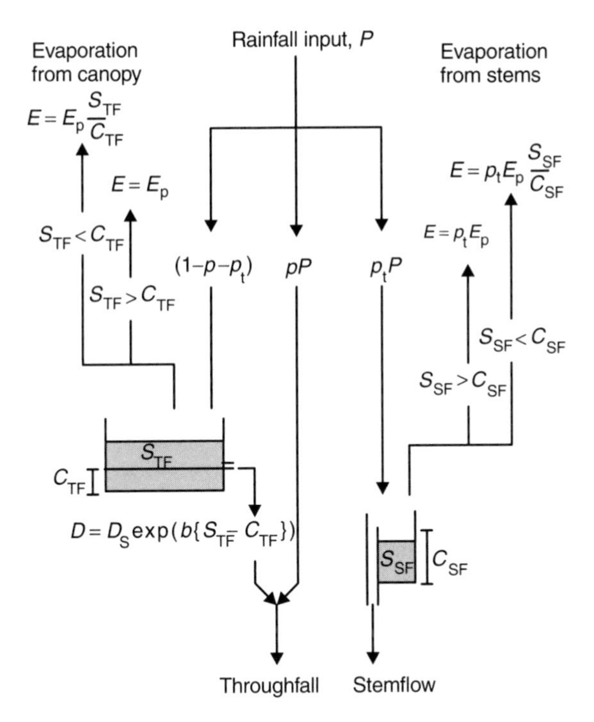

Figure B3.2.1 *Schematic diagram of the Rutter interception model*

of the capacity C_{SF} is assumed to drain rapidly to the ground. The values of the storage minimum capacities C_{TF} and C_{SF} can be determined by the type of regression analysis on storm volumes described above.

The Rutter model also takes account of evaporation from the two stores based on a potential evapotranspiration rate for a wet canopy calculated using the Penman–Monteith equation described in Box 3.1 with a canopy resistance of zero. If either store is greater than its storage capacity, then evaporation takes place at the potential rate. If either storage is below its minimum capacity then it is assumed that part of the canopy is dry and evaporation is reduced proportionally as

$$E_a = E_p \frac{S_{TF}}{C_{TF}} \tag{3.17}$$

where E_a is the actual evaporation rate and E_p is the estimated potential rate from the interception store. A similar form is used for the stemflow store, allowing that this may account for a proportion of the total potential evaporation (usually taken as equal to p_t) until that store is dry.

The Rutter model therefore requires the specification of six parameters, p, p_t, D_s, b, C_{TF} and C_{SF}, together with estimates of rainfalls and potential evaporation as inputs. In many vegetation types, notable crops and deciduous trees, values of the parameters will not be constant throughout the year but will change with the pattern of leaf development. Care should also be taken with adopting values found in the literature, since the values quoted may depend on the particular form of the model used. Values for different types of tree canopy are given in Calder (1977), Gash and Morton (1978), Dolman (1987), Gash *et al.* (1980) and Lloyd *et al.* (1988). The most important parameter will generally be the total storage capacity, $C_{TF} + C_{SF}$, as this will have the dominant control on the amount of evaporation that takes place at the higher wet canopy rates.

The assumptions of the Rutter model may be summarized as follows:

- A1. Rainfall inputs can be apportioned between direct throughfall, canopy interception and stemflow storage.
- A2. Drainage from the canopy interception store is an exponential function of storage in excess of a minimum storage capacity, once that capacity is exceeded.
- A3. Drainage from the stemflow store takes place as soon as the storage capacity is filled.
- A4. Evaporation from the canopy and stemflow stores is at the wet canopy potential rate if storage is greater than the minimum capacity, but is linearly reduced below the potential rate if the canopy or stems are partially wet. Note that as the canopy dries, if the evaporation from the canopy is not sufficient to satisfy the potential rate, then it is possible that there will be some additional transport of water to the atmosphere due to transpiration or soil evaporation.

A simplified analytical variant of the Rutter model was proposed by Gash (1979) and has met with reasonable success in a number of different environments (e.g. Gash *et al.* 1980; Lloyd *et al.* 1988; Navar and Bryan 1994).

The Calder Stochastic Model of Interception

The way in which net throughfall is related to canopy storage in the original Rutter model (which allowed drainage even for a canopy storage less than the capacity value) means that the canopy reaches maximum storage only after much more rainfall than the capacity volume has fallen, while the exponential drainage function predicts a small amount of throughfall even when the storage is zero. To avoid this problem, Calder (1986) proposed an alternative stochastic model of interception based on the probability of raindrops striking elemental areas making up canopy surfaces. In the original formulation a Poisson distribution was assumed to relate the mean number of drops retained relative to the mean number of drops striking the element. The result was a model in which interception was dependent on both storm volume and the mean drop size of the rainfall with the effective canopy storage also being a function of drop size.

In a later development, Calder (1996) has extended the stochastic model to account for secondary drops, i.e. drops falling from upper parts of the canopy that strike lower elemental areas, and for a kinetic energy dependence effect that is a function of drop size and rainfall intensity. This version of the model has been tested by Calder *et al.* (1996) and Hall *et al.* (1996). This model requires seven parameters to be specified. This is too many to be determined from measurements of throughfall rates alone and calibration requires some detailed measurements during wetting experiments, including drop sizes (Calder *et al.* 1996).

Box 3.3 Estimating Snowmelt by the Degree-Day Method

As noted in the main text, there are many problems and complexities in modelling the accumulation and melting of snowpacks on a complex topography and that while process-based energy budget and pack evolution models are available (e.g. Anderson 1968; Blöschl *et al.* 1991; Morris 1991; Marks and Dozier 1992), it has proven difficult to demonstrate that they can produce generally more accurate operational predictions than simpler empirical models, at least without assuming that parameter values such as albedo of the pack are time variable (e.g. Braun and Lang 1986). Of the empirical models, the most widely used is the *degree-day method,* of which there are many variants (e.g. Bergstrom 1975; Martinec and Rango 1981; Hottelet *et al.* 1993; Moore *et al.* 1999). An

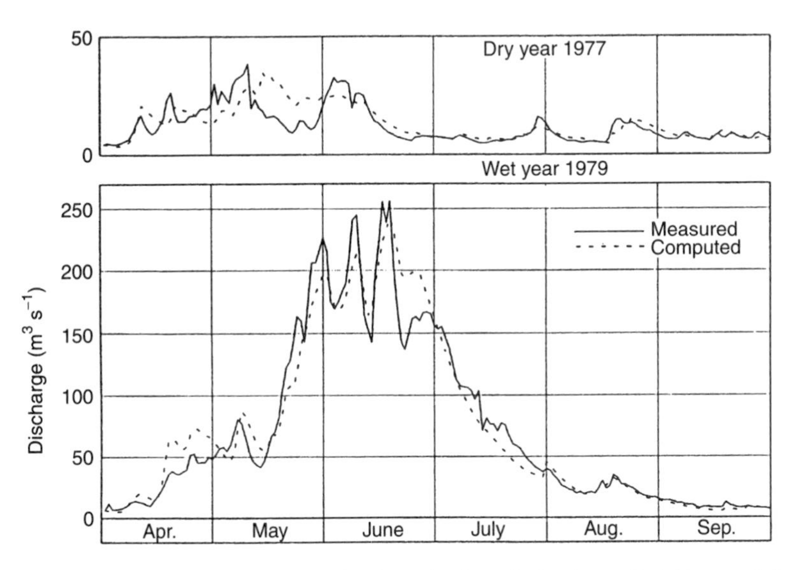

Figure B3.3.1 *Discharge predictions for the Rio Grande basin at Del Norte, Colorado (3419 km²) using the snowmelt runoff model (SRM) based on the degree-day method (after Rango 1995)*

example of the accuracy of snowmelt discharge predictions in a large catchment for two different years using a degree-day snowmelt model with a daily time step is shown in Figure B3.3.1).

In its simplest form, the degree-day method predicts a daily melt rate from

$$M = F \max(0, \overline{T} - T_F) \tag{3.18}$$

where M is melt rate as a water equivalent per unit area [L T^{-1}], F is the degree-day factor [L T^{-1}K^{-1}], $\overline{T}$ is mean daily air temperature [K], and T_F is a threshold temperature [K] close to the freezing point of water. Thus, in its simplest form, the degree-day method takes no account of the temperature of the snowpack, variations in local radiation balance due to slope, aspect, cloudiness, changing albedo, partial snow cover and vegetation projecting above the snow, inputs of heat associated with rainfall, changing snowpack area, etc., except in so far as these can be accounted for by adjusting either F or the temperature values. The degree-day method will work best when the snowpack has ripened to the melting temperature, but even then the diurnal changes in air temperature and day-to-day weather conditions can lead to variations in melt rates in both time and space. It is not expected, however, that the degree-day method will be accurate at all points in space; only that it will produce reasonable estimates of melt and the lifetime of a pack over an area, given some knowledge of the snow water equivalent at the end of the accumulation period.

Various modifications to the degree-day method have been made to try and extend the basic concept of a temperature-dependent melt. In the ETH-4 version of the method (Hottelet *et al.* 1993), a continuous balance of water equivalent and snowpack temperature is maintained throughout the winter period. This version of the degree-day method was used by Ambroise *et al.* (1996a) in modelling the small Ringelbach catchment in the Vosges, France.

In the ETH-4 model, when precipitation occurs, it is added to the pack water equivalent as either liquid water or snow or a mixture of the two depending on the air temperature at the time relative to two threshold temperatures. Above T_{rain}, all precipitation is assumed to be in liquid form; below T_{snow}, all precipitation is assumed to occur as snow. Between the two, the proportion of snowfall is given as the ratio $(T_j^a - T_{snow})/(T_{rain} - T_{snow})$ where T_j^a is the air temperature at time step j. Any precipitation in the form of snow is added directly to the pack.

At each time step the temperature of the pack, T_j^s, is updated according to the equation

$$T_j^s = C_T^1 T_j^a + (1 - C_T^1) T_{j-1}^s \tag{3.19}$$

Rainfall is added to the pack at a rate depending on the pack surface temperature. If T_j^s is less than a threshold temperature T_c (which need not be the same as T_F) all rainfall is assumed to be frozen into the pack. Otherwise, if $T_j^s > T_c$ then the rainfall is added to the pack as liquid water content. Part of this liquid water content may freeze at a rate proportional to the temperature difference between the pack temperature and the threshold temperature. If there is some liquid water content the increase in snowpack water equivalent is calculated as

$$S_j = S_{j-1} + C_T^2 (T_j^s - T_c) \tag{3.20}$$

Various conditions to ensure that there is not more melt or freezing than water equivalent available complete the model.

This simple snowmelt model already contains a number of parameters that must be calibrated. These are the coefficients F, C_T^1 and C_T^2 and the threshold temperatures T_F and T_c. These might vary with location, while it has long been recognized that the melt coefficient F is not generally constant, but will increase during the melt season (e.g. Figure B3.3.2). The World Meteorological Organisation (WMO 1964) recommends values of the degree-day coefficient that increase through the melt season. Hottelet *et al.* (1993) represent this change as a gradually increasing sine curve defined by maximum and minimum values of F, resulting in one additional parameter to be estimated.

Where a catchment covers a wide range of elevations, then it is usual to divide the area up into a number of elevation bands. Rango (1995) reports using bands of 500 m in the snowmelt runoff model (SRM). Air temperatures may then be adjusted for elevation using a simple lapse rate multiplied by the elevation difference from the temperature station so that for band k

$$T_k^a = T_s^a + \gamma (E_k - E_s)/100 \tag{3.21}$$

where γ is the lapse rate (K per 1000 m), T_s^a is the temperature at the recording station, E_k is the mean elevation of band k, and E_s is the elevation of the recording station. A typical lapse rate used would be of the order of 6.5 K per 1000 m. A further important adjustment is to account for the change in snow-covered area within each elevation band. This can be done, either by using standardized depletion curves for the changing average water equivalent of the pack in each elevation band (e.g. Figure B3.3.3), or by updating the current snow-covered area using remote sensing (Rango 1995).

The SRM is a particularly interesting implementation of the degree-day method because of the way in which it will also make use of satellite imaging to determine snow-covered areas in predicting snowmelt runoff in large catchments (for more detail, see Rango 1995).

In some cases, shorter than daily time steps are required for runoff prediction. Rango and Martinec (1995) report a test of a modification to the degree-day method for use in the SRM that incorporates a radiation component such that

$$M = F^* \max(0, \overline{T} - T_F) + R_n/\lambda_m \tag{3.22}$$

where M is now an hourly melt-water equivalent, F^* is now a new degree-day coefficient, R_n is the net radiation, and λ_m is the latent heat of melting. This modification is clearly more

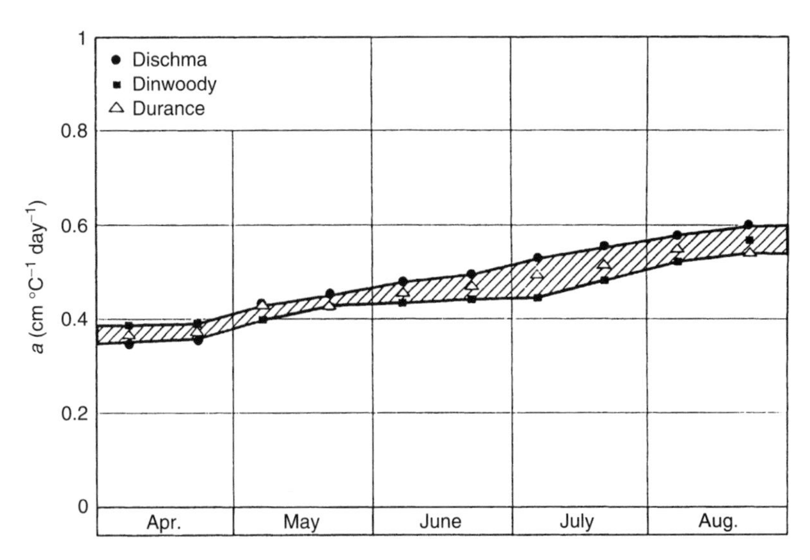

Figure B3.3.2 Variation in average degree-day factor, F, over the melt season used in discharge predictions in three large basins: the Dischma in Switzerland (43.3 km^2, 1668–3146 m elevation range); the Dinwoody in Wyoming, USA (228 km^2, 1981–4202 m elevation range); and the Durance in France (2170 km^2, 786–4105 m elevation range) (after Rango 1995)

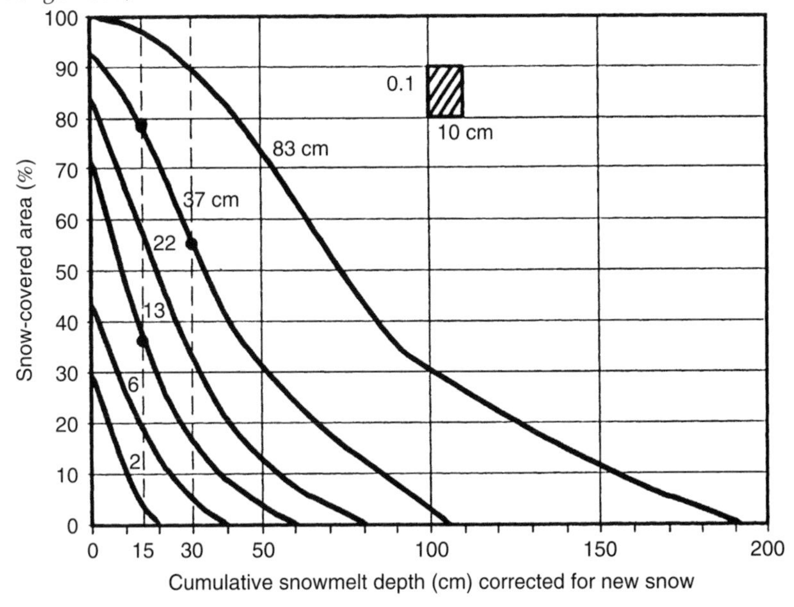

Figure B3.3.3 Depletion curves of snow-covered area for different mean snowpack water equivalent in a single elevation zone (2926–3353 m elevation range, 1284 km^2) of the Rio Grande basin (after Rango 1995)

demanding in terms of data. Net radiation, in particular, may not be measured directly and may need the albedo and long-wave outgoing radiation of the pack to be estimated. Differences in the radiation regime for different slope aspects can be accommodated within this modification. It does, however, take account of the changing radiation regime through the melt season so that the coefficient F^* shows much less variation than F over the season.

Kustas *et al.* (1994) have extended this further to a more complete energy balance melt model for use within SRM. Rango and Martinec (1995) point out that the degree-day method will continue to be used 'not so much because of its simplicity but because of its modest data requirements'. They suggest that the results, averaged over periods of a few days or routed through a runoff model on a large catchment, are comparable in accuracy to the more complex energy budget models available.

The assumptions of the degree-day method tend to vary with the implementation but the most important may be summarized as follows:

- A1. Predicted snowmelt from a ripe snowpack is a linear function of the difference between local mean daily temperature and a threshold temperature.
- A2. The degree-day factor will tend to increase as the melt season progresses.
- A3. Variation in snow-covered area can be taken into account by either using local depletion curves or remote sensing.

4

Predicting Hydrographs Using Models Based on Data

Some experts are fond of saying that the simplest methods are the best. This really confuses the issue. All other things being equal this is clearly true. Unfortunately, all other things are not usually equal. The prime criterion should be accuracy, and until equivalent accuracy is demonstrated, simplicity should be a second-order criterion.

Ray K. Linsley, 1986

4.1 Data Availability and Empirical Modelling

There are two very different, but widely held, views of modelling. The first holds that all models, however physically based their theory, are essentially tools for the extrapolation of available data in time (to different periods) and space (to different catchments). This view of modelling as induction will be the subject of this chapter. The second view holds that models should as far as possible reflect our physical understanding of the processes involved. Only in this way, it is suggested, will we be able to have faith in predictions that lie outside of the range of data available in time (e.g. in the future) and space (in different catchments). This view, modelling almost as deduction, will be the subject of the next chapter. It is modelling *almost* as deduction because, unfortunately, we cannot yet get away without some empiricism in the description of hydrological processes and in estimating model parameters, and may, in fact, never be able to do so (see Beven 2000).

In the first view, the approach is unashamedly empirical. The modelling problem becomes one of trying to make the most of the data that are available; indeed, to learn from the data about how the system works by trying to relate a series of inputs to a series of outputs. This is data-based modelling, usually at the catchment scale, without making much physical argument or theory about process. Another name that is used is *black box modelling*. If we can successfully relate the inputs to the outputs, why worry about what is going on inside the catchment box? The black box analogy, however, is not necessarily a good one. An approach based on an input–output analysis may, in some circumstances, lead to very different conclusions about the operation of the system than an accepted theoretical analysis would suggest. The data are then suggesting that the theoretical analysis may be wrong and that some other physical interpretation may be necessary.

Young and Beven (1994; see also Young *et al.* 1997) have suggested an empirical approach which they call data-based mechanistic modelling, i.e. letting the data suggest an appropriate model structure but then evaluating the resulting model to see if there is a mechanistic interpretation that might lead to insights not otherwise gained from modelling based on theoretical reasoning. They give examples from rainfall–runoff modelling based on transfer functions that are discussed below in Section 4.5. This type of empirical data-based modelling necessarily depends on the availability of data. In the rainfall–runoff case, it will not be possible to use such models on an ungauged catchment *unless* the parameters for that catchment can be estimated a priori.

It is also worth noting that the inductive or empirical approach is very old; indeed it is the oldest approach to hydrological modelling. The Mulvaney *rational method* described at the beginning of Chapter 2, and the early coaxial correlation or graphical technique of Figure 2.1 are essentially examples of empirical modelling based on data; attempts to find some rational similarity in behaviour between different storms and different catchments. I suspect that in the new millennium there will be a return to forms of empirical model building and inductive reasoning in hydrology as the limitations of more theoretical approaches when applied to catchments with their own unique characteristics become increasingly appreciated (for a more detailed discussion, see Beven 2000).

4.2 Empirical Regression Approaches

Empirical regression is a basic statistical technique for the extrapolation of a data set to other situations in either time or space. Data are collated for a *dependent* variable and one or more *independent* or *explanatory* variables and statistical calculations are used to derive an equation relating variations in the dependent variable to values of the independent variables. A typical recent example, taken from Sefton and Howarth (1998), is a relationship between a parameter of the IHACRES model (see Section 4.3.1) and various physical catchment descriptor variables (PDC). The relationship, determined by multiple linear regression, takes the form

$$\frac{1}{c} = 7.7UPLAND + 12.9DECID + 7.4TILLED + 9.0URBAN + 7.3OTHER - 0.4CLASSB - 0.8REHPMN \quad (4.1)$$

where c is the proportion of rainfall contributing to catchment storage; *UPLAND* is the percentage area of heath, moor and bracken land use types; *DECID* is the percentage area of deciduous and mixed forest land use types; *TILLED* is the percentage area of arable crops; *OTHER* is the percentage area of the remaining three land use types; *CLASSB* is the percentage area of semi-permeable mineral soils (no groundwater), and *REHPMN* is the mean relative humidity averaged over the period 1961–1991. The relationship was developed from calibrating the model to 60 different catchments in England and Wales. The correlation coefficient for this relationship was 0.61

(explaining only 37 percent of the variance of the data), implying significant uncertainty in estimating the value of $1/c$ for any individual catchment. Similar equations for other parameters of the model had both better and worse fits to the data.

The independent variables in equation (4.1) were chosen from a list of 10 morphometric variables, 5 soil descriptors, 8 land use types; and 7 climate variables. The statistical fitting techniques used allow the identification of which independent variables add a significant contribution to the explanation of the observed variations in the dependent variable. Any variables that do not provide a significant contribution are not included in the equation. The technique used to derive this type of equation is a linear regression analysis. Any nonlinear relationships must be achieved by transforming either the dependent or independent variables. In one of the other predictive equations in the Sefton and Howarth paper, log transforms on all the variables were used before performing the analysis. This type of analysis is used widely in the estimation of model parameters or discharge characteristics within a region for ungauged catchments. What is often not recorded is what other transforms or other independent variables were tried and rejected in favour of the final published equation.

It would be nice if such relationships always made good hydrological sense. This is sometimes hard to extract from this type of regression equation. In equation (4.1), one is led to wonder about what processes an additive linear function of the mean relative humidity variable is acting as a surrogate for?

Several points are worth making about this type of regression analysis. One is that in an equation such as equation (4.1) which aims to predict the parameters of a model, the analysis is already one step removed from the original catchment data. Thus the values entered into the regression will depend on how well the model was fitted to the original data. There may be a particular problem in this study because two of the parameters show a very strong correlation across the 60 different catchments. Secondly, all such predictions should be associated with a standard error of estimation that allows an expression of the uncertainty associated with the prediction. This is not always done. A third point is that the standard errors of estimation will get larger as the predictions become more extreme in relation to the data set on which the regression was based. Care should be taken in using the resulting estimates, particularly where the correlation of the predictive equation is relatively low, as in equation (4.1). Individual catchments may then depart significantly from the regression estimate.

Similar regression equations are replete throughout the hydrological literature. Studies such as the UK *Flood Studies Report* (NERC 1975) provide many examples relevant to rainfall–runoff modelling, including equations for the peak flow and time-to-peak of a unit hydrograph and for calculating a percentage runoff in estimating an effective rainfall for use with the unit hydrograph. Similar analyses will be part of the new *Flood Estimation Handbook* (IH, 1999) and have been carried out in other parts of the world (e.g. Institution of Engineers, Australia 1977).

In any such empirical regression approach the explanatory variables are not chosen by chance. They will reflect the perceptions and physical reasoning of the analyst. The search is for variables that have explanatory power in this particular statistical context, so it is clearly sensible to try variables that are expected to be related to the dependent variable, whatever that might be. Sometimes the results might be surprising. Hewlett *et al.* (1984), for example, present an analysis of 4094 hydrograph events from 15 small drainage basins, over a range of climates and topographies, which suggests that rainfall intensity variables make only a small contribution to regression equations for peak flow and storm discharge volumes. The marginal coefficient of determination accounted for by maximum hourly rainfall intensity was of the order of 1 percent for storm volumes and 10 percent for the hydrograph peaks. Their conclusions reinforced an earlier study of Hewlett *et al.* (1977). Total storm rainfall has a much more significant effect on both dependent variables. Hewlett *et al.* conclude that these results are not consistent with the basis of many models of runoff generation in which hourly, or shorter period, rainfall intensities play an important role, particularly those models based on a Hortonian infiltration excess runoff generation mechanism. The conclusion is now starting to be supported by detailed modelling studies (see the R-5 catchment case study in Section 5.6).

4.3 Transfer Function Models

The type of empirical regressions described above are primarily aimed at extrapolating information measured at one site to other sites where the type of hydrological measurements required are not available but where the other variables required for the regression equations might be more easily measured. In some situations, we do have hydrological measurements available at a site and the problem is much more one of using a model to extrapolate that information for that site to other conditions.

In this section we will look at the case of having some rainfall–runoff time series data available for a catchment and using that information to model the response of the catchment. A modern empirical approach to this problem is the use of transfer functions relating an input to an output. The traditional unit hydrograph technique is a form of transfer function, as are the triangular and Nash cascade representations of the unit hydrograph shown in Figure 2.6. Modern approaches stem from work in linear systems analysis in which a general linear model is used to suggest an appropriate model structure compatible with the input–output data available (see Box 4.1). The resulting models may, however, have useful mechanistic interpretations. For example, evaluation of catchment transfer functions of the type described in Box 4.1 has frequently suggested that a parallel model structure is appropriate, with a proportion of the runoff being routed through a fast pathway and the remainder through a slow pathway (Figure 4.1). This does not allow any firm conclusions to be drawn about whether surface or subsurface flow processes are involved; it does allow some characteristic time constants for the catchment to be defined in terms of the mean residence times for the fast and slow flow pathways (see Box 4.1).

The problem in applying such methods to the rainfall–runoff system is that rainfall is related to stream discharge in a very nonlinear way. Many years of experience with the use of the unit hydrograph method in hydrological prediction have shown that storm runoff may be more linearly related to an 'effective' rainfall, but here we wish to avoid any need to carry out any prior separations of the rainfall and runoff time series since, as discussed in Section 2.2, hydrograph separation is a pretty desperate analysis technique. However, we can interpret this experience to suggest that it may be possible to use a linear transfer function model for calculating the time distribution of the total runoff

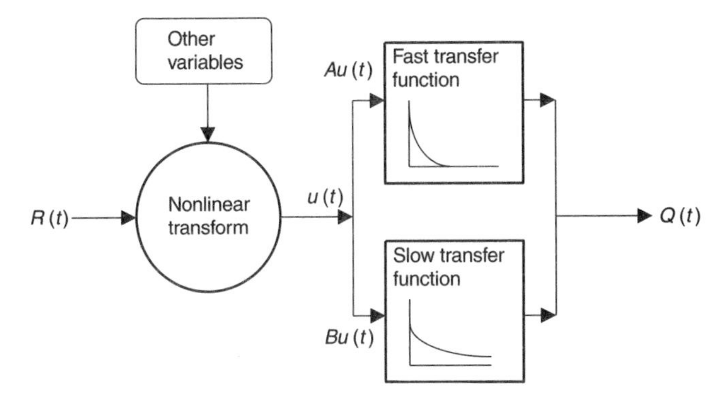

Figure 4.1 *A parallel transfer function structure and separation of a predicted hydrograph into fast and slow responses*

if we can find an appropriate nonlinear filter on the rainfalls to represent the runoff generation processes. The question then is how to find the appropriate form of filter.

One way is to simply assume that a certain form is physically reasonable and that constant parameter values can be found that give a good fit to the data throughout the calibration period. Early work on linear transfer functions of this type is reported in Dooge (1959). If there is truly a linear relationship between the transformed inputs and the measured output data then this should, in fact, be the case. It is the traditional approach used with unit hydrograph theory where the transformation from total rainfall to effective rainfall is based on an infiltration equation, or a method such as the Φ-index method (see Section 2.2). The transfer function based IHACRES model, described below, also adopts this strategy. However, if an estimate of a unit hydrograph is available for a catchment, it can be used in an inverse sense to estimate a pattern of effective rainfalls given a (previously separated) storm runoff hydrograph (see Box 4.2). Young and Beven (1994) extend this idea to the case of a time variable analysis of the transfer function parameters to allow the data to suggest what the form of the model should be (see Section 4.5 and Box 4.3).

4.3.1 The IHACRES Model

The IHACRES model (Identification of unit Hydrographs and Component flows from Rainfall, Evaporation and Streamflow data) of Jakeman *et al.* (1990) derives from the work of Young (1975) and Whitehead *et al.* (1979) which attempted to avoid the problem of hydrograph separation in classical unit hydrograph models by relating total rainfall to total discharge. Recent developments have been the result of a collaboration between the UK Institute of Hydrology (IH) at Wallingford and the Centre for Resource and Environmental Studies (CRES) in Canberra, Australia, a product of which is an IHACRES package for PC computers. The model uses a particular set of functions to filter the rainfall to produce an effective rainfall that is then related to total discharge using a generalized linear transfer function. The rainfall filter introduces a soil storage variable and also, for longer period simulations, uses temperature as an index of evapotranspiration. A number of different forms of rainfall filter have been used in different IHACRES applications (see Jakeman *et al.* 1993; Jakeman and Hornberger 1993; Post and Jakeman 1996; Sefton and Howarth 1998). One form, used in the Sefton and Howarth study discussed in Section 4.2 above, is as follows. If the rainfall input at time step t is denoted as R_t, while the effective rainfall is denoted as u_t, then

$$u_t = R_t(S_t + S_{t-1})/2 \tag{4.1}$$

$$S_t = cR_t + \left[1 - \frac{1}{\tau(T_i)}\right] S_{t-1} \tag{4.2}$$

$$\tau(T_i) = \tau_w \exp(10f - T_i f) \tag{4.3}$$

where S_t is the storage variable at time t, $\tau(T_i)$ is a mean residence time for the soil storage depending on mean daily temperature T_i, c controls the proportion of rainfall contributing to catchment storage, τ_w is the mean residence time for soil storage at 10°C and f is a scaling parameter to allow for the relationship of evapotranspiration effects to this temperature difference. In many respects, this part of the IHACRES model represents a simplified form of ESMA (explicit soil moisture accounting) model (see Section 2.4). The effective rainfall, u_t, then forms the input to a transfer function analysis based on the generalized linear models of Box 4.1 with the total discharge as the output. The parameters of the complete model are calibrated by fitting transfer functions to different values of c, τ_w and f until the best results are achieved. Standard errors and covariances for the transfer function model parameters can be estimated, but uncertainty in the c, τ_w and f parameters has not generally been considered.

The IHACRES model has now been applied to a wide variety of catchments (Jakeman *et al.* 1990, 1993a; Jakeman and Hornberger 1993; see for example Figure 4.2), including catchments subject to significant snowmelt inputs (Schreider *et al.* 1997; Steel *et al.* 1999) and in predictions of the impacts of climate change on catchment hydrology (Jakeman *et al.* 1993b; Schreider *et al.* 1996). The model has also been linked to erosion and water quality components (Jakeman *et al.* 1999). The results generally show that the parallel transfer function of Figure 4.1 is a suitable structure for rainfall–runoff simulation at the catchment scale, with one fast flow pathway and one slower flow pathway (see Box 4.1). The fast flow pathway will provide the major part of the predicted storm hydrograph, the slower pathway the major part of the recession discharge between storm periods. Note again that this does not imply anything about whether surface or subsurface flow processes, old or new water, direct flow or displacement are involved in the fast and slow flow responses. The fast flow pathway should not be interpreted as a surface flow pathway; it could equally be a response controlled by the time-scale of displacement of old water from subsurface storage. This type of parallel pathway routing is also used in other models (such as the Xinanjiang/Arno/VIC model described in Box 2.2, and the Institute of Hydrology PDM model described in Section 6.2); the advantage of the IHACRES approach is that the data are allowed to suggest the form of the transfer function used rather than specifying a fixed structure beforehand.

The IHACRES model has the right sort of functionality to reproduce hydrological responses at the catchment scale with about the right number of parameters for those parameters to be identifiable given a period of calibration data, at least for some environments. The parameters required to apply the model are essentially the two time constants of the fast and slow pathways in the parallel transfer function, the proportion

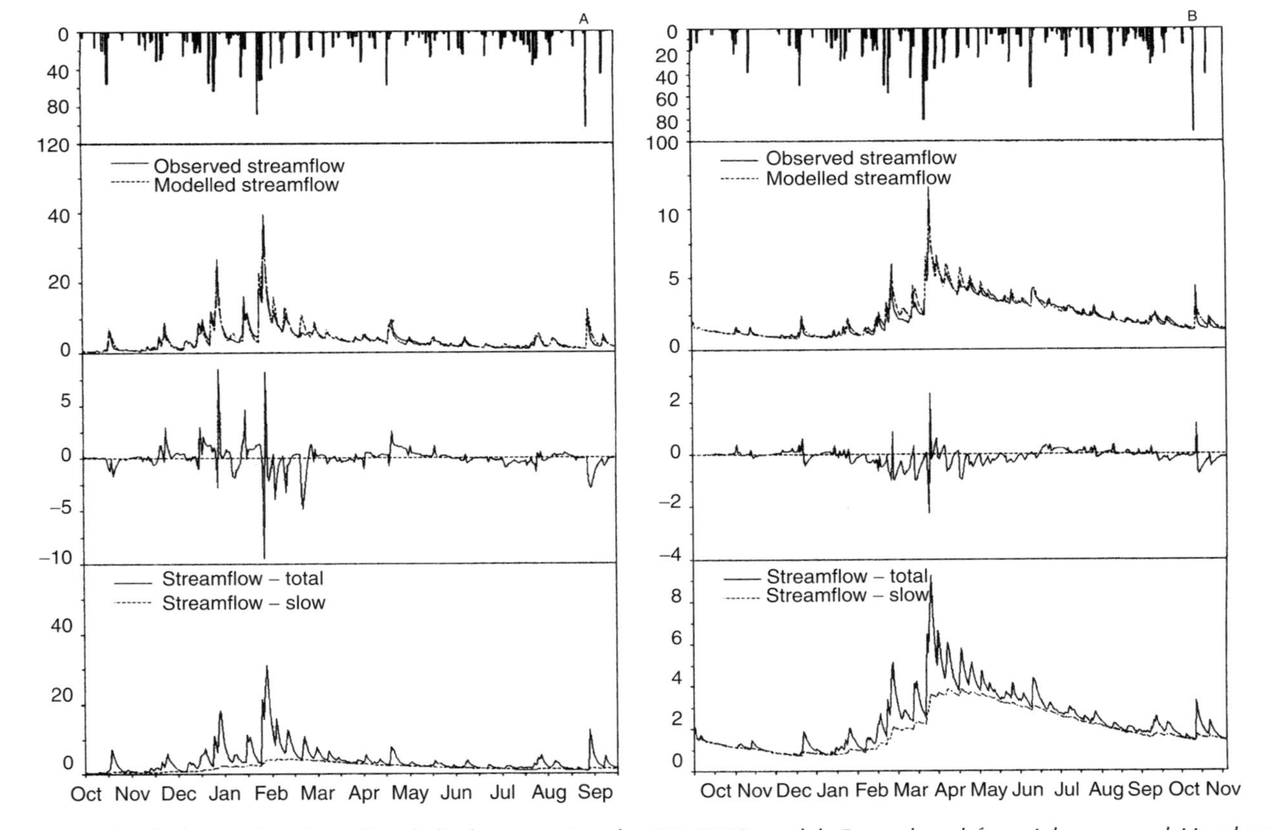

Figure 4.2 *Example of observed and predicted discharges using the IHACRES model. Reproduced from Jakeman and Hornberger, Water Resources Research 30(12): 3567, 1994, copyright by the American Geophysical Union*

of the effective rainfall following each pathway, and the c, τ_w and f parameters of the effective rainfall filter. Jakeman and Hornberger (1993) suggest that these parameters can be considered as the dynamic response characteristics (DRCs) of a catchment and that it might be possible to relate these DRCs to physical catchment descriptors.

This was partially successful in the study of Post and Jakeman (1996), who derived DRC parameters for 16 small catchments in Victoria, Australia. They found that five of the six parameters required by the model could be reasonably well related to physical characteristics of the catchments. The time constant of the fast runoff component was related to drainage density and catchment area; that of the slow component to the slope and shape of the catchment; the temperature modulation parameter, f, to gradient and vegetation type; and the c and τ_w parameters to drainage density and gradient. The one parameter that showed no clear relationship to catchment characteristics was the proportion of effective rainfall going to the fast flow pathway. In their particular study this had, in any case, a very limited range of optimized values.

Sefton and Howarth (1998) used different catchment descriptors and produced different regression relationships for the IHACRES parameters for catchments in England and Wales (including equation (4.1) above). In their data, it was the mean residence time of the slow flow pathway that was least well defined by the DRC variables, while f and the proportion of the effective rainfall going to the fast flow pathway were best defined. It seems that it may be difficult to generalize about such relationships. In Section 4.2, a dependence on the original calibration of the parameters to hydrograph data was also noted and it may also be that there is sufficient interaction in calibrating the different DRC parameter values to limit the extent to which they can truly be considered as unique hydrologically significant descriptors of a catchment. There is also the question of whether the nonlinear rainfall filter used is, in fact, hydrologically appropriate, especially since Jakeman and his co-workers have used more than one in their series of papers on the IHACRES approach. It might be better to let the data suggest an appropriate form for the nonlinearity, as discussed in the next section.

4.3.2 Data-Based Mechanistic Models Using Transfer Functions

The data-based mechanistic (DBM) approach of Young and Beven (1994), as far as possible, makes no prior assumptions about the form of the model other than that a general linear transfer function approach can be used to relate an effective rainfall input to the total discharge. In the spirit of letting the data determine what the structure of the model should be, rather than making *ad hoc* prior decisions about model structure, they make use of time variable parameter estimation to determine the form of the effective rainfall nonlinearity. Their results suggest a nonlinear rainfall filter of the form

$$u_t = Q_t^n R_t \tag{4.4}$$

where u_t is the effective rainfall, R_t is the rainfall input, Q_t is discharge, n is a parameter and t is time. Here discharge is being used very much as a surrogate variable for the antecedent moisture status of the catchment. In general, the measured discharge is the best readily available index of antecedent conditions in a catchment, but its use in this way does mean that discharge is being used in the prediction of discharge. This is not a problem in model calibration at gauged sites; it will be a problem in prediction or real-time forecasting but does not, in fact, turn out to be a difficult problem to overcome (see below). In the original application of Young and Beven (1991) the bilinear model ($n = 1$) was used, but more recent results using time variable parameter estimation (see Box 4.3) have suggested values of n of between 0 and 1 (a bilinear power law model). The application of the DBM approach is demonstrated in the case study of the next section.

In a further application of these techniques to a longer period of data for a catchment in the eastern United States, Young and Beven (1994) have demonstrated that there is a seasonal pattern in the time variable estimates of the transfer function a parameter (and hence the mean time residence time) as well as a similar correlation of the time variable estimates of the gain parameter b with discharge. It was found that the seasonal pattern in a could be correlated with daily mean temperatures. The correlation for this site is, however, interesting in that it suggests that the a parameter changes in an inverse relationship with temperature. This implies that the higher the temperature, the longer the mean residence time. This is also physically reasonable if higher temperatures are an indication of greater summertime evapotranspiration and therefore lower levels of moisture storage and consequently slower response time.

This approach results in a minimal model relating total rainfall to total discharge, with gain and power parameters for the nonlinear filter and one or two time constants for the transfer function. These parameters may be considered to be bulk 'physical' characteristics at the catchment scale but, as in the study of the dynamic response characteristics of the IHACRES model by Post and Jakeman (1996), it is not yet clear how well these parameters might be related to catchment characteristics nor the range of catchments for which such a simple model might be appropriate. However, it is clear that good estimates of the parameters can be obtained from only short periods of rainfall–runoff data so that a period of field measurement of rainfalls and discharges at a site of interest might be the best way of calibrating the parameters.

It is worth noting that one of the features of the linear time series analysis used to derive the transfer function is that standard errors can be estimated for the parameters. These standard errors can be used to evaluate the physical interpretation of the model. Young (1992), for example, has investigated the sensitivity of the proportion of effective rainfall going through the fast and slow flow pathways to error in the estimated parameters. His results suggest that there will be considerable uncertainty in these proportions, limiting any interpretation in terms of fast runoff contributing areas. At least in this type of model these uncertainties can be made explicit; there are similar implications for the more complex models that will be discussed in Chapters 5 and 6, but such sensitivities are then seldom evaluated.

4.4 Case Study: DBM Modelling of the CI6 Catchment at Llyn Briane, Wales

The data-based mechanistic modelling approach will be described in relation to a specific application, the CI6 catchment at Llyn Briane in Wales (Young and Beven 1994).

The steps are as follows:

1. Fit a preliminary transfer function to the input–output data using the techniques described in Box 4.1. One or more rainfall filtering models might be tried at this

stage; the important thing is to derive an estimate of the parameters of a low-order transfer function.

2. Examine the residuals, including tests for nonlinearity. If there is no nonlinear behaviour apparent then the fitted model can be accepted; otherwise go on to the next step.
3. Calibrate a low-order transfer function (first or second order will normally be adequate for rainfall–runoff modelling) using a time variable parameter estimation technique, such as the fixed interval smoothing (FIS) described in Box 4.3. Examine the variation in the parameter values, including the standard errors, through the period of the calibration (e.g. Figure B4.3.1 for the CI6 catchment).
4. Examine the nature of the time variation in the parameters, taking account of the accuracy with which the parameters are estimated. For the CI6 application, it is clear that there is only information added in estimating the b parameter when there is rainfall. The standard error of estimation on this parameter increases rapidly during recession periods.
5. Evaluate the nature of the variation in relation to other variables. In this case, a relation between this parameter and discharge can be found if only the most significant values of the b parameter estimates are used (Figure 4.3). This is physically reasonable, as with the simple bilinear rainfall filter previously used by Young and Beven (1991), in that we would expect higher discharges to suggest wetter antecedent conditions, resulting in a greater proportion of the rainfall becoming discharge. This analysis suggests, however, that the rainfall filter should not be the simple multiplication by discharge used in Young and Beven (1991), but a multiplication by discharge raised to a power, here of the order of 0.65. A block diagram of the resulting model is shown in Figure 4.4.
6. Optimize the resulting model. The results are presented in Figure 4.5 as output from the TFM program (see Section 4.5). The best fit for this catchment is obtained with a power of about 0.63. This is a slightly better model with different time constants (3.95 h and 80.2 h) to those of the earlier bilinear (power = 1.0) model fit for this catchment.
7. Examine the residuals for evidence of further structure or nonlinearity. Here Young and Beven (1994) show that although the resulting nonlinear filter and transfer function explain more than 98 percent of the variance for this period of observed discharges, a third-order autoregressive model of the residuals can be used to explain more than half of the remaining error variance, but there is no evidence of further nonlinearity. Such a model of the residuals might be useful in flood forecasting (see Chapter 8).

The bilinear power law filter with parallel transfer function model has an interesting interpretation. With this model the proportion of the rainfall that goes through the fast flow pathway is $b'Q^n$. Without implying too much about the processes involved, this could be considered as a representation of a contributing area function for the fast responses in the catchment, whether the fast responses be due to surface or subsurface flow processes. The contributing area increases with discharge for all $n > 0$.

Although of some interest, this approach still requires further evaluation since it has no explicit account of seasonality of responses except in so far as seasonality is reflected in the contributing area filtering based on discharge. In another application of the

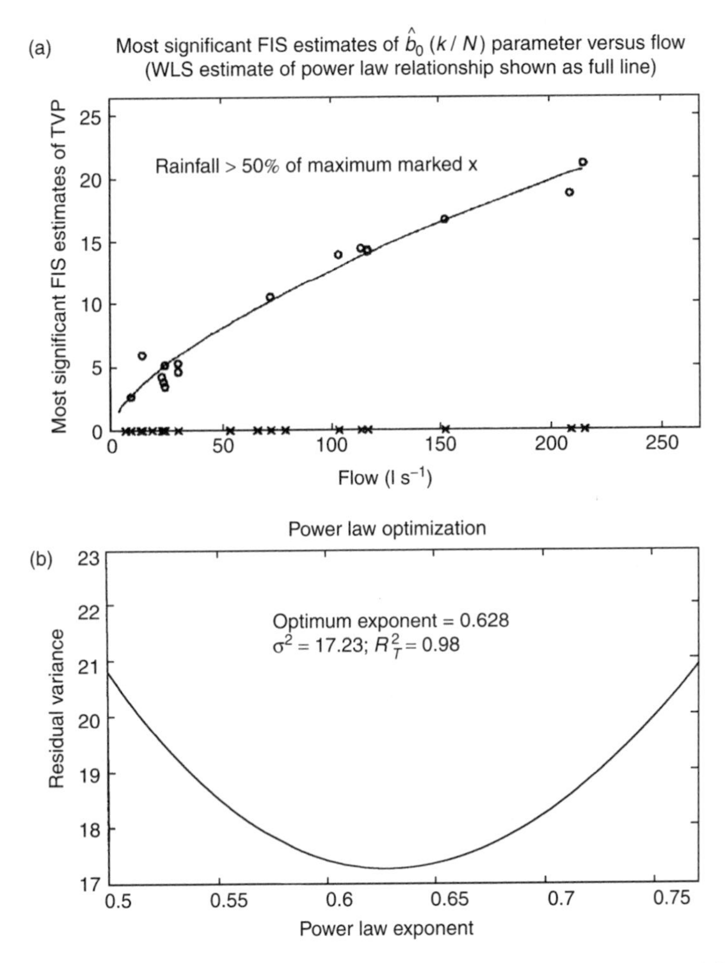

Figure 4.3 *(a) Time variable estimates of the gain coefficient in the bilinear model for the CI6 catchment plotted against the discharge at the same time step. (b) Optimization of the power law coefficient in fitting the observed discharges (after Young and Beven 1994). Reproduced with permission of John Wiley & Sons Limited*

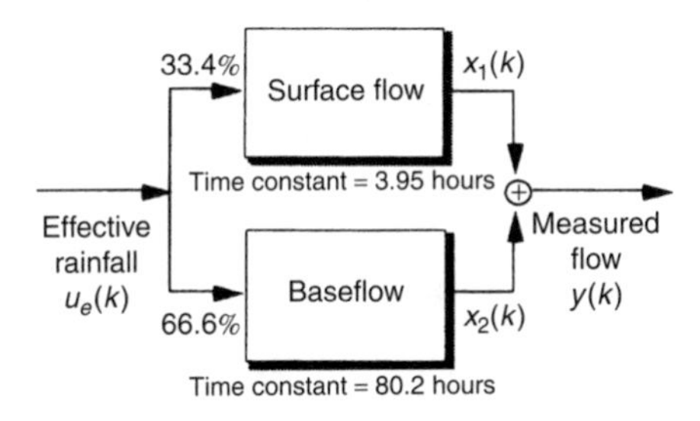

Figure 4.4 *Final block diagram of the CI6 bilinear power law model used in the predictions of Figure 4.3 (after Young and Beven 1994). Reproduced with permission of John Wiley & Sons Limited*

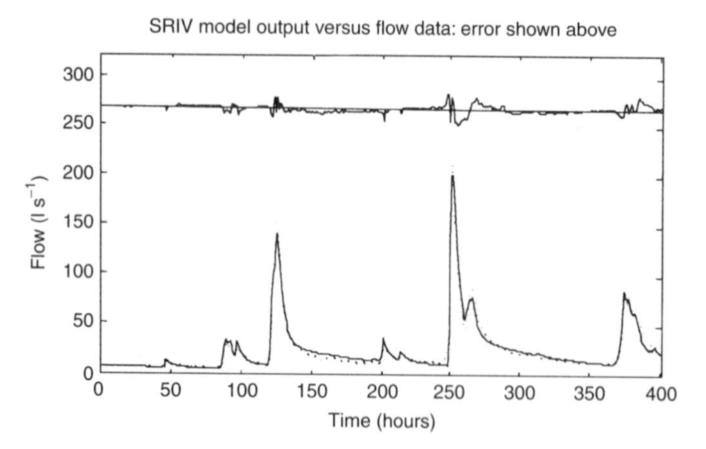

Figure 4.5 *Example of observed and predicted discharges for the CI6 catchment at Llyn Briane, Wales, using the bilinear power law model with* n = *0.628 (after Young and Beven 1994). Reproduced with permission of John Wiley & Sons Limited*

bilinear power law model to the Canning River in Australia, Young *et al.* (1997) have shown that this approach can explain 95.8 percent of the variance of daily discharges over a two-year period, including dry summer periods. A split record test of the fitted model over two further years of data produced almost as good fits (88.9 and 92.4 percent). The IHACRES model does take account of seasonal wetting and drying by including a temperature variable as an input to the nonlinear effective rainfall filter component. Young and Beven (1994), in a second example of modelling daily discharges for one of the Coweeta catchments, also show that temperature can be used as a surrogate variable to improve the accuracy of longer period simulations.

4.5 The TFM Software

One of the software packages introduced in Appendix A is the TFM package which provides a simple implementation of the transfer function modelling methods described above. It has been set up specifically for rainfall–runoff modelling for short periods of data and includes two different types of effective rainfall filtering, as well as an algorithm for fitting different linear transfer function structures. The two filtering methods are a soil storage approach, similar to the IHACRES model but without any temperature compensation; and the bilinear power law approach of the Young and Beven data-based mechanistic approach. The test data set supplied is the data for the CI6 catchment case study.

4.6 Nonlinear and Multiple Input Transfer Functions

The single effective rainfall input linear transfer function models described above generally work very well if some data are available for calibration, although if the form of the transfer function is allowed to be flexible in calibration, the identified model does not always have the parallel form noted above. There is also clearly some uncertainty about the appropriate form of the nonlinear filtering that is used to create the effective rainfall to be routed through the transfer function. Early concerns about the linearity of hydrological responses resulted in a number of attempts to formulate a nonlinear transfer function (e.g. Amorocho and Brandstetter 1971; Diskin and Boneh 1973). These early attempts were based on the use of Volterra series (see also the recent studies of Ahsan and O'Connor 1994; G. C. Liang *et al.* 1994). More recently, a new methodology based on an extension of the class of generalized linear models to the nonlinear case using NARMAX (nonlinear autoregressive moving average with exogenous inputs) models has been applied to hydrological problems. Tabrizi *et al.* (1998) demonstrate how NARMAX models can be used for both single input and multiple input cases.

Another way of addressing the nonlinearity problem is in terms of nonstationary or time variable transfer function parameters (see Box 4.3). This is essentially the approach used in the DBM modelling strategy described above, where the power law function was used to encapsulate the parameter variations revealed by the time variable analysis. Recent unpublished work by Peter Young based on applying the fixed interval smoothing to sorted discharge ordinates suggests that it may be possible to make the nonlinear filtering of the rainfall even more flexible in future.

Another ongoing issue in TFM modelling is the use of multiple rainfall input series in modelling discharges, derived either from multiple raingauges in a catchment area or from radar rainfall data. Multiple-input transfer functions have been proposed, for example, by G. C. Liang *et al.* (1994), Tabrizi *et al.* (1998) and Kothyari and Singh (1999). A major problem with these approaches is the correlation to be expected among the multiple inputs. In the general case there may be no unique solution to the multiple-input single-output problem (Cooper and Wood 1982) and a robust identification of the parameter values may be difficult. One solution to this problem has been proposed by Cooper and Wood (1982) using canonical correlation to identify an appropriate model structure and maximum likelihood estimation to identify the required parameters.

4.7 Physical Derivation of Transfer Functions

In the previous two sections the transfer functions have been fitted to the data using the generalized linear model developed in Box 4.1. It is possible, however, to develop a transfer function based on the form of a catchment, in a way similar to the Clark (1945) time–area diagram interpretation of the unit hydrograph. We will consider two more recent types of transfer function based on catchment form, one based on the network width function, the other the geomorphological unit hydrograph. Note, however, that both of these approaches address only the routing problem and not how much of the rainfall to route. Thus, both require prior estimation of effective rainfalls but can be used with a variety of effective rainfall models, including those of Chapter 2 and the nonlinear filters described earlier in this chapter. In this respect they are in the classical unit hydrograph tradition.

4.7.1 Using the Network Width Function

The routing of runoff in catchments is a function of both the hillslope and channel responses. Work by Kirkby (1976) and Beven and Wood (1993) has demonstrated how the time delays in small catchments tend to be dominated by the routing of surface and subsurface flows on the hillslopes, while in large catchments routing in the channel network will play the dominant role in shaping the hydrograph, especially during overbank flow conditions. If the hillslope runoff inputs to the channel network are distributed along the reaches of the network, then at least in large catchments the shape of the hydrograph should reflect the form of the network. This is the idea behind using the network width function to derive a transfer function for runoff in the network. The width function is formed by counting the number of channel reaches at a given distance away from the outlet (see Figure 4.6). Different network shapes will give different width functions. Under assumptions of a constant wave velocity in the network (which does not imply that the flow velocity must be everywhere constant; see Beven (1979) and the section on kinematic wave models of surface flows in Chapter 5), the width function can be used directly as a transfer function for routing runoff inputs into the channel.

This type of routing algorithm has been used, for example, in the TOPMODEL software of Chapter 6. It has the advantage that it requires only the network width function, which can be derived directly from maps or digital terrain data, and a single parameter, the wave speed in the channels. It has the disadvantage that it does not deal explicitly with routing on the hillslopes, and that there is no dispersion of the form of the width function in routing the flow to the outlet since, for a unit input everywhere along the channel, the shape of the resulting hydrograph will directly reflect the width function. This would be obscured, of course, if different patterns of runoff were being produced in different parts of the catchment. Both of these limitations can be relaxed. It is not difficult in an analysis of digital terrain data to derive a distance to the nearest channel for every point on the hillslopes of a catchment, so that this type of approach can be extended to routing surface runoff (at least) to the channel, perhaps using a different velocity. Secondly, Mesa and Mifflin (1986) and Naden (1992) have shown that a diffusive routing algorithm in each reach can be implemented relatively easily at the expense of introducing an additional parameter.

Diffusive network width function algorithms have also gained some recent popularity within the macroscale hydrology field to allow the routing of runoff at the continental scales in a computationally efficient way. The model of Naden (1992) was applied to the Thames and Severn catchments in Naden (1993) and to the Amazon and Arkansas–Red River catchments using GCM-generated rainfall inputs in Naden *et al.* (1999). It is also included as a component of the UP macroscale model discussed in Chapter 9. At this scale, it will generally be necessary to estimate the routing parameters required by the model. For a uniform channel, the effective wave velocity, c, and the dispersion parameter, D, may be approximately related to the characteristics of the flow at a site as follows:

$$c = \tfrac{3}{2}v_o \tag{4.5}$$

$$D = \frac{q_o}{2S_o}\left(1 - \frac{{F_o}^2}{4}\right) \tag{4.6}$$

where v_o is the mean velocity at a reference discharge q_o in a channel of bed slope S_o and Froude number F_o. In a large basin, the discharge characteristics and

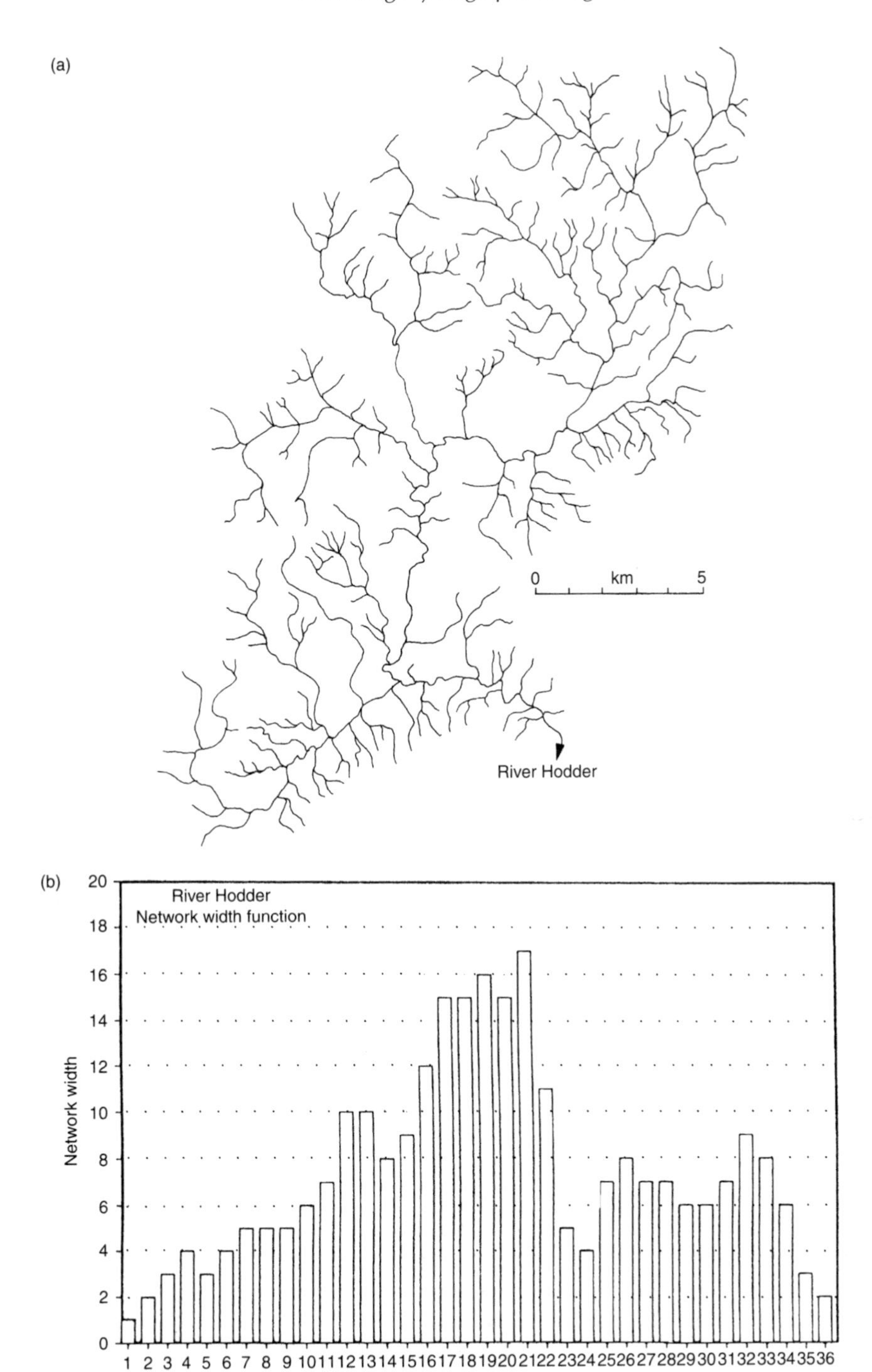

Figure 4.6 *(a) Network and (b) Network width function for River Hodder catchment (250 km², UK (after Beven and Wood 1993). Reproduced with permission of John Wiley & Sons Limited*

channel dimensions will change downstream. Work by Snell and Sivapalan (1995) and Robinson *et al.* (1995) showed how effective values for these parameters over the whole network could be related to the at-a-site and downstream hydraulic geometry of the channels. The parameter estimates have to be considered very approximate and further work needs to be done to both evaluate the routing procedures and improve them by taking full account of the variation in the parameters within these large networks.

4.7.2 The Geomorphological Unit Hydrograph (GUH)

Related to the use of the network width function as a transfer function for routing runoff is the geomorphological unit hydrograph concept. This idea was initiated by Ignacio Rodriguez-Iturbe in a series of papers (summarized in Rodriguez-Iturbe 1993) that explored the cause–effect linkages between hillslope form, runoff production, channel growth and network development. The network is a reflection of the runoff-producing mechanisms of the hillslopes operating over a long period of time, but development of the network has a feedback effect on the form of hillslopes and consequent runoff production. These geomorphological linkages result in structural regularities in the form of catchments and it should be possible to take advantage of these regularities in making hydrological predictions. The regularities have been studied by geomorphologists for a long time and are summarized in Horton's laws that express the expected relationships between channel numbers, upstream areas, lengths and slopes, for different orders of channel where, in the Strahler ordering system, a first-order stream is a stream with no upstream junctions (an exterior link on the network); a second-order stream is formed by the junction of two first-order streams (creating an interior network link), and so on (Figure 4.7).

The geomorphological unit hydrograph is then developed by considering the probability of a 'raindrop' contributing to a stream of a given order, and considering a 'holding time distribution' for both the hillslopes and the streams of each order. The important part of the theory is the way in which it uses Horton's laws to determine the probability of a drop contributing to each order of stream and the links between the streams of different orders. A third-order stream, for example, will have some direct contribution from local hillslopes, a contribution from the second-order streams feeding it from upstream, and the possibility of a contribution from additional first-order streams. The complete theory is complex and depends on making some simplifying assumptions about the holding time distributions for mathematical tractability. Thus, starting with Rodriguez-Iturbe and Valdes (1979), there have been attempts to simplify the resulting predictions in terms of functional forms of the unit hydrograph. They derived expressions for the time-to-peak and peak discharge of a triangular unit hydrograph; Rosso (1984) did the same for the gamma probability density function (i.e. the mathematical form of the Nash cascade discussed in Section 2.3). It was later shown by Chutha and Dooge (1990) that all forms of GUH that assume an exponential holding time in each reach of the network must, under reasonable geomorphological assumptions, produce a GUH that is close to a gamma distribution in shape. Recall that the gamma probability distribution unit hydrograph is defined by two parameters, N and K (equation 2.2) where in the Nash cascade, N is the number of linear stores in series (which does not have to be an integer number), each of time constant K. Rosso (1984)

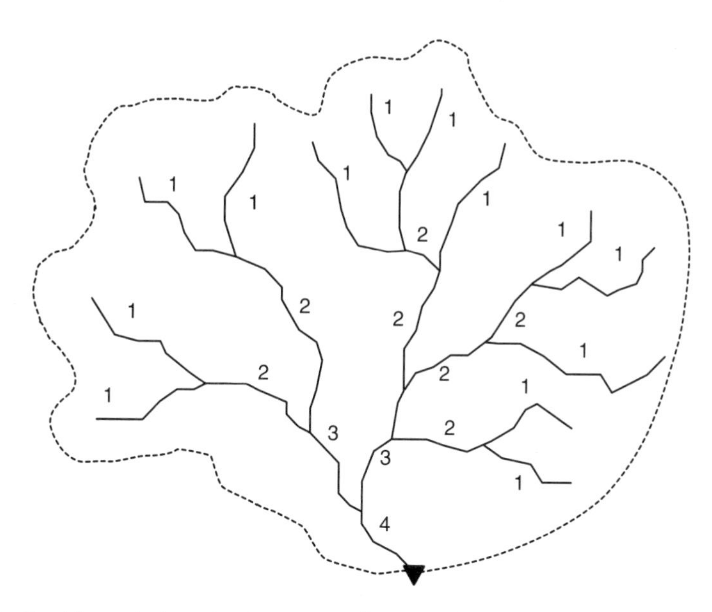

Figure 4.7 *Strahler ordering of a river network as used in the derivation of the geomorphological unit hydrograph*

showed that the two parameters N and K could be related to the geomorphological structure of the channel network as

$$N = 3.29 \left[\frac{R_A}{R_B} \right]^{0.78} R_L^{0.07} \tag{4.7}$$

and

$$K = 0.70 \left[\frac{R_A}{R_B R_L} \right]^{0.48} L_\Omega v^{-1} \tag{4.8}$$

where R_B is called the bifurcation ratio of the network and is equal to the ratio of the number of streams of order $\overline{\omega}$ to that of order $\overline{\omega}+1$; R_A is the area ratio of the network and is equal to the ratio of the average catchment area of streams of order $\overline{\omega}$ to that of order $\overline{\omega}+1$; R_L is the length ratio of the network and is equal to the ratio of the mean length of streams of order $\overline{\omega}$ to that of order $\overline{\omega}+1$; L_Ω is the length of the highest order stream in the network and v is an average stream velocity.

There have now been a large number of studies that have used the GUH in various predictive contexts, from hydrograph prediction, flood frequency predictions, solute transport predictions, predicting the impacts of climate change, and predicting catchment sediment yields (useful summaries can be found in Rodriguez-Iturbe 1993; Rodriguez-Iturbe and Rinaldo 1997). Note that, as well as the bifurcation, area and length ratios of the network, there is still a velocity parameter to be calibrated, in the same way that the constant velocity network width function approach of the last section also requires a velocity parameter to be specified. This is commonly calibrated using effective rainfall and storm hydrograph data, but the values derived will depend on the formulation of the GUH used. Al-Wagdany and Rao (1998), for example, have shown that calibration of three different formulations of the GUH results in different, but correlated, velocity values. In fact, given the ease with which a full channel network can now be derived from GIS or digital terrain data, there does not seem to be all that much advantage in representing the network in terms of its geomorphological ratios over using the full network structure. Some of the detailed information about the network will be lost in the GUH ratios since Horton's laws are only approximate relationships. Beven (1986a) has discussed the use of the network width function approach as an alternative to the geomorphological unit hydrograph (see also Naden 1992; Naden *et al.* 1999) while Gandolfi *et al.* (1999) have used a triangular approximation to the distribution of contributing area to the channel network as the basis for a routing model. Nash and Shamseldin (1998) have recently suggested that the holding time assumptions of the GUH approach may be overly restrictive in shaping the form of the unit hydrograph and that little is added by the geomorphological ratios to the original Nash cascade unit hydrograph. They suggest that the GUH theory must still be considered as a hypothesis that has not been adequately tested.

4.8 Using Transfer Function Models in Flood Forecasting

Flood forecasting is one of the most important applications of rainfall–runoff modelling. Flood forecasting requires decisions to be made as to whether flood warnings should be issued on the basis of the data coming in from raingauges, radar rainfall images and stream gauges, and the model predictions as the event happens in 'real time'. The requirement is for those forecasts and warnings to be made as accurately as possible and as far ahead or with the greatest *lead time* as possible. We can use a model for these predictions that has been calibrated on historical data sets but we will often find that during a flood event the model predictions of river stage or discharge will start to deviate from those values being received online at a flood forecasting office from the telemetry outstations. Thus, in any flood forecasting situation it is useful to have methods that both allow for *real-time updating* of the forecasts as the event proceeds, and give an indication of how uncertain the forecasts might be.

Forecasting methods based on transfer function modelling are ideally suited to these requirements and have been quite widely implemented (e.g. Moore *et al.* 1990; Sempere Torres *et al.* 1992; Cluckie 1993). An example application is provided by Lees *et al.* (1994) who describe the implementation of a flood forecasting system for the town of Dumfries on the River Nith in Scotland, which is described in more detail as a case study in Chapter 8.

4.9 Empirical Rainfall–Runoff Models Based on Neural Network Concepts

An alternative and recent approach to data-based rainfall–runoff modelling is the use of artificial neural networks. Neural networks stem from research in artificial intelligence

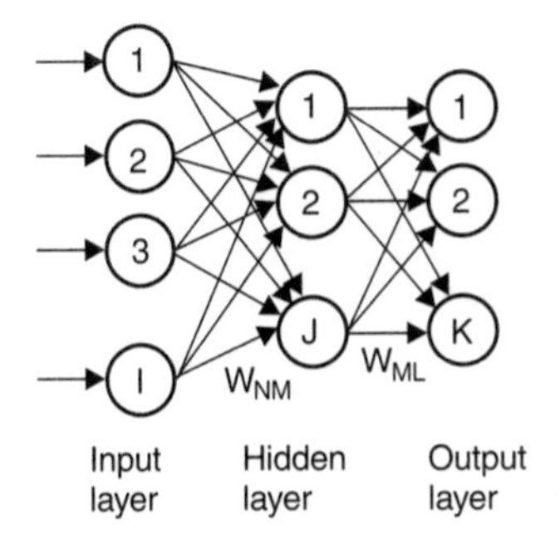

Figure 4.8 *The structure of a neural network showing input nodes, output nodes and a single layer of hidden nodes. Each link is associated with at least one coefficient (which may be zero)*

as a simple attempt to mimic the workings of the brain in terms of nodes connected by neurones. The simplest neural networks relate an input signal (here we are interested in inputs comprising rainfall and past discharge measurements) to an output signal (current or future discharge) by means of a series of weighting functions that may involve a number of layers of interconnected nodes, including intermediate 'hidden layers' (Figure 4.8). Some applications have used additional filtering functions (essentially simple transfer functions) for each node in a hidden layer, so that the outputs will also depend on the form and parametrization of these functions. A variety of techniques are available for determining the appropriate model structure and weights given a *learning set* of input and output data.

There is a clear analogy between the neural network weights and the parameters of other modelling approaches, and between the learning set and what we have before called a period of calibration data. Work in neural networks often does not draw this analogy but it is a useful one in that just as an increase in the number of parameters gives a model more degrees of freedom in calibration but may result in overparametrization with respect to information in the data set, so in a neural network an increase in the number of layers, nodes and interconnections will also result in more degrees of freedom in fitting the learning set, also with the possibility of overparametrization.

A number of studies of the rainfall–runoff problem using neural network techniques have been published (e.g. Lek *et al.* 1996; Minns and Hall 1996; Dawson and Wilby 1998; Fernando and Jayawardena 1998; Tokar and Johnson 1999). For the most part these models have been produced for the purposes of N-step ahead forecasting rather than simulation over long periods. The availability of previous water level or discharge data as an input to the neural net is generally important to the success of such modelling since it allows some of the nonlinearity of the rainfall–runoff process to be reflected in the net for short-term forecasting. Campolo *et al.* (1999), using a neural net based on multiple rainfall and discharge inputs, demonstrate a good performance of up to 5 hour ahead forecasts for the 2480 km^2 Tagliamento catchment in Italy (Figure 4.9). In forecasting, neural network models can also be used in the same way as transfer

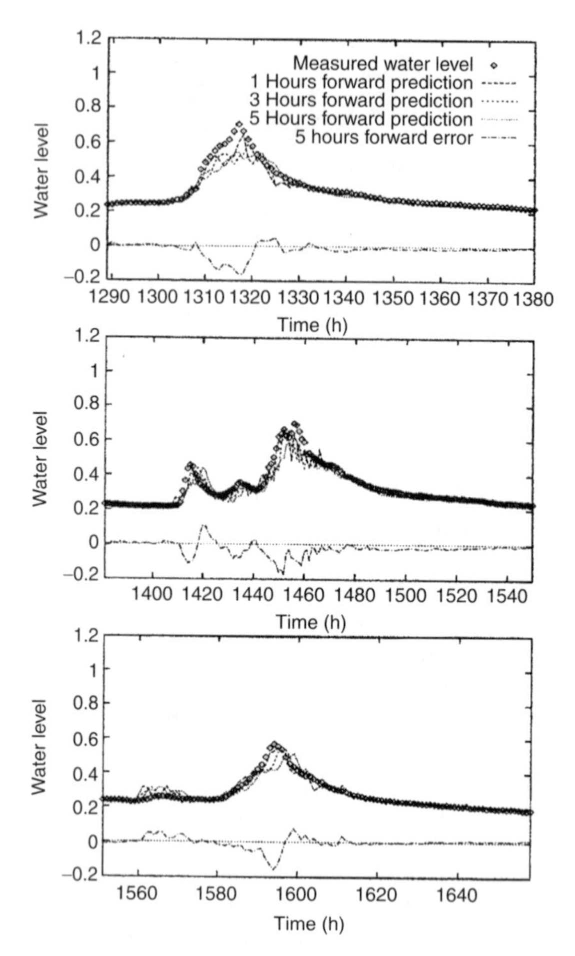

Figure 4.9 Application of a neural network to rainfall–runoff modelling for flood forecasting in the River Tagliamento catchment, northern Italy (1950 km^2). 1, 3 and 5 hour ahead forecasts are based on input data of lagged rainfalls from five sites over the previous 15 time steps and discharges from the previous five time steps (after Campolo et al. 1999). Reproduced from Water Resources Research 35: 1191–1197, 1999, copyright by the American Geophysical Union

functions to forecast downstream river levels or discharges given some upstream data (e.g. Thirumalaiah and Deo 1998). One study has used neural networks to predict the two-year return period peak discharge for ungauged basins as an alternative to a multiple regression approach (Muttiah *et al.* 1997).

Overparametrization of neural net models relative to the information in the learning set is an issue with this type of model (as in any empirical model). The danger of overparametrization is that in general it will lead to greater uncertainty in prediction or extrapolation, particularly in prediction or extrapolation beyond the range of the learning or calibration set. A good performance in fitting the learning set does not guarantee a good performance in prediction when the conditions go outside the range seen in the learning set.

4.10 Key Points from Chapter 4

- This chapter has dealt with models for the rainfall–runoff process derived directly from data, without explicit consideration of the processes involved.
- Empirical relationships based on regression analysis have been widely used in hydrological prediction. Regression relationships should always be associated with an estimate of the uncertainty associated with the predictions of the dependent variables.
- Modern transfer function techniques, an extension of the unit hydrograph approach, can be used to derive catchment-scale parameters based directly on the analysis of the observations. Transfer function techniques require a nonlinear transformation of the rainfall and the results will depend on the form of transformation adopted. In the data-based mechanistic approach of Young and Beven (1994), a flexible approach to model structure is adopted in which time variable parameter estimation is used to suggest a form for the nonlinear transformation.
- The resulting transfer functions are often parallel in form, with one fast flow pathway and one slow pathway. This does not directly imply any interpretation in terms of flow processes, and sensitivity analysis shows that estimates of the proportion of effective rainfall following each pathway may be subject to significant uncertainty.
- Transfer functions may also be derived directly from the structure of the channel network in the catchment. The use of the network width function and the geomorphological unit hydrograph are discussed. Both only address the problem of routing an estimate of effective rainfall.
- Other data-based rainfall–runoff modelling methods include the use of neural networks to relate inputs to outputs. A neural network model can sometimes be interpreted in terms of a transfer function, but because of the possibility of large numbers of weights (parameters) to be identified for all the linkages in the network, such methods may not be accurate when predicting outside the range for which they have been developed.

Box 4.1 Linear Transfer Function Models

The Building Block: the First-Order Linear Store

A linear store is a model element for which the predicted output, Q [$L^3 T^{-1}$], is directly proportional to the storage, S [L^3](see Figure B4.1.1). Thus, we assume

$$Q = S/T \tag{4.9}$$

where T [T] is a parameter equivalent to the mean residence time of the store. For water, the linear store is physically equivalent to a straight-sided bucket with a hole in the bottom, allowing the storage in the bucket to escape as an output Q.

The mass balance equation for the linear store (or bucket) can be written as

$$\frac{dS}{dt} = u - Q \tag{4.10}$$

where the differential dS/dt is the rate of change of storage with time and u [$L^3 T^{-1}$] is an input rate (here an effective rainfall). To obtain an equation in the outflow Q, since $dS/dQ = T$, we can modify this equation to

$$T\frac{dQ}{dt} = u - Q \tag{4.11}$$

We assume here that the input sequence has already been suitably transformed to an effective input that can be related linearly to the outputs.

For simple patterns of the effective input this equation can be solved analytically. For example, for a sudden input of effective rainfall u^* into an initially dry store at time t_o,

$$Q_t = \frac{u^*}{T}\exp\{-(t - t_o)/T\} \tag{4.12}$$

This is the impulse response or transfer function of the linear store expressed in continuous time. It has the form of an initial step rise followed by an exponential decline in the outflow (Figure B4.1.1(b)).

In hydrology and many other modelling applications it is often usual to have measurements of inputs and outputs at discrete time increments (e.g. every hour) rather

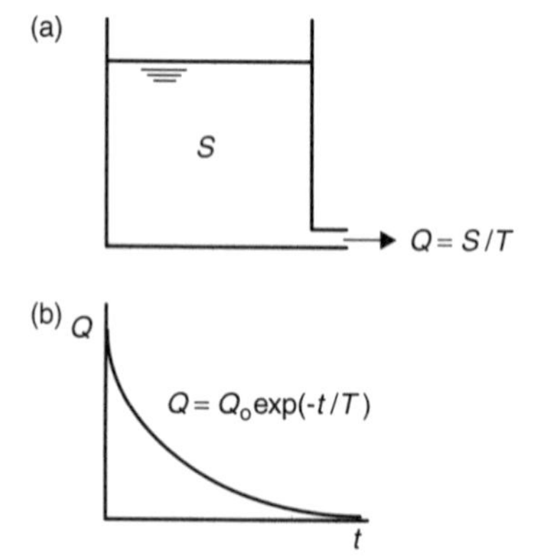

Figure B4.1.1 *The linear store*

than in continuous time. Thus, using a simple explicit finite difference form of the mass balance equation of the linear store over a discrete time step of length Δt,

$$\frac{Q_t - Q_{t-\Delta t}}{\Delta t} = \frac{u_t - Q_{t-\Delta t}}{T} \tag{4.13}$$

or

$$Q_t = \frac{\Delta t}{T} u_t + \left\{ 1 - \frac{\Delta t}{T} Q_{t-\Delta t} \right\} \tag{4.14}$$

or

$$Q_t = aQ_{t-\Delta t} + bu_t \tag{4.15}$$

where $a = 1 - \Delta t/T$; $b = \Delta t/T$; and to ensure mass balance between total effective rainfall inputs and total discharge outputs, $a + b = 1$.

In hydrological systems there is sometimes a delay after the start of a rainfall, before the discharge starts to rise. Assuming, for the moment, that this delay can be considered a characteristic of the system under study, it can be introduced as

$$Q_t = aQ_{t-\Delta t} + bu_{t-\delta} \tag{4.16}$$

where in this discrete time equation, the delay δ must be expressed as an integer number of time steps.

In considering more complex linear transfer function models it is convenient to introduce the backward difference operator, z. This is defined as

$$z^{-1} u_t = u_{t-1} \tag{4.17}$$

so that an input delayed by δ time steps may be written as

$$u_{t-\delta} = z^{-\delta} u_t \tag{4.18}$$

and the discrete time mass balance equation can be written as

$$Q_t = az^{-1} Q_t + bz^{-\delta} u_t \tag{4.19}$$

or rearranging this

$$Q_t = \frac{bz^{-\delta}}{1 - az^{-1}} u_t \tag{4.20}$$

Higher-Order Transfer Function Models

Higher-order transfer function models may then be easily constructed from the basic linear store building block components in either series or parallel structures (feedback structures are also possible but less amenable to a hydrological interpretation). For two stores in series, the individual components are multiplied together so that

$$Q_t = \left(\frac{b' z^{-\delta'}}{1 - a' z^{-1}} \right) \left(\frac{b'' z^{-\delta''}}{1 - a'' z^{-1}} \right) u_t \tag{4.21}$$

or

$$Q_t = \frac{b_o}{1 - a_1 z^{-1} - a_2 z^{-2}} u_{t-\delta} \tag{4.22}$$

where $b_o = b'b''$; $a_1 = a' + a''$; $a_2 = -a'a''$; and $\delta = \delta' + \delta''$.

For two stores in parallel, two first-order components are added such that

$$Q_t = \left(\frac{b' z^{-\delta'}}{1 - a' z^{-1}} + \frac{b'' z^{-\delta''}}{1 - a'' z^{-1}} \right) u_t \tag{4.23}$$

or, if in this case, both components have the same time delay δ

$$Q_t = \frac{b_o + b_1 z^{-1}}{1 - a_1 z^{-1} - a_2 z^{-2}} u_{t-\delta} \tag{4.24}$$

where $b_o = b' + b''$; $b_1 = -\left(b'a'' + b''a'\right)$; $a_1 = a' + a''$; $a_2 = -a'a''$.

By extension, a general higher-order linear transfer function model may be written

$$Q_t = \frac{b_o + b_1 z^{-1} + \cdots + b_m z^{-m}}{1 - a_1 z^{-1} - a_2 z^{-2} + \cdots + a_n z^{-n}} u_{t-\delta} \tag{4.25}$$

This is the general form of linear transfer function that forms the basis of the TFM package described in Section 4.5 and Appendix A. The Nash cascade discussed in Section 2.3 is one specific form of the general linear model. Combinations of linear storage elements were also investigated for rainfall–runoff modelling in the 1960s by Diskin and Kulandaiswamy (see Chow 1964; Chow and Kulandaiswamy 1971; Diskin and Boneh 1973). The specific case of a second-order parallel model in hydrological applications has been discussed by Young (1992).

Model Structure Identification

Given an input–output data set, there is then the problem of identifying an appropriate transfer function model structure, i.e. finding the best values of (m, n, δ) and the corresponding coefficients. There may be no unique answer to this problem, partly because the model will always be an approximation to what may really be a complex nonlinear relationship between the inputs and outputs, partly because there will always be some degree of error associated with the input and output data sets, and partly because the real-time delay might not be a constant or might not be precisely an integer number of time steps. It is often found that there are several model structures that give acceptably accurate simulations after calibration of the a and b coefficients, so some way of evaluating these structures is required.

There are two considerations in making such an evaluation. One is how well the model fits the data, i.e. the goodness of fit. A commonly used goodness of fit index is the coefficient of determination or proportion of the variance in the data explained by the model (introduced as the efficiency measure for rainfall–runoff modelling by Nash and Sutcliffe 1970). This is defined as

$$R_t^2 = 1 - \frac{\sigma_e^2}{\sigma_o^2} \tag{4.26}$$

where σ_o^2 is the variance of the observed output data calculated over all time steps used in fitting the model, and σ_e^2 is the variance of the residual differences between observed and simulated outputs at each time step. As the model fit improves, the value of R_t^2 will approach 1. If the model is no better than fitting the mean of the observed outputs (when $\sigma_e^2 = \sigma_o^2$), the value of R_t^2 will be zero or less. In general, the higher the model order, the better the fit will be, because there are more coefficients or degrees of freedom in the model that can be adjusted to fit the data. The aim is to find a model structure that gives a good fit but is *parsimonious* in having a small number of coefficients.

There is, however, a danger of allowing too many coefficients or too high a model order. This might result in good values of R_t^2, but the model may then be over-fitted or over-parameterized with the result that the transfer function may not be physically realistic (perhaps with negative ordinates or oscillations), or that, when used in simulation, its predictions are very sensitive to the input sequence used. One way of checking for over-parametrization is by looking at how well the coefficients of the model are estimated. For example, in the TFM package (see Appendix A), the transfer function model coefficients are fitted using the Simplified Recursive Instrumental Variable algorithm (Young 1984). This algorithm allows the standard errors associated with each of the *a* and *b* coefficients to be estimated every time a model structure is calibrated to a data set. The higher the model order (i.e. the greater the number of *a* and *b* coefficients), the larger the standard errors will tend to be. Very large standard errors on the coefficients are an indication that a model is over-parametrized. Young has suggested a criterion for model structure evaluation that combines elements of goodness of fit and standard errors on the coefficients. This Young information criterion (YIC) is defined as

$$YIC = \ln\left(\frac{\sigma_e^2}{\sigma_o^2}\right) + \ln\left(\frac{1}{N}\sum_N \sigma_e^2 \frac{P_{ii}}{\alpha_i}\right) \tag{4.27}$$

where α_i, $i = 1 \ldots N$ are the model coefficients, and P_{ii} is the ith diagonal element of a scaled parameter covariance matrix. The YIC value should be as negative as possible, indicating on its logarithmic scale that the variances of both the model residuals and coefficient values are as small as possible. Over-parametrization leads to a very rapid increase in the value of YIC. However, before accepting a transfer function model of this type, the user should always check the shape of the transfer function to ensure that it is physically reasonable for the system being studied. In the case of rainfall–runoff model or flow routing, for example, a transfer function that has negative ordinates would not normally be considered acceptable.

Factorizing a Second-Order Model

If a model is identified as second order ($n = 2$) and has one or two b parameters ($m = 1, 2$), then it is possible to factorize the model into two first-order components by treating the denominator as follows. From the equations above, for both a serial or a parallel model,

$$a_1 = a' + a''$$
$$a_2 = -a'a''$$

Combining these two equations, having fitted values of the two coefficients a_1 and a_2, gives

$$a'^2 - a_1 a' - a_2 = 0 \tag{4.28}$$

that is, a quadratic equation in a'. For cases with real roots, this may be solved using the standard quadratic formula ($x = (-B \pm \sqrt{B^2 - 4AC})/2A$ with $A = 1$, $B = -a_1$, $C = -a_2$). The two solutions will be the two coefficients a' and a''.

Once a' and a'' are known, for a parallel model ($m = 2$) the relative amounts of flow passing through each component can also then be determined since it is easy to show from the expressions $b_o = b' + b''$; $b_1 = -(b'a'' + b''a')$ that

$$b' = -(b_1 + a'b_o)/(a' - a'')$$
$$b'' = b_o - b'$$

Factorization is also sometimes possible for higher-order models with real roots, but the number of possible combinations of serial and parallel connections increases rapidly with model order.

Determination of the Time Constant for a First-Order Model Component

By setting the impulse response of the continuous time and discrete time models to be equivalent it can be shown that the a coefficient of a first-order model is related to the mean residence time of the continuous time equivalent approximately as

$$T = -\Delta t / \ln(a) \tag{4.29}$$

where Δt is the time step of the discrete time model. The mean residence time, with units of time, has a direct physical interpretation. Young and Beven (1994), for example, demonstrate that for a discrete time model fitted to hourly data for the CI6 catchment at Llyn Briane, Wales, a transfer function model suggests two parallel pathways (see Section 4.5), with mean residence times of 3.95 h and 80.2 h, though it must be remembered that there will be some uncertainty associated with these values. Determination of the b coefficients, b' and b'', for the two components suggests that some 33 percent of the effective rainfall takes the faster pathway in the parallel model and the remaining 67 percent the slower pathway. The fit of the model for this case is shown in Figure 4.5 ($R_t^2 = 0.98$). This application used the bilinear power filter applied to the total rainfall to create an effective rainfall series for use as an input to the transfer function modelling.

Summary of Model Assumptions

The assumptions made in applying the generalized linear transfer function model are as follows:

- *A1.* The transfer function uses an effective input (such as a suitably transformed *effective rainfall*) to predict an output such as stream discharge, although differences in total volumes of input and output can be scaled by the calibrated coefficients of the numerator (b_o, b_1 ...).
- *A2.* A transfer function is made up of one or more linear storage elements with different time constants combined in series or parallel, choosing the simplest structure that is consistent with the observations.

Box 4.2 Use of Transfer Functions to Infer Effective Rainfalls

The normal use of transfer function models in rainfall–runoff modelling is to allow the transformation of a time series of effective rainfalls into a time series of predicted discharges. Thus, assuming for demonstration purposes that a simple first-order linear transfer function model of Box 4.1 is appropriate:

$$Q_t = \frac{b_o}{1 - a_1 z^{-1}} u_{t-\delta} \tag{4.30}$$

where $u_{t-\delta}$ is an effective rainfall input at time $t - \delta$, δ is a time delay, the a and b values are coefficients and z is the backward difference operator.

Because the transfer function is linear, however, once the coefficients and time delay for a particular catchment have been identified, this equation may be rewritten as:

$$u_{t-\delta} = \frac{1 - a_1 z^{-1}}{b_o} Q_t \tag{4.31}$$

Thus, given the observed discharges Q_t and an estimate of the transfer function, a time series of effective rainfalls may be derived. There is, however, a problem in that a time series of effective rainfalls is needed to calibrate the transfer function coefficients. Thus, the identification of effective rainfalls in this way tends to be implemented as an iterative deconvolution procedure as follows:

1. Use a simple model (such as the Φ-index approach of Section 2.2 or a bilinear power law model of Section 4.3.2) to transform the rainfall inputs into a first estimate of the series of effective rainfalls.
2. Use this series of effective rainfalls to calibrate a transfer function model.
3. Invert the transfer function model to derive a series of calculated effective rainfalls.
4. Repeat the last two steps until both the series of effective rainfalls and the calibrated transfer function converge to stable values over successive iterations of the process.

There have been several studies of this type of procedure within the context of unit hydrograph theory. The success of the iterative procedure may depend on the initial model of effective rainfalls used and the form of transfer function assumed. Where the transfer function is represented as a series of ordinate values, rather than the functional form of the equations above, the inverse problem is not mathematically well posed. However, Olivera and Maidment (1999) have even implemented a deconvolution procedure to determine spatially varied inputs to a channel network transfer function, by initially identifying a mean catchment effective rainfall and then redistributing this to source areas on the basis of a relative runoff coefficient taken from the rational method.

Duband *et al.* (1993) have demonstrated some success in achieving convergence using this type of deconvolution procedure. The success of their FDTF-ERUHDIT (First Differenced Transfer Function–Excess Rainfall and Unit Hydrograph by a Deconvolution Iterative Identification Technique) algorithm is almost certainly helped by two features. One is that they apply the process to a whole series of rainfall events, rather than separately to individual rainfall events. The second is the fact that they write the equations in terms of the changes in discharge at each time step as the dependent variable. This neatly avoids the need for any initial hydrograph separation and helps reduce the effects of correlation in the errors in calibrating the transfer function.

Other studies of this type include Chapman (1996) who has developed a method for the estimation of unit hydrographs directly from a collection of streamflow hydrographs that can be used without knowledge of the original rainfall records. This involves the inference of a pattern of effective rainfalls for each individual event by an iterative procedure. The method still involves an initial 'baseflow separation' for each event, and the resulting unit hydrographs are sensitive to the method of separation chosen. He shows that the resulting average unit hydrographs tend to have higher peaks, shorter times to peak and somewhat shorter durations than those derived by conventional methods (Chapman 1996). The implied effective rainfalls tend to lag behind corresponding peaks in the measured rainfalls and continue after rainfall has stopped. He suggests that effective rainfalls might best be viewed as the output from a highly nonlinear storage process rather than as the result of a 'loss function' as in the more conventional methods for the calculation of effective rainfalls described in Box 2.2.

There is one further method of inferring a pattern of effective rainfalls that can be used without the need for an initial hydrograph separation. This is based on using a preliminary identification of a transfer function for a series of hydrographs followed by the estimation of a time variable gain parameter to track the nonlinear way in which total rainfall is related to total discharge. The time variation in the gain identified in this way may be useful in suggesting a nonlinear function for estimating an effective rainfall for prediction of further storm responses (see Young and Beven 1994). More details of this technique are given in Box 4.3.

Box 4.3 Time Variable Estimation of Transfer Function Parameters

There are a number of situations in which allowing the parameters of a transfer function to vary in time can be a useful analysis tool. The two main ones are in investigating the nature of the rainfall–flow nonlinearity (as in the data-based mechanistic modelling approach discussed in Section 4.5) and in adaptive real-time flow forecasting (see Section 8.4). We will consider only the simplest (but very useful) case of time variable transfer function parameter estimation here; that of estimating a time variable gain.

Consider the simple first-order, discrete time, transfer function of Box 4.1:

$$Q_t = \frac{b_o}{1 - a_1 z^{-1}} u_{t-\delta} \tag{4.32}$$

where Q_t is the discharge output, u_t is the time series of effective rainfall inputs, a_1 and b_o are parameters, δ is a time delay and z^{-1} is the backward difference operator. To allow that the b_o or a_1 parameters might be time variable, it is necessary to make some assumptions about the nature of the time variability. Since we are primarily interested in the broad range and trends of changes in the parameters, we do not wish to allow them to vary too rapidly. One set of assumptions is based on the idea that any time variable parameter, such as the gain, can be modelled as a stochastic generalized random walk (GRW) process, which effectively acts to smooth the changes in the time step to time step values of the parameter. Changes in the parameter values can then be analysed within a framework called fixed interval smoothing (e.g. Young 1993, 2000; Young and Beven 1994; Young *et al.* 1997).

Fixed interval smoothing is based on a linearization of the general stochastic model

$$Q_t = \Im(\Omega_t) + \xi_t \tag{4.33}$$

Here $\Im()$ is a nonlinear function of the variables

$$\Omega_t = [Q_{t-1}, Q_{t-2}, \ldots, u_{t-\delta}, u_{t-\delta-1}, \ldots, U_t, U_{t-1}, \ldots] \tag{4.34}$$

where the u values are past values of effective rainfall inputs and, for generality, we include values of other exogenous variables such as temperature, U, at this and previous time steps. The variable ξ_t represents the stochastic part of the model and is assumed to be a zero mean stochastic variable, independent of the variables u and U.

The first-order transfer function model above may then be written as

$$Q_t = x_t + \xi_t \tag{4.35}$$

where x_t is the noise-free output of the model defined as

$$x_t = \frac{b_o}{1 - a_1 z^{-1}} u_{t-\delta} \tag{4.36}$$

Let $\mathbf{p}$ represent the vector of parameter values $[b_o, a_1, \ldots]$. In the time variable parameter case, changes in any individual parameter, p_i, are described by the generalized random walk of the form

$$\mathbf{p}_{it} = \mathbf{F}_i \mathbf{p}_{it-1} + \mathbf{G}_i \eta_t \tag{4.37}$$

where the elements of $\mathbf{p}_i$ are made up of two components for each parameter value, a changing level and a changing slope, and η_t is a vector of zero mean white noise inputs.

$$\mathbf{F}_i = \begin{bmatrix} \alpha & \beta \\ 0 & \gamma \end{bmatrix}, \quad G_i \begin{bmatrix} \zeta \\ 1 \end{bmatrix} \tag{4.38}$$

Some specific examples of this general model are the Random Walk (RW: $\alpha = 0$; $\beta = \gamma = 1$; $\zeta = 0$), the Smoothed Random Walk (SRW: $0 < \alpha < 1$; $\beta = \gamma = 1$; $\eta_t = 0$) and the Integrated Random Walk (IRW: $\alpha = \beta = \gamma = 1$; $\zeta = 0$).

Then, given some measurements with which to compare the model outputs, the transfer function model may be written in what is called a state space form as the two equations:

$$\text{Observation equation:} \qquad Q_t = \mathbf{H}\mathbf{p}_t + \xi_t$$

$$\text{State equation:} \qquad \mathbf{p}_t = \mathbf{F}\mathbf{p}_{t-1} + \mathbf{G}\eta_t$$

where for the simple first-order transfer function above the parameter vector **p** contains the time varying values and slopes for the changing transfer function parameters a_1 and b_o.

The time variable state space equations are readily solved using a form of Kalman filtering (see Young 1984) that involves both forward filtering and backward smoothing passes through the data. The algorithm is recursive, in that the calculations are carried out time step by time step so that estimates of the updated parameters are available at every time step. In matrix notation the time variable parameter estimation algorithm has the following form:

1. Forward pass filtering
   (a) Prediction:

$$\widehat{\mathbf{p}}_{t|t-1} = \mathbf{F}\widehat{\mathbf{p}}_{t-1} \tag{4.39}$$

$$\mathbf{P}_{t|t-1} = \mathbf{F}\mathbf{P}_{t-1}\mathbf{F}^T + \mathbf{G}\mathbf{N}_r\mathbf{G}^T \tag{4.40}$$

   (b) Correction

$$\widehat{\mathbf{p}}_t = \widehat{\mathbf{p}}_{t|t-1} + \mathbf{P}_{t|t-1}\mathbf{H}_t^T[1 + \mathbf{H}_t\mathbf{P}_{t|t-1}\mathbf{H}_t^T]^{-1}\{Q_t - \mathbf{H}_t\widehat{\mathbf{p}}_{t|t-1}\} \tag{4.41}$$

2. Backward pass smoothing

$$\widehat{\mathbf{p}}_{t|N} = \mathbf{F}^{-1}[\widehat{\mathbf{p}}_{t+1|N} + \mathbf{G}\mathbf{N}_r\mathbf{G}^T\mathbf{L}_t] \tag{4.42}$$

$$\mathbf{L}_t = [\mathbf{I} - \mathbf{P}_{t+1}\mathbf{H}_{t+1}^T\mathbf{H}_{t+1}]^T[\mathbf{F}^T\mathbf{L}_{t+1} - \mathbf{H}_{t+1}^T\{Q_{t+1} - \mathbf{H}_{t+1}\widehat{\mathbf{p}}_{t+1}\}] \tag{4.43}$$

$$\mathbf{P}_{t|N} = \mathbf{P}_t + \mathbf{P}_t\mathbf{F}^T\mathbf{P}_{t+1|t}[\mathbf{P}_{t+1|N} - \mathbf{P}_{t+1|t}]\mathbf{P}_{t+1|t}^{-1}\mathbf{F}\mathbf{P}_t \tag{4.44}$$

with the starting condition $L_N = 0$.

In these equations, the matrices **F** and **G** are as defined above, **I** is the identity matrix, $\mathbf{N}_r$ is the noise variance ratio matrix and **P** is defined as

$$\mathbf{P}_t = \frac{\mathbf{P}_t^*}{\sigma_\xi^2} \tag{4.45}$$

where σ_ξ^2 is the variance of the observation noise and $\mathbf{P}_t^*$ is the error covariance matrix of the state estimates (the states here being the parameter values). The estimation of a covariance matrix for the parameters, updated recursively at each time step, is an important part of this algorithm since the user can then follow how well the parameter values are estimated in different parts of the study period, as well as the changes in the parameters themselves. The noise variance ratio (NVR) matrix, $\mathbf{N}_r$, is normally assumed to be a diagonal matrix with elements for each of the parameters to be estimated. The NVR controls the effective memory of the algorithm. Large values will allow rapid changes in the parameter values at each time step; small values will mean that those changes are damped over a larger number of time steps. The NVR may be optimized to achieve the best overall forecasting performance.

An example of the application of this fixed interval smoothing algorithm is demonstrated in Figure B4.3.1 which shows the time variable estimation of a single gain parameter b_o

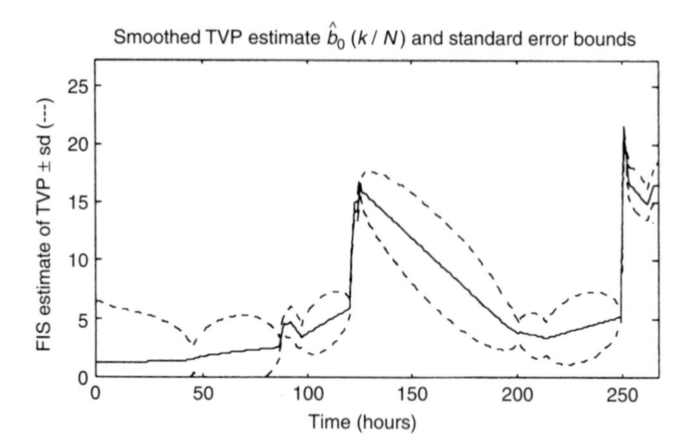

Figure B4.3.1 *Time variable parameter estimation of the gain coefficient in the bilinear model for the CI6 catchment. The solid line represents the best estimate; the dotted lines represent 90 percent confidence intervals (after Young and Beven 1994). Reproduced with permission of John Wiley & Sons Limited*

for the first-order transfer function in modelling the rainfall–runoff relationship for the CI6 catchment at Llyn Briane in Wales. This is the example on which the analysis of nonlinear changes in the parameters discussed in Section 4.4 is based. In this application, the value of the a_1 parameter was fixed on the basis of an initial estimation of the transfer function using a simple bilinear effective rainfall input. The FIS algorithm was then used to examine changes in the gain parameter using the original rainfall record, rather than an estimate of an effective rainfall, as input. In estimating the time variable changes in the gain parameter, b_o, we can then get a good idea of the nonlinear nature of the rainfall–runoff process which might then be used to formulate an approximate runoff generation model (see Section 4.4).

A further important use of time variable parameter estimation is in adaptive real-time flood forecasting (see Section 8.4). In this case, errors due to the nonlinearities in the runoff generation processes and limitations in the measurements of both rainfall and flow can be allowed for by using an adaptive gain parameter (Lees *et al.* 1994). In this case, only the forward pass Kalman filtering step is normally used, since observations are only available up to time t. The parameter estimates $\widehat{\mathbf{p}}_t$ are then used to make predictions into the future, before being updated as new data are received at the next time step. The NVR is normally chosen so as to give a relatively long estimator memory, so that the parameters (and consequent forecasts) do not change too rapidly from time step to time step.

5

Predicting Hydrographs Using Distributed Models Based on Process Descriptions

Until the knowledge gained through research in the sciences concerned with hydrologic phenomena becomes adequate enough to permit good quantitative descriptions of these phenomena and their functional relationships, a very strong subjective influence will pervade the entire endeavor. As long as this is true, the efforts of practising hydrologists will remain largely in the realm of the arts. It is natural to expect that under such conditions there may arise numerous emotional controversies regarding the relative merits of apparently conflicting lines of attack to specific problems.

J. Amorocho and W. E. Hart, 1964

But now we have to ask: is this really such a bad state of affairs after all? For is it not usually better to live with an unformed grammar than with a fully-formed grammar in which one does not truly believe, in which one does not have total faith? Better to be prepared *for conversion to the true faith than to rush prematurely into a false one. We have only to look at the examples of present-day politics, education and medicine to assure ourselves that an inappropriate grammar, and so an inauthentic theory, may do more harm than good in many places as it arrives at the level of its dialectic, at the level of its practical implementation. Surely, with such salutary examples in mind, hydrology should not be in too much hurry in this respect.*

Mike Abbott, 1992

5.1 The Physical Basis of Distributed Models

A general model of rainfall–runoff processes requires representations of the interacting surface and subsurface processes. As noted in Section 2.5, an outline of the physics underlying such a description was first published by Freeze and Harlan (1969), although the individual process descriptions had all been established well before then. Most physically based models today are still based on the Freeze and Harlan 'blueprint', and many are, in fact, simplifications of that blueprint. Even so, their blueprint is not complete. Relative to the perceptual model of Section 1.4, several elements are missing, including the effects of macropores and other heterogeneities of the flow processes (although attempts to introduce the effects of macropores into hillslope- and catchment-scale models have been made by Zuidema (1985) and more recently by Bronstert and Plate (1997) and Faeh *et al.* (1997). In this chapter some models based on the Freeze and Harlan blueprint will be outlined, concentrating on the assumptions made, so that other distributed modelling strategies can later be evaluated in the light of these most 'physically based' models. Initially, subsurface (soil water and groundwater) and surface (overland and channel flow) flow components will be considered separately, followed by their interactions.

5.1.1 Subsurface Flows

The basis of all descriptions of subsurface flow used in distributed models is Darcy's law. Darcy's law assumes that there is a linear relationship between flow velocity and hydraulic gradient, with a coefficient of proportionality which is now called the hydraulic conductivity. Thus,

$$v_x = -K\frac{d\Phi}{dx} \tag{5.1}$$

where v_x is velocity in the x direction [L T^{-1}], Φ is the total potential or hydraulic head [L], and K is the hydraulic conductivity [L T^{-1}]. Originally established empirically by Henri Darcy (1856) for flow through saturated sands, Richards (1931) generalized the use of Darcy's law for the case of unsaturated flow by making the assumption that the same linear relationship holds but that the constant of proportionality should be allowed to vary with soil moisture content or capillary potential. Thus,

$$v_x = -K(\theta)\frac{d\Phi}{dx} \tag{5.2}$$

where θ is volumetric soil moisture. The notation $K(\theta)$ is used to indicate that K is now a function of θ. The total potential, Φ, is often approximated as the sum of a capillary potential ψ [L], and an elevation above some datum, z[L]($\Phi = \psi + z$), ignoring other terms such as osmotic potentials associated with differences in solute concentrations. For an unsaturated soil, the capillary potential ψ will become increasingly negative as the water content decreases and the water is held in smaller and smaller capillary pores. This is because there is a pressure drop across an air/water interface that is inversely related to the radius of curvature of the interface. This radius will be smaller for smaller pores.

Darcy's law can be derived from more fundamental equations of flow, called the Navier–Stokes equations, if some assumptions are made about the nature of the pore space through which the flow is taking place, and if the flow is slow enough to stay in the laminar regime (e.g. Hassanizadeh 1986). This is usually a good assumption for flow in a porous matrix but may break down for flow in soils with heterogeneous characteristics (where it may be difficult to define a gradient of potential except at very small scales) and macropores (where flow in large pores and flow in the porous matrix may be responding to different local gradients). Thus Darcy's law will be strictly valid only over a limited range of scales. There has been a recent initiative towards the development of appropriate flux equations to replace Darcy's law for application at larger hillslope element scales (e.g. Reggiani *et al.* 1999) but the theory has not reached a stage where it is useful in rainfall–runoff modelling. However, it should be noted that the use of Darcy's law to represent fluxes in large elements of a distributed model is a gross approximation and may mean that the effective value of the hydraulic conductivity required may be different from anything that could be measured in the field (see Section 5.2 below).

The other important equation in this description of subsurface flow is the continuity or mass balance equation (see Box 2.3). The combination of Darcy's law with the continuity or mass balance equation results in a flow equation (generally called the Richards equation) which may be written with capillary potential ψ as the dependent variable as follows:

$$C(\psi)\frac{\partial\psi}{\partial t} = \nabla[K(\psi)\nabla\psi] + \frac{\partial K(\psi)}{\partial z} - E_T(x, y, z, t) \quad (5.3)$$

where ψ is the local *capillary potential*, $K(\psi)$ is unsaturated hydraulic conductivity, which is now expressed as a function of ψ rather than θ, $C(\psi)$ is a function of ψ defined as the rate of change of moisture content θ with change in ψ called the *specific moisture capacity*, and $E_T(x, y, z, t)$ is a local uptake of water by plant roots to satisfy evapotranspiration (see Section 5.1.3 below). An equivalent equation with soil moisture content θ as the dependent variable can also be written. The derivation of both forms of the Richards equation is given in Box 5.1. The Richards equation is a partial differential equation (see Box 2.3), and because of the nonlinear change of hydraulic conductivity with moisture content, it is a nonlinear partial differential equation. Such equations tend to be very difficult to solve analytically, except for some very simple cases of initial and boundary conditions and simple forms for the nonlinear relationships relating moisture content, capillary potential and hydraulic conductivity – the so-called *soil moisture characteristic curves*. Some solutions for the special case of infiltration at the soil surface are given in Box 5.2.

For most cases of interest to hydrologists it will be necessary to use an approximate numerical solution of the equation. A number of different techniques are available, including finite-difference, finite-element, boundary element, integrated finite-difference, and finite-volume techniques (e.g. Pinder and Gray 1977). All these methods involve discretizing the flow domain into a network of grids or elements (as shown for a finite-element discretization in Figure 5.1) and solving for values of moisture content θ or capillary potential ψ at a large number of nodes, either on the edges or at the centre of the elements (see Box 5.3).

This type of model is very demanding in its data requirements. Model parameters must be provided for every grid element in the flow domain and boundary conditions must be specified for every discrete length or area of the domain. Figure 5.1 shows a two-dimensional section through a hillslope with a discretization into a finite-element grid and an indication of the boundary conditions that might be applied. Specified flux boundaries are called Cauchy-type boundary conditions; zero flux (impermeable) boundaries are called Neumann-type boundary conditions; specified pressure boundaries are called Dirichlet-type boundary conditions.

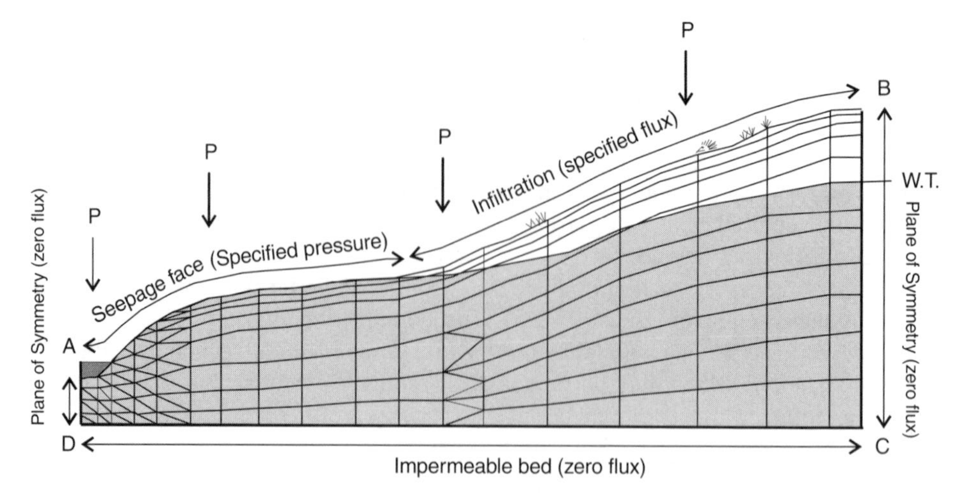

Figure 5.1 *Finite-element discretization of a vertical slice through a hillslope using a mixed grid of triangular and quadrilateral elements with a typical specification of boundary conditions for the flow domain. The shaded area represents the saturated zone which has risen to intersect the soil surface on the lower part of the slope*

Along the boundaries AD and BC a symmetry boundary is usually implemented under the assumption that flow conditions are identical for the hillslope at the other side of the divide at B or the channel at D. A symmetry boundary is equivalent to having a no-flow condition along a direction normal to the boundary. Along boundary CD, a no-flow boundary is also generally assumed on the basis that the hillslope is underlain by an impermeable layer or *aquiclude*. The boundary condition along AB may be time variable. While it is raining and the soil surface has not ponded, there will be a specified flow rate equal to the net rainfall rate at the ground surface. If the soil surface reaches saturation and infiltration into the soil starts to fall below the rainfall rate then part of this boundary may be controlled as a fixed head boundary, with a potential equal to the depth of ponded water. Under dry conditions, again there may be a specified flow rate equal to the estimated loss of water from the surface as evaporation. Time-varying solutions also require the initial conditions at the start of the simulation period to be specified. The initial conditions are values of θ or ψ for every node in the flow domain at the start of the simulation.

One major problem in applying the Richards equation is specifying the nonlinear soil moisture characteristic curves for a particular location or solution grid element. Most models of this type use functional relationships between the moisture content, capillary potential and hydraulic conductivity variables (see Box 5.4). All such relationships are defined by a number of parameter values. Parameter values will need to be specified for every element in the mesh. A variety of functional forms for describing the soil moisture characteristics have been proposed. All require a number of different parameters to be specified. Most are single valued relationships, i.e. each value of θ is associated with a unique value of ψ, $K(\theta)$ and $C(\psi)$. Not all soils show such single valued relationships, however. The appropriate curves for a soil that is wetting may be different to the appropriate curves for a soil that is drying. This is known as soil moisture *hysteresis*. The appropriate values of ψ, $K(\theta)$ and $C(\psi)$ then depend on the changes in θ over time. There are some models of hysteretic soil moisture characteristics available (for a review, see Jaynes 1990), but they tend to rely on idealized representations of the soil to simplify the problem of keeping track of the wetting and drying history of each node.

Measuring the soil moisture characteristics, whether in the field or on samples brought back to the laboratory, is time-consuming and expensive and, in heterogeneous soils, the values obtained on one sample may not be representative of the effective grid element values needed in the model. One technique that has been developed to solve this problem is the use of what have been called *pedotransfer functions*, which attempt to provide estimates of the parameter values in mathematical descriptions of the soil moisture characteristics curves in terms of variables, such as soil texture variables, that are more easily measured (see Box 5.5).

The pedotransfer function concept is fine in principle but may need to be applied with some circumspection in practice since the pedotransfer functions currently available have generally been developed from data obtained from small sample experiments. The parameter values estimated in this way may then not necessarily be appropriate at the model grid scale. Most pedotransfer functions are based on regression analysis of soil characteristic parameters against texture and other variables. The resulting estimates can therefore be associated with a standard error of estimation as a measure of the uncertainty associated with the estimates. One set of pedotransfer function procedures has been incorporated into the STATSGO soil characterization computer package of the US Department of Agriculture (USDASCS 1992) which contains soil texture information for all major soil groups in the US, from which soil moisture characteristic parameter estimates can be derived.

A further method for deriving the parameter values for soil moisture characteristic functions is to calibrate a model of the functions within a Richards equation solution algorithm, so as to best simulate a set of soil moisture and capillary potential data. Where this method is applied to a laboratory soil column, discharge from the column can also be used in calibration. This is also called the inverse method. There is a huge literature on inverse methods for groundwater problems involving only saturated flows (e.g. McLaughlin and Townley 1996) and one of the most widely used groundwater flow packages, MODFLOW, is now available from the USGS with a parameter optimization routine known as MODFLOWP (Poeter and Hill 1997). For unsaturated flows, the inverse method has been reviewed by Kool *et al.* (1987) and an application to the determination of hysteretic soil moisture characteristics has been made by Simunek *et al.* (1999). In general, the calibration of subsurface flow parameters is not a well-posed inverse problem; there is often inadequate information about the flow domain and estimated parameter values may be sensitive to errors in model structure, boundary conditions and observations. Particularly in the nonlinear case of unsaturated flow, it may be difficult to obtain a clear optimum set of parameter values (e.g. Abeliuk and Wheater 1990; Hollenbeck and Jensen 1998). Parameter calibration is discussed more generally in Chapter 7.

A number of different computer packages are available for solving the Richards equation in one, two or three dimensions under the assumption that effective parameter values for the Darcy flow law can be specified at the element scale, such as the

HYDRUS-2D finite-element program for flow and transport calculations of Simunek *et al.* (1996). These also include models that use a two-domain approach to the prediction of preferential flow (e.g. the CHAIN-2D model of Mohanty *et al.* (1998) and the MACRO model of Jarvis *et al.* (1991)). The complexities of obtaining accurate solutions are such that developing solution techniques is best left to the specialist numerical analyst, but the following points are worth remembering in assessing any package:

- All the solution techniques for this type of nonlinear problem are approximate and it is difficult to generalize about whether one method will be more accurate than another for a particular problem.
- For any solution algorithm that is consistent with the differential equation, accuracy will depend on the time and space discretization used. The finer the space increments (or the smaller the elements used), the shorter the time steps will have to be.
- It may be necessary to use a very large number of nodes to represent the flow domain, especially in three dimensions. This will then require the solution of a very large sparse matrix equation at least once at each time step, together with the calculation of the nonlinear functions at each node. The computer time required therefore mounts rapidly with the number of nodes.
- Problems where steep hydraulic gradients are expected, such as a wetting front during infiltration or around a pumped well, will require small elements (and consequently small time steps) to represent the gradient and flow velocities adequately in that part of the flow domain. This may seem obvious but has not always been evident in published applications of distributed models.
- Solutions with soil moisture θ as the dependent variable tend to be better for dry soil conditions; solutions with capillary potential ψ as the dependent variable tend to be better for wet soil conditions.
- Some solutions, particularly those known as *explicit solutions* when the solution at time step t depends only on values of the nonlinear functions calculated at time step $t - 1$ (see Box 5.3), will produce unstable solutions if the time step is too big. Instability in the solution will normally be seen as increasingly large oscillations in the solutions at some nodes. A good implementation of an explicit solution scheme should check for stability, albeit at the expense of additional calculations, and adjust the time step accordingly.
- *Implicit solutions*, which use variable and function values at both t and $t - 1$ (see Box 5.3), will generally be more stable and can use much longer time steps, but may involve a number of iterations at each time step to converge on the solution at time t.
- There are some stability problems inherent in the solution of the Richards' equation, even using an implicit time stepping scheme. This is because the nonlinear soil moisture capacity function ($C(\psi)$ in equation 5.1) peaks at a certain value of ψ. Thus it is possible for the solution at a node to oscillate either side of that value of ψ, while still calculating an appropriate value of $C(\psi)$.
- In heterogeneous flow domains, the values of the parameters required at the grid or element scale may be dependent on the scale of the elements. It is not necessarily the case that a Darcian description of the flow using effective parameter values at the model element scale will be an adequate representation of the flow processes.

The effects of spatial heterogeneity of soil characteristics, time variability due to crusting and other processes, and preferential flow in structured soils, remain topics of research with no generally accepted descriptive models.

- Always test the model, using different time and space steps, against some simple test cases. There is no absolute guarantee that an approximate solution of a nonlinear partial differential equation will be stable and accurate under all circumstances. Validation against some test cases will at best (but also at least) give a guide.

5.1.2 Surface Runoff and Channel Routing

The physical basis of models of overland and channel flow is essentially the same. In both cases, in catchment-scale rainfall–runoff modelling one-dimensional flows downslope or downstream are usually considered as a convenient approximation to the full three-dimensional flows. One-dimensional solutions must use average cross-sectional velocities as solution variables, even for the case of overbank flow for the channel and variable-depth overland flows (Figure 5.2). The one-dimensional case may be described by the equations developed by the Barré de St Venant (1797–1886). These equations assume that the flow may be described in terms of average cross-sectional velocities and depths and are developed from the balances of both mass and momentum in the flow. Thus for a flow of average velocity v, of average depth h, in a cross-sectional area A, and wetted perimeter P, on a bed slope of S_o, and lateral inflow

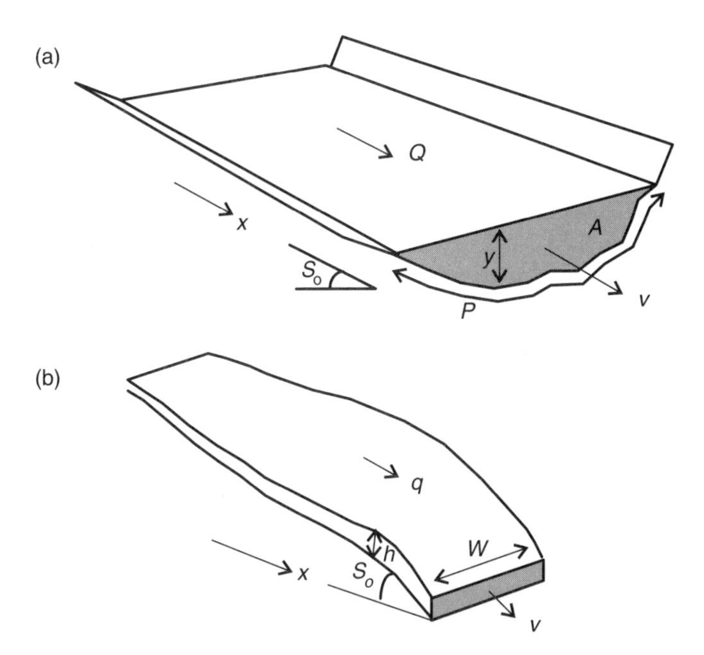

Figure 5.2 Schematic diagram for surface flows. (a) One-dimensional representation of open channel flow with discharge Q, cross-sectional area A, wetted perimeter P, average velocity v and average depth y. (b) one-dimensional representation of overland flow as a sheet flow with specific discharge q, width W, average velocity v and average depth h. In both cases the slope is S_o and distance x is measured along the slope

per unit length of slope or channel of i, the mass balance equation may be written as:

$$\frac{\partial A}{\partial t} = -A\frac{\partial v}{\partial x} - v\frac{\partial A}{\partial x} + i \tag{5.4}$$

and the momentum balance equation, assuming that water is incompressible, as:

$$\frac{\partial Av}{\partial t} + \frac{\partial Av^2}{\partial x} + \frac{\partial Agh}{\partial x} = gAS_o - gP\frac{f}{2g}v^2 \tag{5.5}$$

where f is the Darcy–Weisbach uniform roughness coefficient.

The St Venant equations are a fully dynamic or dynamic wave description of the flow that can be used in routing flood waves and hydrographs down a channel or in a reach of a channel network. The derivation of these equations is given in Box 5.6, together with explanations of simplified versions known as the diffusion wave and kinematic wave approximations, resulting from neglecting different terms in equation (5.5). As in the case of the Richards equation of the last section, solution of the St Venant equations for cases of practical interest generally requires an approximate numerical solution algorithm. The first attempts at an explicit finite-difference solution of the St Venant equations date back to Stoker in 1957. There are now well-established finite-difference schemes such as the four-point implicit method described by Fread (1973, 1985), which has been used even under the extreme conditions of routing the flood wave due to a dam break (e.g. Fread 1985). Routing for rainfall–runoff modelling is generally not so extreme.

The other requirements to apply such a model are information about the geometry of the channel, a specification of the initial velocities and depths of flow at the start of a simulation, and boundary conditions at both the upstream and downstream boundaries of a channel reach. All of these – geometry, roughness coefficients, initial and boundary conditions – are only ever known imprecisely and some simplifying assumptions are usually necessary.

In respect of the channel geometry, it has usually been assumed that the shape of the channel can be interpolated between surveyed cross-sectional profiles measured at different distances along the channel. At low flows this will not give a good representation of the effects of the pool–riffle geometry of the channel; at high overbank flows it may not take proper account of the effects of embankments, field boundaries and other obstructions to the flow. These will also affect the appropriate values of the effective roughness parameter that must reflect all the causes of momentum loss in a reach of channel. Thus, the effective values might be different from that inferred from a measured velocity profile at any single point in the channel and may also require a parametric description of how the roughness coefficient changes with depth of flow, especially for overbank flow under flood conditions.

The boundary conditions will have an important effect on the solution. The St Venant equations require boundary conditions to be specified for both upstream and downstream boundaries (in contrast to the kinematic wave approximation discussed below). In fact, since there are two unknowns in the solution, velocity and depth for each cross-section, two upstream boundary conditions and two downstream boundary conditions are required for every simulated reach. Junctions between reaches will require some special conditions to ensure consistency in the solutions for the upstream and downstream reaches. It is, in fact, rare for the boundary conditions to be specified directly in terms of velocity and depths at each boundary. These are not generally available. Water surface elevation or stream stage is more generally available, at least at the gauging sites which will often mark the boundary points for a solution in a larger river. A measurement of stage can be used with an appropriate rating curve and cross-sectional survey to get approximate estimates of discharge and cross-sectional area, from which a mean velocity can be derived. The rating curve may be measured, may be a theoretical rating of a gauging structure, or may be derived by assuming that there is a locally *uniform flow* at the boundary. In the last case, the relationship between velocity and stage can be described by one of the uniform flow equations such as the Manning or Darcy–Weisbach equations, (see Box 5.6), given knowledge of the appropriate roughness coefficient.

A little bit of care is necessary here however. Use of such a uniform flow rating curve implies that the water surface is always parallel to the bed. The fully dynamic equations, however, imply that the water surface should be steeper than the bed slope on the rising limb of a hydrograph and less steep on the falling limb, resulting in a hysteretic or looped rating curve. The uniform flow assumption can then only provide an approximate boundary condition for the solution. Again, it is important to make an assessment of what assumptions are being made in the description and solution for each process.

It must be remembered that this description of channel flow is a one-dimensional description, with the solution variables being average velocities and depths of flow. This type of description will not be as accurate during flood conditions, when local roughness coefficients, velocities and depths of flow may vary dramatically in the cross-section. Recent developments have seen two-dimensional models coming into more widespread use. Such models can predict a pattern of depth-averaged velocities within the channel and across a flood plain, although there has been little validation of such predictions against observations to date. Examples of such models are TELEMAC_2D and RMA2 (see e.g. Bates *et al.* 1992, 1995). Three-dimensional models of surface flows using general computational fluid dynamics packages are just starting to be used (see the review of Lane 1998), but as yet are computationally feasible for only small-scale problems. Even then, much remains to be learned about appropriate representations of turbulence and momentum losses in natural channels for such models.

As in the case of subsurface flow solutions, there are a number of points that should be borne in mind when evaluating a hydraulic model of the surface flow processes:

- All numerical solutions of the flow equations will be approximate and may be subject to numerical diffusion. Again, a large number of nodes may be necessary to represent the flow domain, and explicit solutions in particular may require very short time steps.
- An important control on accuracy in representing the real flow processes will be the specification of the geometry of the flow domain. Surveys are expensive and compromises are generally necessary in applications.
- The specification of boundary conditions and roughness coefficients will also be important. Effective roughness coefficients may be flow depth dependent and may have to take account of momentum losses associated with obstructions (trees, walls, hedges) as well as the surface roughness of the bed and banks.

- Always test the model, using different time and space steps against some simple test cases to evaluate the convergence and stability properties of the solution algorithm.

5.1.3 Interception, Evapotranspiration and Snowmelt

Any physically based catchment model also requires components for interception, evapotranspiration and snowmelt. These, in conjunction with primarily the subsurface flow component, will control the simulation of the antecedent conditions prior to an event, and the inputs during an event, which will be important in predicting the runoff from that event. A typical set of components, as in the SHE model discussed in the next section, uses a Penman–Monteith actual evapotranspiration calculation (see Box 3.1), an interception storage model such as the Rutter model (see Box 3.2) and either a full energy balance or degree-day snowmelt model (see Box 3.3).

5.2 Physically Based Rainfall–Runoff Models at the Catchment Scale

5.2.1 Coupling the Surface and Subsurface Process Descriptions: Towards a Fully Three-Dimensional Description

The distributed model defined by Freeze and Harlan (1969) was a fully three-dimensional saturated–unsaturated subsurface flow description coupled to a two-dimensional overland flow description and one-dimensional channel flow description. The coupling of the different process descriptions can be achieved through common boundary conditions. For example, the depth of ponding of water on the soil surface predicted by an overland flow solution can be used to define a local head boundary for the subsurface flow solution in simulating infiltration rates. Similarly, the depth of flow predicted in the channel might provide a local head boundary condition for the prediction of fluxes from the saturated zone through the bed of the channel. In principle, therefore, the whole system of processes could be solved in one system of equations, taking proper account of all the common boundary conditions. In practice, to apply such a description at the scale of a catchment, or even at the scale of a hillslope, requires prodigious amounts of computer time, even with today's computing power. Most distributed models have therefore attempted to reduce the amount of computing power in some way, although three-dimensional solutions are beginning to be considered as viable tools of the future.

A number of different strategies have been used. The first is to use a coarser mesh, so that there are fewer nodes, a smaller number of equations must be solved at each time step, and fewer parameters need to be specified. There is clearly then a danger of having a model that is not an accurate solution to the original equations. This is a very real danger; it applies to certainly most of the distributed models that have been used in representing the rainfall–runoff process at the catchment scale to date.

A second strategy has been to reduce the dimensionality of the problem, i.e. break it down into smaller pieces. One way to do this has been to treat the unsaturated zone where flows are predominantly vertical as a one-dimensional problem, and the saturated zone where flows are predominantly lateral as a two-dimensional problem.

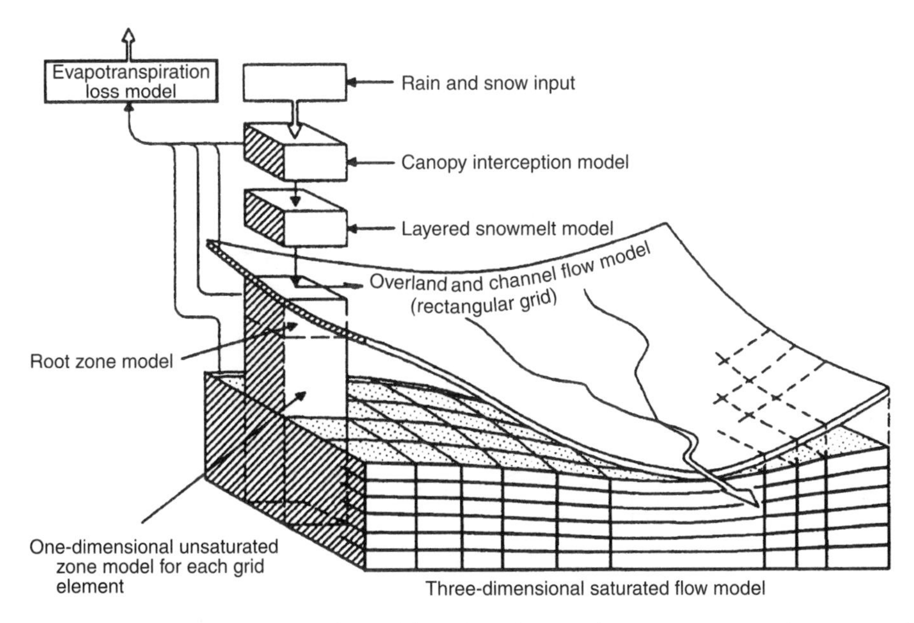

Figure 5.3 *Schematic diagram of a grid-based catchment discretization as in the SHE model (after Refsgaard and Storm 1995). Reprinted with permission of Water Resources Publications LLC*

This is the approach adopted by the SHE model (Figure 5.3; see Abbott *et al.* 1986a). This can give rise to numerical problems in coupling the solutions at a boundary which moves up and down with the water table as the soil wets and dries. The solution of the saturated zone problem depends on the profile of water content in the unsaturated zone of each grid element and vice versa. Normally some iterations are required to achieve convergence of the two solutions. Similar iterations may be required to achieve convergence at nodes on the soil surface, where the boundary may be changing from infiltration to ponded conditions during a storm.

An alternative strategy is to avoid decoupling the unsaturated and saturated zones and instead make the split along the lines of greatest slope in the catchment to make a number of hillslope planes which are then solved separately 'in parallel'. The vertical section along each plane is then discretized in two dimensions, assuming that conditions across each plane can be considered uniform. This is the approach adopted by the IHDM model (Figure 5.4; see Calver and Wood 1995). Then there are models such as TOPOG (Vertessy *et al.* 1993) which uses variable-width hillslope planes but separates the unsaturated zone and saturated zone; VSAS2 which uses variable-width hillslope planes but has a time variable separation of a saturated contributing area to ease the numerical problems of solving the Richards equation when part of the flow domain is fully saturated (Bernier 1985; Prevost *et al.* 1990; Davie 1996); and the model of Duffy (1996) which also uses hillslope planes but solves for the moisture storage at each point, integrated over the profile of both saturated and unsaturated zones.

The major point to be made here is that this is a further level of approximation introduced by computer and data limitations. In time, with the advent of more and more powerful desktop parallel computers, these limitations will become less

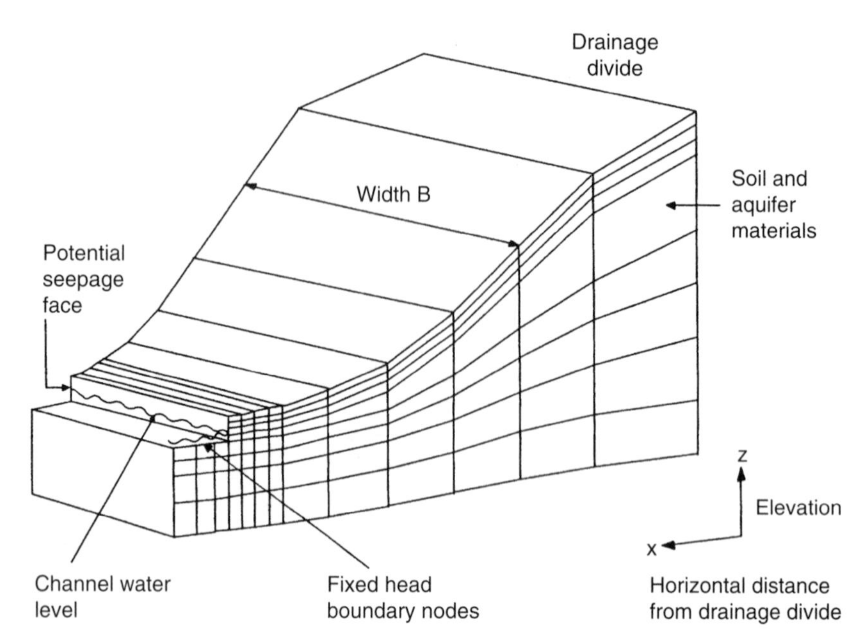

Figure 5.4 *Schematic diagram of a hillslope plane catchment discretization as in the IHDM model (after Calver and Wood 1995)*

constraining. Fully three-dimensional models will become more feasible to implement and, as computer power increases, finer and finer grid or element sizes will be possible. Already subsurface flow models with millions of nodes are being used for hypothetical simulations and fully three-dimensional catchment models have been applied at relatively small scales (Binley *et al.* 1989; Paniconi and Wood 1993).

Specifying all the parameters will still, however, be a problem. The finer the elements, the more parameter values must be specified. A minimal list of the parameters that might be required for a full catchment-scale model is given in Table 5.1. Note that many of these parameters, while often assumed as constant for a particular model run, may in fact be dependent on other variables. Canopy resistance, for example, may require a more fundamental parametrization to account for its variation with soil moisture, solar radiation and surface temperature (see Box 3.1); interception storage may vary with crop growth; channel flow resistance may vary with flow depth; a degree-day factor may increase during the melt season (Box 3.3). All these dependencies would need to be specified in a complete model of the processes and since a solution with thousands of elements requires many thousands of parameter values, this will only be feasible by linking such models to databases for the preparation and storage of parameter values, effectively a Geographical Information System (GIS). Similar software will be required for the post-processing of the results. The finer the elements the more parameter values that must be specified and the more data that are produced by each simulation. The only way to assess such information easily will be visually, in the form of computer graphics. Some of these developments can already be seen in the applications of current-day models (e.g. Abbott and Refsgaard 1996).

Table 5.1 *Minimal parameters required for a catchment-scale process-based model*

Parameter	Symbol	SI units
Subsurface flow parameters (for each soil type/horizon element)		
Saturated hydraulic conductivity matrix	$\underline{K}_s$	$m\,s^{-1}$
Porosity	θ_s	–
Soil moisture characteristic parameters (see Box 5.4)		
Vegetation parameters (for each vegetation type)		
Proportion of direct throughfall	p	–
Interception storage capacity	C	m
Interception drainage parameter		–
Aerodynamic resistance (may vary with wind speed)	r_a	$s\,m^{-1}$
Canopy resistance (may vary with other variables)	r_c	$s\,m^{-1}$
Albedo	α	–
Proportion of active roots to distribute $E_T(x, y, z, t)$ over elements		–
Overland flow parameters (for each slope element)		
Overland flow roughness (may vary with flow depth)	f	–
Local surface slope angle	S	–
Channel flow parameters (for each reach element)		
Channel flow roughness (may vary with flow depth)	f	–
Overbank flow roughness (may vary with flow depth)	f	–
Local channel bedslope	S_o	–
Snow parameters (degree-day model)		
Threshold temperature	T_o	K
Degree-day factor (may vary in time)	F	$mm\,day^{-1}K^{-1}$

5.2.2 Grid-Element-Based Models: The SHE Model

The Système Hydrologique Européen or SHE model is the most widely known model of this type. The development of SHE was started in 1977 as a joint collaboration, between the UK Institute of Hydrology, the Danish Hydraulics Institute (DHI) and SOGREAH of Grenoble in France. An early description of the model was published by Beven *et al.* (1980), an explanation of modelling philosophy was provided by Abbott *et al.* (1986a,b), while the first full application to the Institute of Hydrology River Wye experimental catchments at Plynlimon, Wales (10 km^2), was published in the series of articles by Bathurst (1986a,b). Other applications have been published, ranging from the 1.4 km^2 Rimbaud catchment in the south of France (Parkin *et al.* 1996) to the 820 km^2 Kolar and 4955 km^2 Narmada catchment in India (Jain *et al.* 1992; Refsgaard *et al.* 1992). Summaries of various applications of SHE are given in Refsgaard and Storm (1995), Abbott and Refsgaard (1996), Bathurst *et al.* (1995) and Bathurst and Cooley (1996).

The SHE is a grid-based model, splitting the catchment into a number of square or rectangular grid elements, linked to channel reaches that run along the boundaries of the hillslope grid (Figure 5.3). The size of grid used has varied in different applications, ranging from 50 m on a side for the small 40 ha Upper Sheep Creek catchment in Idaho, up to 2 km on a side for the Kolar and Narmada catchments in India. Note that in the latter case, as the authors acknowledge, the grid size is so large that the model cannot be considered to be representing flow on the hillslopes or in the smaller channels of the catchment in any meaningful way. Each hillslope grid element has a specified surface elevation and model components for interception, evapotranspiration, snowmelt, and one-dimensional vertical unsaturated zone flow where appropriate. The grid elements are linked by two-dimensional surface runoff and groundwater components. Internal boundary conditions allow the coupling of surface flow and infiltration into the unsaturated zone, the unsaturated and saturated zones at the local water table, and groundwater and channel flows. Great effort has been made to ensure that the processes are properly coupled and that the numerical solutions are stable for a wide range of conditions although, because of the nonlinearity of the unsaturated zone equations and the coupling of different processes, stability is not guaranteed. The model can predict a variety of runoff generation processes on each grid element, including both infiltration excess and saturation excess runoff, and the groundwater flow component can be used to simulate subsurface contributions to the hydrograph under suitable conditions. The descriptions of the unsaturated and saturated zones are based on Darcy's law; overland and channel flows are described by a diffusion wave approximation to the St Venant equations, and various options are included for simulating interception and evapotranspiration, including the Penman–Monteith equation of Box 3.1. Snowmelt is simulated, using either a degree-day method or a full energy balance (see Bathurst and Cooley (1996) who make a comparison of both implementations).

The types of parameter values required are similar to those listed in Table 5.1 and there is the potential to have different parameters for every grid element and within each grid element for different layers in the vertical. Any application of the SHE model will therefore require the specification of thousands of parameter values. The parameter values required are effective values at the grid element scale, which may not be the same as values that might be measured locally. There is also the potential to specify fully distributed precipitation and meteorological data across the model grid elements, if the data are available. The predictions will, however, be dependent on the grid scale used. Refsgaard (1997), using the SHE model, is one of the few studies to have looked at the effect of the grid scale on the model predictions. His study, on the Karup catchment in Denmark, compared predictions using a finest grid of 500 m, to those with degraded grids of 1000, 2000 and 4000 m. His conclusion was that above 1000 m, it might still be possible to obtain reasonable simulations of catchment discharge but that this would require recalibration of parameters and possibly reformulation of some model components. He infers that not much improvement in accuracy would be gained by using finer scale grids than 500 m, but this conclusion may be conditional on the nature of the Karup catchment which is dominated by groundwater flows. Xevi *et al.* (1997) have also demonstrated that the results of the SHE model are sensitive to grid size.

The different SHE development teams have implemented impressive pre- and post-processing packages for preparing model applications and visualizing the distributed predictions, including graphical animations of the predicted responses. The distributed predictions of the SHE model have also allowed other model components to be developed within the most recent versions which are now being developed independently by the original partners. The UK version, SHETRAN, now based within the Water Resource Systems Research Unit at the University of Newcastle, has added contaminant and sediment transport components (Bathurst *et al.* 1995). The DHI version, MIKE SHE, has also added a contaminant transport component (Refsgaard and Storm 1995). In both cases the predictions of contaminant transport are based on the advection–dispersion equation. Both DHI and University of Newcastle now have versions of SHE which make fully 3-D solutions for the unsaturated–saturated flow domain. MIKE SHE has also added an option to predict a preferential recharge to the saturated zone as a simple proportion of the infiltration rate (Refsgaard and Storm 1995), although there is no real physical justification for such a conceptual description.

There are other models based on grid elements available. The fully 3-D models of Binley *et al.* (1989) and Paniconi and Wood (1993) use a grid-based spatial discretization. The ANSWERS model (see Beasley *et al.* 1980; Silburn and Connolly 1995; Connolly *et al.* 1997), which has its origins in one of the very first fully distributed grid-based models of Huggins and Monke (1968), essentially considers only an infiltration excess runoff generation mechanism, using the Green–Ampt infiltration equation (see Box 5.2) to predict excess rainfall on each grid element. The runoff generated is then routed towards the stream channel in the direction of steepest descent from each grid element. The CASC2D model of Doe *et al.* (1996) is similar in that it also uses a Green–Ampt infiltration equation, but uses a 2-D diffusion wave approximation to model overland flow on the hillslopes and a 1-D diffusion wave model for the channel reaches. The 3-D version of HILLFLOW of Bronstert and Plate (1997) is a grid-based model, with the interesting option of modelling the Richards equation using the fuzzy rules methodology of Bárdossy *et al.* (1995). HILLFLOW also has a 2-D option for modelling individual hillslope elements in a way similar to the models of the next section, and a 1-D version for individual soil profiles. All the HILLFLOW versions have a component for modelling preferential flow in macropores, at the expense of introducing additional parameters. Bronstert (1999) provides a review of experience in using HILLFLOW in a variety of applications.

5.2.3 Hillslope-Element-Based Models: IHDM, TOPOG

The main alternative catchment discretization strategy is to make a subdivision into hillslope planes (Figure 5.4). Such a subdivision is ideally made along flow lines, such that any lateral exchanges of water between adjacent hillslope elements can be neglected. Some early physically based distributed models which only attempted solutions for a single hillslope were essentially of this type (e.g. Freeze (1972) using a finite-difference solution and Beven (1977) using a finite-element solution). It is, of course, much easier to determine flow lines if the flow follows the form of the surface topography. Hillslope elements may then be determined on the basis of a topographic analysis of the catchment. Thus this type of model will work best where the hydrologically active layer is near to the soil surface, and no deeper regional aquifer flows are involved. For deeper systems, a two-dimensional in plan (as in SHE) or fully three-dimensional solution of the subsurface flow domain will be more appropriate.

However, there are many catchments for which the hillslope element discretization based on surface topography will be a reasonable approximation to the flow directions. In some early catchment models of this type, the variable width, variable depth and variable slope hillslope elements were represented by 'equivalent' planes of uniform width, uniform depth and uniform slope (and usually uniform soil and surface parameters). Early versions of the Institute of Hydrology Distributed Model (IHDM) were of this type, as well as some models based on Hortonian infiltration excess runoff generation that did not include a full subsurface flow solution, but treated infiltration as a 'loss' (e.g. the model of Smith and Woolhiser (1971) which later developed into the KINEROS package described by Smith *et al.* (1995); see also Section 5.3.2).

The model of Beven (1977) showed that using finite elements it was relatively easy to follow the actual shape of the hillslope and allow the depths of different horizons within a hillslope to vary (as in the vertical plane discretization of Figure 5.1). This study also introduced the simple idea of including slope width into the equations so that convergent and divergent hillslopes could be represented (see also Box 5.7). This was also introduced into Version 4 of the IHDM (Beven *et al.* 1987) and there have been further numerical improvements since (Calver and Wood 1995). Applications are reported by Calver (1988) and Binley *et al.* (1991) for the Wye catchment at Plynlimon in Wales, and by Calver and Cammeraat (1993) for an experimental hillslope in Luxembourg.

As a result of this form of discretization, there is an implicit assumption that the soil and surface parameters must be considered to be constant across the width of the hillslope (in the same way as for the SHE model effective values are required for each grid element). Variations in parameter values between different soil horizons, or for individual elements in a discretization such as that of Figure 5.1, can be represented but these must be effective values integrating over any heterogeneity across the slope. Thus, it may be difficult to measure such values in the field and Calver and Wood (1995), for example, report that their experience in using the model is that measured values of hydraulic conductivity tend to underestimate the values required to represent fast subsurface stormflow in the model.

In Australia, two similar models, THALES and TOPOG, have been developed based on the TAPES-C topographic analysis package which identifies one-dimensional downslope sequences of hillslope elements from contour data without any intervening interpolation onto a raster elevation grid (see, for example, Figure 3.5). Both models use a kinematic wave approximation of downslope flows in the saturated zone and are described in more detail in Section 5.5.3 below.

5.3 Case Study: Modelling Flow Processes at Reynolds Creek, Idaho

Reynolds Creek is a 234 km^2 rangeland catchment in the Owyhee Mountains of Idaho, managed by the USDA North West Watershed Research Centre. It was the site of one of the very first attempts to evaluate the predictions of a distributed process-based hydrological model. Stephenson and Freeze (1974) used a two-dimensional finite-difference partially saturated Darcian subsurface flow model in an application to a vertical slice through a complex hillslope within the Reynolds Creek catchment. Model predictions were checked against field measurements made during a snowmelt season. After making initial estimates of parameter values for the soil and rock layers in the slope, they carried out a trial and error calibration of the model, adjusting the parameter values to try and improve the fit to the observations. At that time, computer constraints severely limited the number of calibration runs that could be made. The results of the best simulation are shown in Figure 5.5.

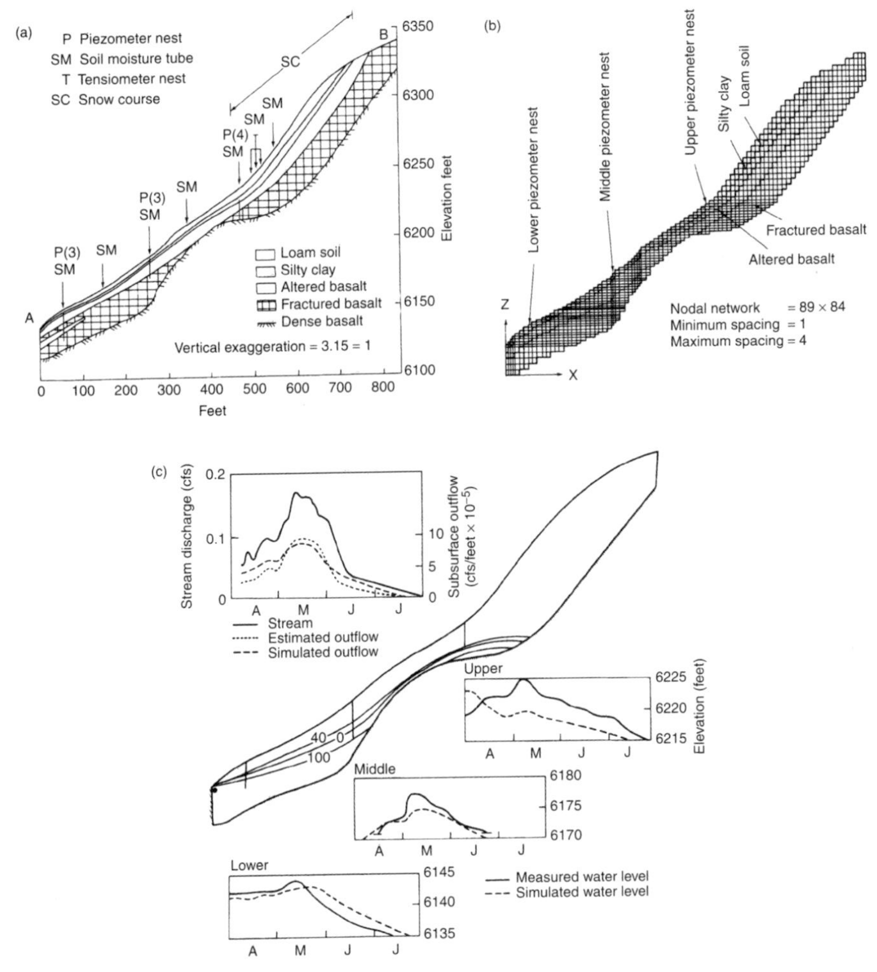

Figure 5.5 Process-based modelling of the Reynolds Creek hillslope. (a) Topography, geology and instrumentation. (b) Discretization of the hillslope for the finite-difference model. (c) Calibrated transient simulation results for the 5 April–13 July 1971 melt season (after Stephenson and Freeze 1974). Reproduced from Water Resources Research 10(2): 284–298, 1974, copyright by the American Geophysical Union

This study is interesting, even after 25 years, because it is one of the first studies to recognize that there might be limitations to the application and validation of this type of model at particular sites. They conclude, 'we recognize that our calibration is less than perfect but it is probably representative of what can be attained when a fully deterministic mathematical model is applied to a field site with a fairly complete, but as always imperfect, set of field measurements' (Stephenson and Freeze 1974, p. 293).

They also noted that validating such models was a particularly difficult problem since it presupposes perfect knowledge of all the boundary conditions, parameter values and initial conditions required. Imperfect knowledge will always introduce a degree of flexibility (or uncertainty) in any attempt at validation of a model.

More recently, the SHE model has been applied to the 40.4 ha Upper Sheep Creek subcatchment by Bathurst and Cooley (1996). At this site, at an elevation of over 2000 m in the headwaters of Reynolds Creek, average annual precipitation is of the order of 1016 mm, with more than 70 percent falling as snow. Snow accumulation is highly variable, with a deeper pack building up in the lee of a ridge where the pack may reach depths of more than 5 m.

Bathurst and Cooley (1996) have simulated a single snowmelt period using both energy budget and degree-day snowmelt models within the SHE model framework. A model discretization based on 161 grid cells (50 m × 50 m) was used. All parameters for the model were specified on the basis of knowledge of the catchment soils and vegetation. Initial snowpack characteristics were assigned on the basis of snow course information and photographs; initial saturated zone thickness was assigned to reproduce the initial flow at the start of the simulation. It is not clear from the paper how the initial unsaturated soil moisture profiles were defined, and only a 12 hour period was allowed for the model to 'run in' before the predictions were compared with observations, so some sensitivity to the definition of the initial pattern of moisture on the hillslope should be expected.

The study aimed to test four different hypotheses about the processes in the catchment based on how well the model reproduces the stream discharge during the simulation period. These hypotheses varied in assumptions about the extent of frozen soil and depth of the effective impermeable layer. Unlike the Stephenson and Freeze (1974) paper, no attempt was made to validate the model predictions against internal state measurements.

The authors state that 'the traditional calibration approach of adjusting the parameter values (within each hypothesis) to improve the agreement played a secondary role and was carried out under the constraint that the [parameter] values must reflect the field measurements where these exist or should otherwise lie within physically realistic limits' (Bathurst and Cooley 1996, p. 194). The earliest run number reported in the paper is 69, the last 107. It is clear that this application, like that of Stephenson and Freeze 20 years earlier, was still limited by computer run times.

The best discharge predictions (shown in Figure 5.6(a)) were found for the hypothesis which assumed that the majority of the runoff is generated by a near-surface subsurface flow mechanism close to the stream while, on the rest of the slope, snowmelt infiltrates the soil surface and percolates vertically to a deep saturated zone in the porous weathered basalt. This is consistent with the earlier, much more limited, study of Stephenson and Freeze (1974) but in a sensitivity analysis it was found that different parameter sets gave equally acceptable results within the limitations of the data available for model evaluation.

The degree-day snowmelt calculations could also produce acceptable discharge predictions (Figure 5.6(b)) but only after calibration of the degree-day coefficient to a value that was high relative to those reported in the literature.

5.4 Case Study: Blind Validation Test of the SHE Model on the Rimbaud Catchment, France

Ewen and Parkin (1996) have outlined a methodology for the *blind validation* of a hydrological model that involves the specification of tests and criteria of success before

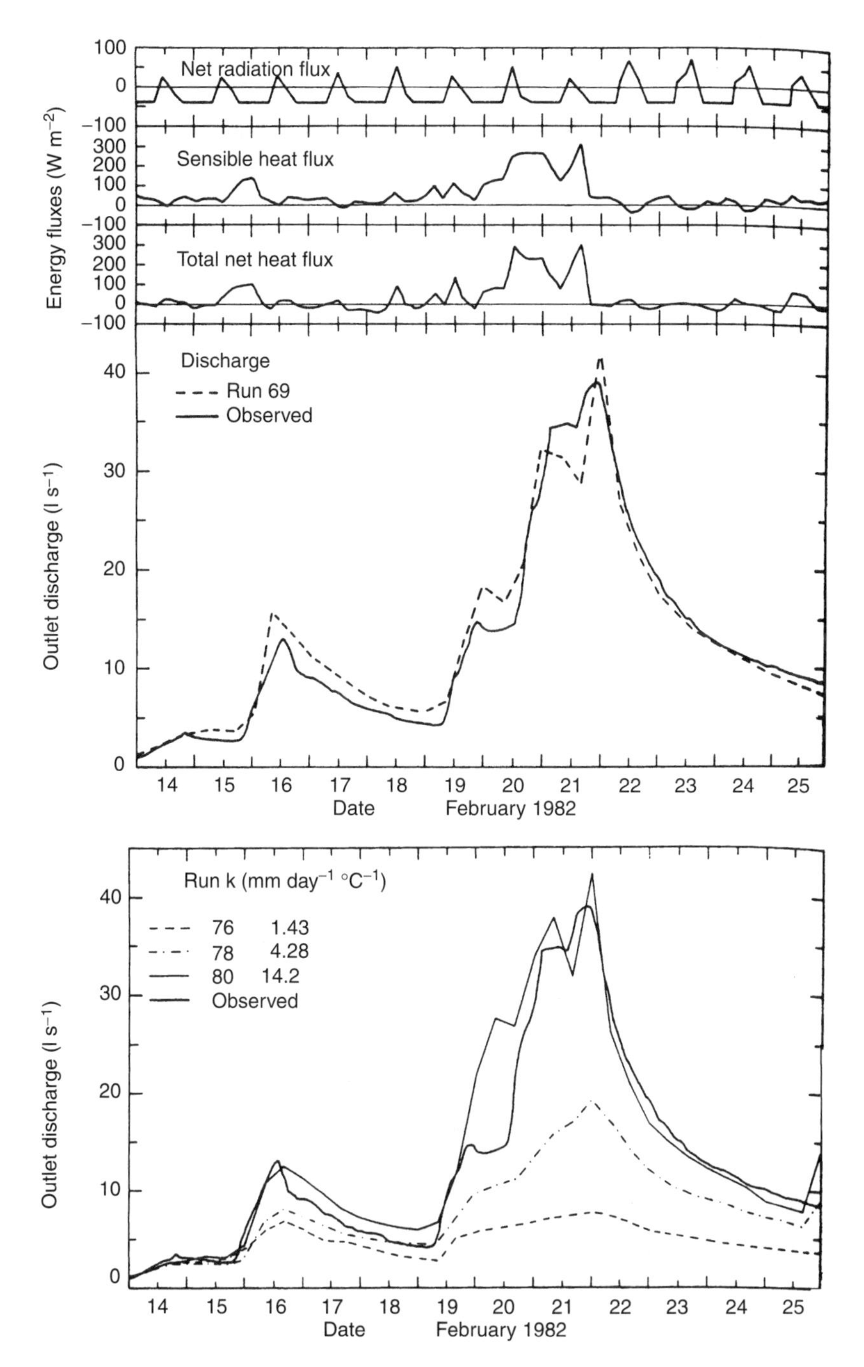

Figure 5.6 *Results of the Bathurst and Cooley (1996) SHE modelling of the Upper Sheep Creek subcatchment of Reynolds Creek: (a) Using the best-fit energy budget snowmelt model; (b) Using different coefficients in a degree-day snowmelt model. Reprinted from Journal of Hydrology 175, Bathurst and Cooley, 181–211, copyright (1996), with permission of Elsevier Science*

the model simulations are compared with observed discharges or other observations. In an application of this methodology to the 1.4 km^2 Rimbaud catchment in the Maures Massif near Toulon in southern France, Parkin *et al.* (1996) tested the SHETRAN version of SHE using only prior estimates of the parameter values. The Rimbaud catchment is one of a number of nested subcatchments in the Real Collobrier basin managed by CEMAGREF.

The catchment was represented by 144 grid squares of dimensions 100 m × 100 m (Figure 5.7). The parameters required by the model (similar to those specified in Table 5.1) were estimated from information about the soils and vegetation. Uncertainty in these estimates was allowed by specifying a range for each parameter. Some general information about the runoff responses was used to set the criteria for success in the model evaluation. The evaluation was 'blind' in that the modellers did not have access to the observed discharge record from the catchment before making the model runs.

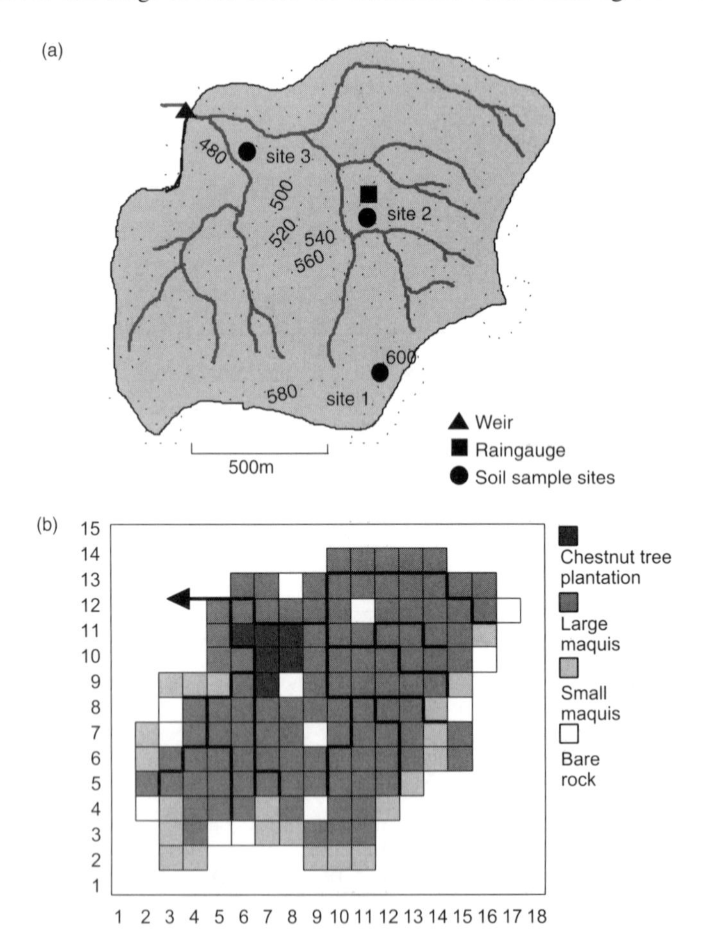

Figure 5.7 *The Rimbaud catchment used in the SHE model blind evaluation tests (after Parkin et al. 1996). (a) Topography and instrumentation. (b) SHE discretization showing channel reaches and modelled vegetation pattern. Reprinted from Journal of Hydrology 175, 595–613, Copyright (1996), with permission from Elsevier Science*

The model was run for a 13-month period during 1968–1970. Computer run times were still a limitation in this study. Rather than sample combinations of parameter values across the full prior ranges, a limited number of combinations of values at the range limits was used on the basis that these extremes should bracket the range of feasible model responses. Even so, some of these runs were then rejected on the basis of the judgement of the modellers as unlikely to be feasible, before the evaluation. The bounds of predicted discharges for part of the simulated period are shown in Figure 5.8. These bounds therefore represent a censored range of simulations.

The model was evaluated on four criteria set before the start of the blind test. It was required that the prediction bounds bracket 90 percent of the observed discharges, 90 percent of the peak discharges, 11 out of 13 monthly runoff volumes and the total

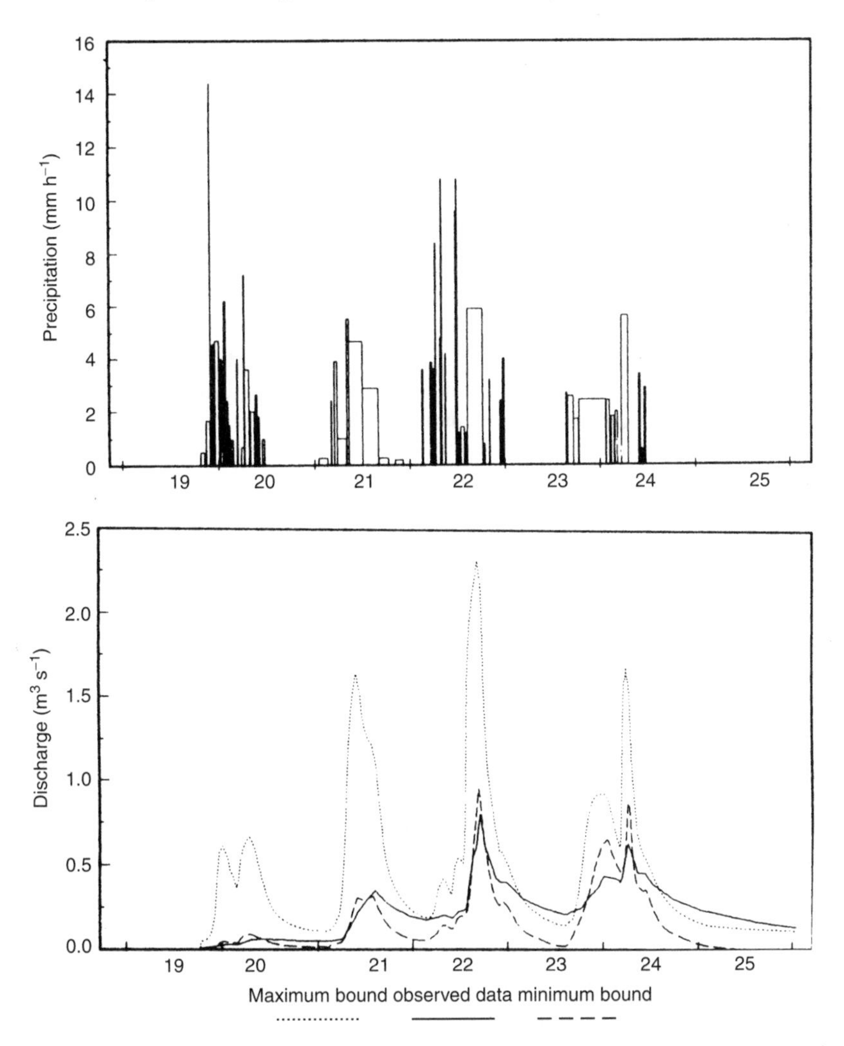

Figure 5.8 *Censored prediction limits using prior estimates of parameter ranges for the Rimbaud catchment, France, 19–25 February 1968 (after Parkin et al. 1996). Reprinted from Journal of Hydrology 175, 595–613, copyright (1996), with permission from Elsevier Science*

runoff volume. In fact, the model was successful in only the last of these criteria. Only 78 percent of the discharge hydrograph was in the prediction bounds, 47 percent of the peak flows (the model tended to generate runoff by an infiltration excess mechanism resulting in an overestimating of the peaks and storm recessions that were too steep), and 10 of the 13 monthly flows were within the prediction bounds. No attempt was made to evaluate any model predictions of internal catchment variables relative to observations.

This study is one of very few studies using such a blind testing framework, although Refsgaard and Knudsen (1996) also report on the use of the MIKE SHE model with prior estimation of parameters (the 'proxy basin' validation test of Klemeš 1986). They also found only limited success, with no apparent advantage of this type of model over more conceptual models in such an application, concluding that all types of models benefit from at least a short period of calibration (see also Section 10.5). Somewhat greater success has been claimed by Lange *et al.* (1999) in predicting peak flows in a semiarid environment using a simpler distributed model based on Hortonian infiltration excess concepts, but with surface runoff generation based directly on field measured plot infiltration experiments.

5.5 Simplified Distributed Models

The latest versions of SHE and similar physically based rainfall–runoff models represent the most complex rainfall–runoff models available. They have the advantage that they are based on physical theory though, as we have seen, they make significant simplifying assumptions to allow a model that is computationally feasible. They also have the important advantage that they make predictions that are distributed in space so that the effects of partial changes to the catchment, and the predicted spatial dynamics of the processes can be assessed. They have important disadvantages in the computational resources required and in the problems of specifying the huge numbers of parameters required over all the spatial elements of the model. These disadvantages have led to the investigation of simplified distributed models of two main types. One type is that based on kinematic wave theory, which is considered in the remainder of this chapter. The second is the type of probability distributed models considered in the next chapter, in which elements with similar characteristics are grouped together to reduce the calculations required.

5.5.1 Kinematic Wave Models

Kinematic wave models are simplified versions of the surface or subsurface flow equations of the previous section, resulting from making additional approximations. The first such model reported in the literature was, in fact, a grid-based model of surface runoff developed by Merrill Bernard in 1937 (see Hjelmfelt and Amerman 1980). In another early study, Keulegan (1945) analysed the magnitude of the various terms of the St Venant equations for shallow surface runoff over a sloping plane and concluded that a simplified equation, essentially the kinematic wave equation, was an adequate approximation. The flow routing of the Huggins and Monke (1968) grid-based model mentioned above was also effectively a kinematic wave solution. There are some problems in applying kinematic wave principles in two-dimensional cases (see

Section 5.5.5) and most models have used a catchment discretization based on one-dimensional hillslope planes (as in Figure 5.4). Early models used fixed width planes or planes with radial symmetry to allow analytical solutions to be made, but it is now easy to implement variable width planes in numerical solutions (e.g. Li *et al.* 1975).

All kinematic wave models are combinations of the continuity equation with a storage–flow relationship (Box 5.7). Generally some simple mathematical function is used for the storage–flow relationship but this is not strictly necessary. Numerical solutions could use any function represented as a look-up table, even hysteretic functions, although to my knowledge hysteretic functions have never been tried. However, the resulting models are flexible and relatively easy to implement, with many analytical solutions available for simple boundary conditions. A comprehensive coverage of kinematic wave theory and its application in surface hydrology has been published by Singh (1996). The general kinematic wave equation for a variable-width flow domain has the following form:

$$W_x \frac{\partial h}{\partial t} = -c \frac{\partial W_x h}{\partial x} + W_x r \tag{5.6}$$

where h is a depth of flow, W_x is the width of the slope or channel, r is the inflow rate per unit area of slope or channel, and c is the kinematic wave velocity or *celerity* which will in general be a function of flow depth (but may be a constant in some special cases). The form of that function will vary with the relationship between downslope flow rate and depth of flow (see Box 5.7 for surface and subsurface flow examples).

There is one important limitation of using kinematic wave models, even in applications to one-dimensional systems. Unlike both the Richards equation for subsurface flow and the St Venant equations and diffusion wave analogy for surface flow, the kinematic wave equation cannot reproduce the effects of a downstream boundary condition on the flow. Essentially the effects of any disturbance to the flow will generate a kinematic wave, but the equation can only predict the downslope or downstream movement of these waves. Thus a kinematic wave description cannot predict the effects of the drawdown of a water table due to an incised channel at the base of a hillslope, or the backwater effects of an obstruction to the flow for a surface flow. This has led to a number of theoretical studies of the conditions under which the kinematic approximation is a valid approximation to a more complete description (see Box 5.7), but it is worth noting that these are theoretical studies, comparing one mathematical description with another. The problems of parameter estimation and uncertain knowledge of the subsurface geometry and values of recharge or lateral inflows will often mean that these differences are not so important in real applications, and that the kinematic wave approximation may be a useful predictive model. This has been demonstrated, for example, in the study of Zoppou and O'Neill (1982) in a comparison of methods for routing flood waves on the River Yarra in Australia (Figure 5.9).

5.5.2 Kinematic Wave Models for Surface Runoff

An early exposition of the mathematics of kinematic wave theory by Lighthill and Whitham (1955) used traffic routing and flow routing in channels as example applications. This work was later developed by Eagleson (1970) for the case of overland flow

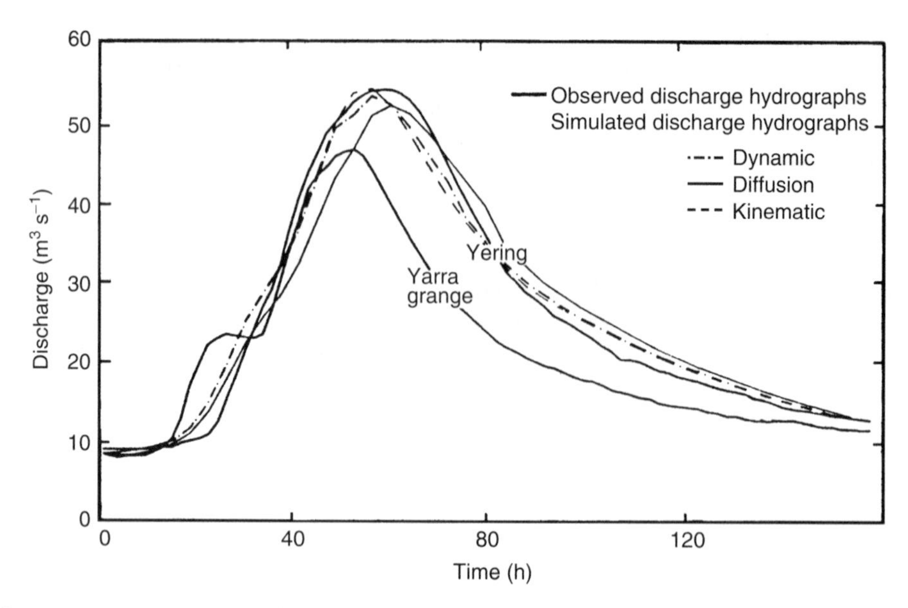

Figure 5.9 *A comparison of different routing methods applied to a reach of the River Yarra, Australia (after Zoppou and O'Neill 1982)*

routing on hillslopes to predict hydrographs. Eagleson gave analytical solutions for the case of a constant effective rainfall input. Later Li *et al.* (1975) provided a simple numerical solution that could be used for arbitrary sequences of inputs. Numerical solutions have been used in a number of catchment rainfall–runoff models based on the infiltration excess overland flow runoff mechanism, the most well known of which are probably KINEROS (Smith *et al.* 1995) and the US Corps of Engineers model HEC1 (Feldman 1995). Both of these models treat a catchment area as a sequence of hillslope segments, bounded by streamlines. Flow is treated as one-dimensional in the downslope direction. Each hillslope may be represented by a single plane, or by a cascade of planes of different widths and slopes. In fact, as shown in equation (5.6) (see also Box 5.7), it is not difficult to include continuous changes in width and slope in the kinematic wave equations. This makes analytical solutions difficult, but is not a problem with numerical solutions. Goodrich *et al.* (1991) describe a finite-element solution of the kinematic wave equation for overland flow for a catchment discretiztion based on a triangular irregular network (TIN).

The one-dimensional description requires an appropriate function for the storage–discharge relationship. This may be different for overland and channel flows. However, it has been common in surface water hydrology to use a uniform flow relationship such as the Manning equation for both overland and channel flows. The Manning equation has the form

$$v = \frac{1}{n} S_o^{0.5} R_{\mathrm{h}}^{0.67} \tag{5.7}$$

where R_{h} is the hydraulic radius, and S_o is the local slope angle. Recall that the hydraulic radius is defined as the cross-sectional area of the flow, A, divided by the wetted perimeter, P. Thus, for flows that are wide compared with their depth $R_h = A/P \approx Wh/W \simeq h$, where W is the width of the flow and h is the local flow depth. The discharge can then be calculated as

$$Q = vhW = \frac{1}{n} W S_o^{0.5} h^{1.67} \tag{5.8}$$

This has the general form of the power law storage–discharge relationship used in Box 5.7 (where an equivalent expression is developed for the Darcy–Weisbach uniform flow equation)

$$q = bh^a \tag{5.9}$$

where the specific discharge $q = Q/W$, and for Manning's equation, $b = S_o^{0.5}/n$ and $a = 1.67$. The kinematic wave velocity or celerity c is equal to the rate of change of discharge with storage (here $\mathrm{d}q/\mathrm{d}h$). It is an expression of the rate at which the effects of a local disturbance will propagate downslope or downstream. For the power law, $c = (abh^{a-1})$, and this wave velocity will always increase with discharge if $a > 1$. For $a = 1$, q is a linear function of h and both the flow velocity and the wave speed c are constant with changing discharge.

Other relationships may show different types of behaviour. Wong and Laurenson (1983) show, for a number of Australian river reaches, how the form of the relationship between wave speed and flow in a river channel may change as the flow approaches bankfull discharge and goes overbank (Figure 5.10). At a much smaller scale, Beven (1979), for example, showed that field measurements in the channels of a small upland catchment suggested a velocity–discharge relationship of the form

$$v = \frac{aQ}{b + Q} \tag{5.10}$$

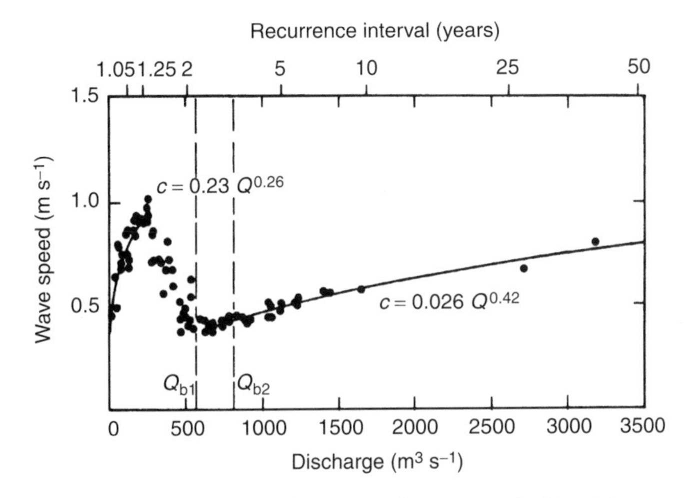

Figure 5.10 *Wave speed–discharge relation on the Murrumbidgee River over a reach of 195 km between Wagga Wagga and Narrandera (after Wong and Laurenson 1983).* Q_{b1} *is the reach flood warning discharge;* Q_{b2} *is the reach bankfull discharge. Reproduced from Water Resources Research 19: 701–706, 1983, copyright by the American Geophysical Union*

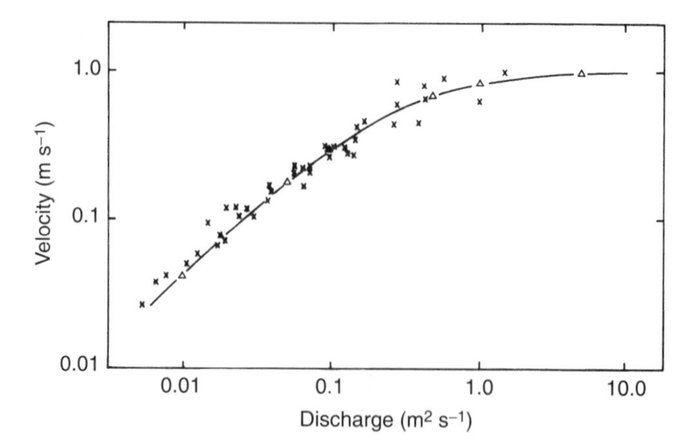

Figure 5.11 Average velocity versus discharge measurements for several reaches in the severn catchment at Plynlimon, Wales, together with a fitted function of the form of equation (5.10) which suggest a constant wave speed of 1 m s^{-1} (after Beven 1979b). Reproduced from Water Resources Research 15: 1238–1242, 1979, copyright by the American Geophysical Union

or, assuming an irregular channel with $Q = vA$

$$Q = a(A - b) \tag{5.11}$$

where b is interpreted as the cross-sectional area at a discharge of zero (not unreasonably in a channel with pools). This relationship gives a constant celerity of $c = a$ for all flows, even though the flow velocity of the water is always increasing with discharge (Figure 5.11). This then allows a further simplification of the routing procedure to the type of constant wave velocity network width based routing discussed in Section 4.6.1. For the particular data set from small channels in upland Wales examined by Beven (1979), the value of a was 1 ms^{-1} and was always faster than the mean flow velocity (Figure 5.11).

5.5.3 Kinematic Wave Models for Subsurface Stormflow

In the application of the kinematic wave model to saturated downslope subsurface flow, a similar representation of the catchment as one-dimensional streamtubes can be made. The important simplification made, relative to more complex descriptions of such flows, is that the hydraulic gradient can be approximated by the bed slope angle $\sin\beta$ (or, as an approximation, the surface slope). It is therefore assumed that the water table is approximately parallel to the bed (or surface). The Darcy velocity (velocity per unit cross-sectional area of the flow) is then given by

$$v_x = K_s \sin\beta \tag{5.12}$$

where v_x is the Darcian velocity (flux per unit cross-sectional area of saturated soil) measured with respect to the downslope distance x (measured along the slope), K_s is the saturated hydraulic conductivity of the soil (for the moment assumed constant with depth of saturation), and $\sin\beta$ is the slope angle. The kinematic wave approximation was first applied to saturated subsurface stormflow by Henderson and Wooding

(1964). Later, Beven (1981) showed that, at least for steeper slopes and high hydraulic conductivities, it could be a useful approximation to a more complete description of shallow saturated flow over an impermeable layer on a hillslope (see Box 5.7). This work was later extended to include delays associated with the propagation of a wetting front into the soil before recharge starts and different profiles of hydraulic conductivity with depth (Beven 1982). The kinematic wave equation will be a better approximation if hydraulic conductivity increases with depth of saturation, as is the case in many soils due to the increased macroporosity expected near the surface (Kirkby 1988).

For the constant hydraulic conductivity case, an examination of the wave velocity is interesting. As shown in Box 5.7, for saturated subsurface flow the wave velocity is given by

$$c = K_s \sin\beta/\varepsilon \tag{5.13}$$

where ε is a storage coefficient representing the 'effective' difference between soil water content and saturation in the region above the water table. The value of ε is always less than 1 so that c is always greater than v_x. If all the three variables controlling the wave velocity were constant, c would be constant, but in practice ε is likely to vary in magnitude both with depth of saturation and distance downslope. For wet soils ε may be very small. In this case, comparing the expression for c with that for the Darcy velocity v_x, the wave velocity c may be very much faster than the Darcy velocity. That is to say, disturbances to the flow, such as new input of recharge, must propagate downslope faster than the Darcy velocity of the flow. The effects of recharge will also propagate downslope faster than mean pore water velocity, i.e. the average flow velocity through the part of the cross-section that is pore space rather than solids. This is given, for saturated conditions, by

$$v_p = K_s \sin\beta/\theta_s \tag{5.14}$$

where θ_s is the porosity. The wave velocity will be faster than v_p since ε will always be smaller than θ_s.

This is one explanation why storm hydrographs that have an important contribution of subsurface flow, tend to show a high proportion of 'old' water, even at peak flows (see Section 1.5). The effects of recharge to the saturated zone will move downslope with the wave velocity, which is faster than the pore water velocity. The effect will be that the discharge downslope rises more quickly than water can flow from significant distances upslope, so that some of the water which flows out of the slope must be displaced from storage in what was the unsaturated zone above the water table prior to the event. The same analysis holds for more complex descriptions of subsurface flows, but the comparison of the v_x, v_p and c velocities in the kinematic wave description illustrates the effect quite nicely.

The THALES and TOPOG models (Grayson *et al.* 1992a, 1995; Vertessy *et al.* 1993; Vertessy and Elsenbeer 1999; Zhang *et al.* 1999) are both models using the kinematic wave approximation on one-dimensional sequences of hillslope elements representing a catchment. Both are based on the TAPES-C digital terrain analysis package (see Section 3.7). THALES allows that each element may have different infiltration characteristics (using either the Green–Ampt or Smith–Parlange infiltration models described in Box 5.2), vertical flow in the unsaturated zone assuming a unit hydraulic gradient and using the Brooks–Corey soil moisture characteristics of Box 5.4, and downslope saturated zone flow using the one-dimensional kinematic wave approximation of Box 5.7. The use of THALES in an application to the Walnut Gulch catchment is described in Section 5.6. The TOPOG-Dynamic model, developed by CSIRO in Australia, uses an analytical solution of the Richards equation to describe vertical flow in the unsaturated zone and an explicit kinematic wave solution for the lateral saturated zone fluxes (Vertessy *et al.* 1993). Latest versions include plant growth and carbon balance components for ecohydrological modelling of the effects of land use change (Dawes *et al.* 1997; Zhang *et al.* 1999).

5.5.4 Kinematic Wave Models for Snowpack Runoff

An early implementation of the kinematic wave equation in hydrology was for modelling flow through a snowpack by Colbeck (1974). The model was later used by Dunne *et al.* (1976) and further analytical solutions have recently been outlined by Singh *et al.* (1997). All of these studies have assumed a snowpack of constant porosity and hydraulic conductivity. The earlier studies assumed that all the specified net melt would percolate downslope through the snowpack; the later study of Singh *et al.* explicitly incorporates the effects of infiltration into underlying soil. In the general case, the time-varying infiltration rate may be a function of the depth of saturation in the pack. Singh *et al.* (1997) provide analytical solutions for the case where the infiltration rate may be assumed to be constant.

5.5.5 Kinematic Shocks and Numerical Solutions

One of the problems of applying kinematic wave theory to hydrological systems is the problem of kinematic shocks (see Singh 1996). The kinematic wave velocity can be thought of as the speed with which a particular storage or depth value is moving downslope. If the wave velocity increases with storage, then waves associated with larger depths will move downslope faster than waves associated with shallower depths. Very often this may not be a problem. In channels, kinematic shocks will be rare (Ponce 1991). On a hillslope plane of fixed width and slope subjected to uniform rainfall, wave velocity will never decrease downslope and there will be no shock. If, however, a concave-upwards hillslope is represented as a cascade of planes and subjected to a uniform rainfall, then faster flow from the steeper part of the slope will tend to accumulate as greater storage as the slope decreases, causing a steepening wave until a kinematic shock front occurs. The paths followed by kinematic waves in a plot of distance against time are known as characteristic curves. A shock front occurs when two characteristic curves intersect on such a plot. The effect of such a shock front reaching the base of a slope would be a sudden jump in the discharge.

This is clearly not realistic. It is a product of the mathematics of the kinematic wave approximation, not of the physics of the system itself which would tend to disperse such sharp fronts. One example of a kinematic shock that can be handled analytically using kinematic wave theory is the movement and redistribution of a wetting front into an unsaturated soil or macropore system (Beven 1982, 1984; Smith 1983; Charbeneau 1984; Germann 1990). In fact, the Green and Ampt (1911) infiltration model outlined in Box 5.2 can be interpreted as the solution of a kinematic wave description of infiltration, with the wetting front acting as a shock wave moving into the soil. Models have been developed that try to take account of such shocks by solving for the position of the characteristic curves (called the method of characteristics; see Borah *et al.* 1980), but most numerical solutions rely on numerical dispersion to take care of such shocks.

Using a finite-difference form of the kinematic wave equation to obtain a solution, the approximation adds some artificial or numerical dispersion to the solution. If numerical dispersion is used to reduce the problem of shocks in a solution of the kinematic wave equation (for cases where shocks might occur), then the solution will not be a true solution of the original kinematic wave equation. There is a certain irony to this problem in that the approximate solution might then be more 'realistic' since, as already noted, nature would tend to disperse such sharp fronts (Ponce 1991). However, numerical dispersion is not well controlled; the effect will depend on the space and time increments used in the solution, together with any other solution parameters. The reader needs be aware, however, of the potential problem posed by shocks and of the fact that the approximate solution may not be consistent with the solution of the original equation in such cases. There is also a possibility that a large shock might lead to instability of the approximate numerical solution.

Shocks occur where a larger wave (a depth moving through the flow system) catches up with or meets another smaller one. This will tend to occur where flow is slowed for some reason (reduction in slope, increase in roughness) or where there is flow convergence. A number of models that purport to use two-dimensional kinematic wave equations (in plan) have been proposed, going all the way back to the original grid-based model of Bernard in 1937 (see Hjelmfelt and Amerman 1980). More recent examples are the catchment surface flow model of Willgoose and Kuczera (1995), and the subsurface stormflow model of Wigmosta *et al.* (1994). Mathematically this is not really a good idea. Wherever there is flow convergence there is the possibility of two different kinematic waves meeting and a shock front developing. This was, in fact, explicitly recognized by Bernard in the 1930s (see Hjelmfelt and Amerman 1980) but there appears to be nothing in most of the more recent models to explicitly deal with shocks. It seems that they generally rely on numerical dispersion to smear out any effects of shock fronts. The models work and have been shown to give good results in some test cases, but if shocks are dispersed in this way they are not true solutions of the original kinematic wave equations. I am aware of this problem because I once spent several months trying to develop a finite-difference two-dimensional kinematic wave model. The solution kept going unstable for some test runs on a real catchment topography. It took a long while for me to realize that this was not just a program bug, or a matter of getting the numerics of the approximate solution right, but a problem inherent in the mathematics of a kinematic wave description. With hindsight, the reason is fairly obvious!

Having raised the issue of kinematic shocks and numerical dispersion, it should be added that it is not necessarily a problem that should worry us unduly. Problems of parameter calibration might well dominate any physical and theoretical approximations inherent in the numerical solution of kinematic wave equations. We can in fact use the numerical dispersion to our advantage, if the resulting model is actually more realistic. An example here is the still widely used Muskingum–Cunge channel flow routing model. The Muskingum method was originally developed as a conceptual flow routing model for the Muskingum River in the 1930s. Cunge (1969) showed that the Muskingum method was equivalent to a four-point explicit finite-difference solution of the kinematic wave approximation for surface flow. It follows that, for this interpretation, any dispersive and peak attenuation effects of the Muskingum–Cunge routing model come from numerical dispersion associated with the finite-difference approximation. In applying the model it is normal to fit its two parameters so as to match the observed peak attenuation, which allows some control over the numerical dispersion by parameter calibration. This is an interesting historical example, but the details of the Muskingum–Cunge model will not be given here as, in applications to long reaches, it suffers another serious defect of not properly allowing for the advective time delays in the channel (i.e. any change in the upstream inflow has an immediate effect on the predicted reach outflow, regardless of the length of the reach). This means that the transfer function of the Muskingum–Cunge model will often show an initially negative response (e.g. Venetis 1969) as a way of producing a time delay. The more general transfer function techniques described in Chapter 4, which include the possibility of an explicit time delay, are a better approach to defining a simple flow routing model where observed hydrographs are available for model calibration (Young and Wallis, 1985). The Muskingum–Cunge model is, in fact, a specific case of the general linear transfer functions outlined in Box 4.1, mathematically equivalent to a first-order (one a coefficient) model, with two b coefficients and zero time delay.

The problems of kinematic shocks can, in fact, be largely avoided in many situations by solving each node separately, as if it was a node on a one-dimensional plane, by taking advantage of the fact that kinematic waves only move downslope. Thus, in theory, there is no dependence of the solution on downslope conditions. If the kinematic wave equation is solved for flow, q, at a node, then all the inputs from upslope can be lumped together as one input, even if they converge from more than one node. The solution is made, and the resulting inputs can then be dispersed to form the inputs for one or more downslope nodes as necessary. This is the approach used in the subsurface kinematic model of Wigmosta *et al.* (1994) which is based on a two-dimensional raster elevation grid, solving for a depth of saturation for every grid element. Palacios-Velez *et al.* (1998) provide an algorithm for constructing such a cascade of kinematic solution elements for both TIN and raster spatial discretizations of a catchment. A cautionary note that should be added here is that this approach will still add numerical dispersion in a way that is not well controlled, and may still be subject to stability problems under rapidly changing conditions. The user should also always remember that a kinematic wave model cannot model backwater effects: in channels, because of weirs or restrictions to the flow, or in subsurface flow, as a result of groundwater ridging or the drawdown effect of an incised channel in low-slope riparian areas. At least a diffusion analogy model is required to simulate such effects.

The simplicity of the kinematic wave equation with its straightforward combination of the continuity equation and a storage–discharge function, makes it very appealing as an approximation of the real physics. It is particularly appealing in cases where some 'effective' storage–discharge function might be required to take account of limited understanding of flow over rilled surfaces or flow through structured soils that might not be well represented by the normal theory used for surface and subsurface flow (e.g Beven and Germann 1981; Faeh *et al.* 1997). This degree of flexibility makes it valuable as a modelling strategy. It remains necessary, however, to be aware of the limitations of the approach to essentially one-dimensional flows and the possible effects of kinematic shocks.

5.6 Case Study: Modelling Runoff Generation at Walnut Gulch, Arizona

One of the real challenges of semiarid hydrology is still to model the extensive data set collected by the USDA Agricultural Research Service on the well-known Walnut Gulch experimental catchment in Arizona. This has been the subject of numerous experimental and modelling studies, a selection of which are discussed in this section. This is a semiarid catchment, with 11 nested subcatchments that range in area from 2.3 to 150 km^2, and an additional 13 small catchment areas ranging from 0.004 to 0.89 km^2. Spatial variability in rainfall is assessed using a network of 92 gauges. The catchment has been the subject of two intensive field campaigns combining field measurements with aircraft platform remote sensing (Kustas and Goodrich 1994; Houser *et al.* 1998). The perception of runoff generation in this environment is that it is almost exclusively by an infiltration excess mechanism (Goodrich *et al.* 1994).

At the hillslope runoff plot scale in Walnut Gulch, Parsons *et al.* (1997) have compared observed and predicted discharges, together with flow depths and velocities, at several cross-sections. The model used the simplified storage-based infiltration model of Box 5.2 with two-dimensional kinematic wave routing downslope. The storage–discharge relationship used was a power law with parameters that varied with the percentage cover of desert pavement in each grid cell of the model. In the first application to a shrubland site, the model underpredicted the runoff generation but was relatively successful in predicting the shape of the experimental hydrograph (Figure 5.12). A second application to a grassland plot, reported in Parsons *et al.* (1997), was less successful, despite a number of modifications to the model, including the introduction of stochastic parameter values.

At a somewhat larger scale, Goodrich *et al.* (1994) and Faurès *et al.* (1995) have applied KINEROS (Smith *et al.* 1995) to the 4.4 ha Lucky Hills LH-104 subcatchment to examine the importance of different antecedent soil moisture estimates and the effects of wind and rainfall pattern on the predicted discharges (Figure 5.13). KINEROS uses a Smith–Parlange infiltration equation (see Box 5.2) coupled to one-dimensional kinematic wave overland flow routing on hillslope planes and in channel reaches. At this scale, both studies conclude that an adequate representation of the rainfall pattern is crucial to accurate runoff prediction in this environment. Using average initial soil moisture contents from different remote sensing and modelling methods had little effect on the predictions, as did trying to take account of the effects of wind direction and velocity on raingauge catch. However, checking the model predictions for different combinations of different numbers of raingauges they showed that combinations of four gauges (i.e. a density of one per hectare) gave a variation in predicted discharges that spanned the observed discharges and had a similar coefficient of variation to that estimated for the discharge measurements (Faurès *et al.* 1995). Goodrich *et al.* (1994) also looked at the sensitivity of runoff production to pattern of initial moisture

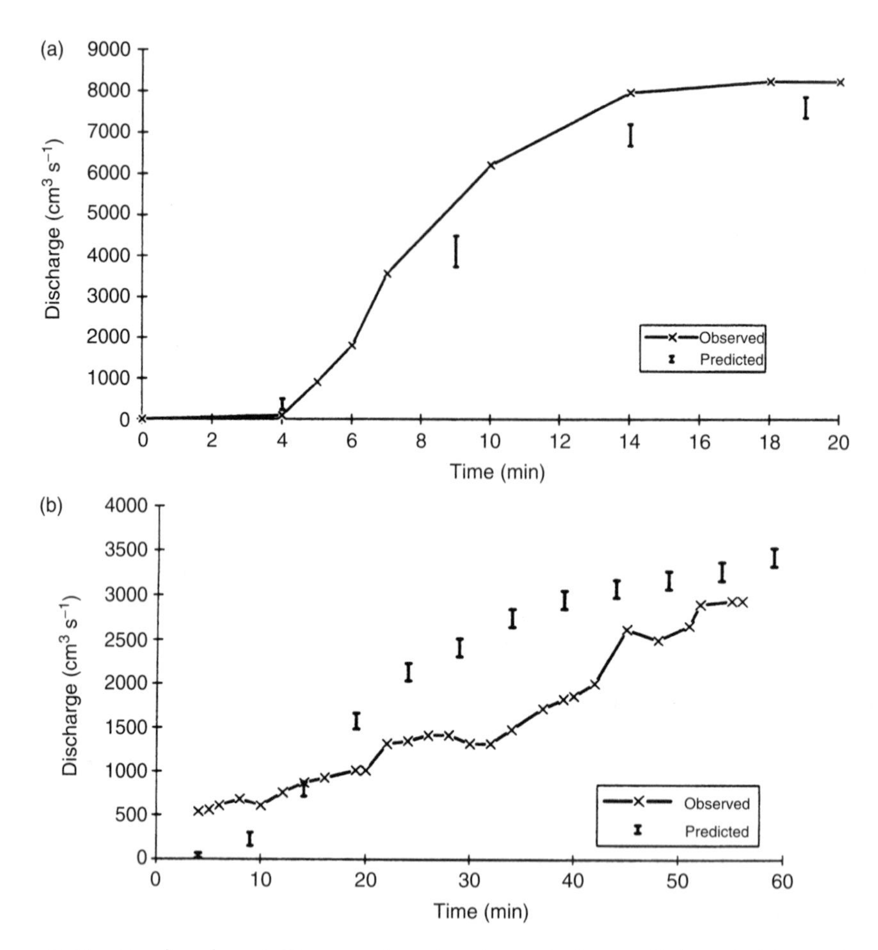

Figure 5.12 *Results of modelling runoff at the plot scale in the Walnut Gulch catchment. Results are for (a) the shrubland plot and (b) the grassland plot. The error bars on the predictions indicate the range of 10 randomly chosen sets of infiltration parameter values (after Parsons et al. 1997). Reproduced with permission of John Wiley & Sons Limited*

content at the larger scale of the WG-11 subcatchment (6.31 km^2). They suggest that a simple basin average of initial moisture content will normally prove adequate and that, again, knowledge of the rainfall patterns is far more important. A water balance model estimate of initial moisture content did as well as the remotely sensed estimates.

Michaud and Sorooshian (1994) compared three different models at the scale of the whole catchment, including a lumped SCS curve number model, a simple distributed SCS curve number model, and the more complex distributed KINEROS model. The modelled events were 24 severe thunderstorms (mean runoff coefficient 11 percent), with a raingauge density of one per 20 km^2. Their results suggested that none of the models could adequately predict peak discharges and runoff volumes, but that the distributed models did somewhat better in predicting time to runoff initiation and time to peak. The lumped model was, in this case, the least successful.

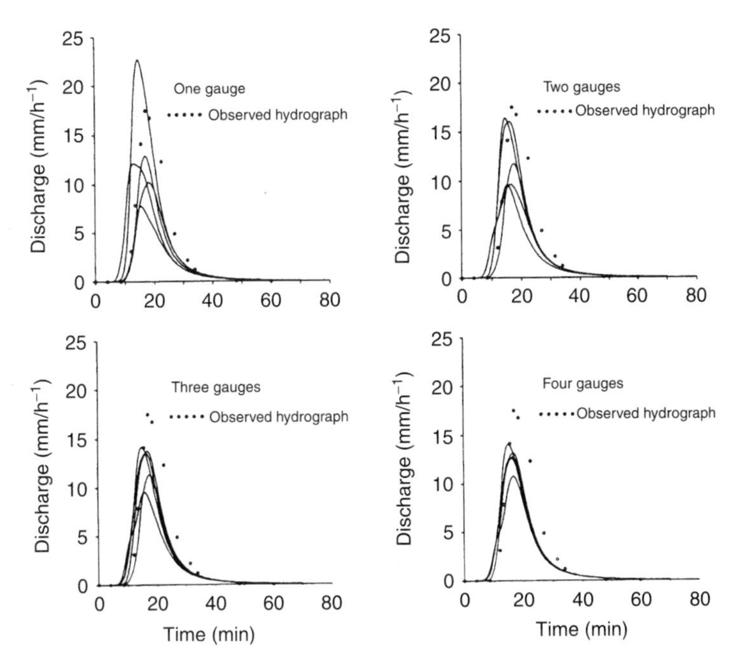

Figure 5.13 *Results of modelling the 4.4 ha Lucky Hills LH-104 catchment using KINEROS with different number of raingauges to determine catchment inputs (after Faurès et al. 1995). Reprinted from Journal of Hydrology 173: 309–326, copyright (1995), with permission from Elsevier Science*

More recently, Goodrich *et al.* (1997) have used the data from all the 29 nested subcatchments within Walnut Gulch, with drainage areas ranging from 0.2 to 13 100 ha, to investigate the effects of storm area and catchment scale on runoff coefficients. They conclude that, unlike humid areas, there is a tendency for runoff responses to become more nonlinear with increasing catchment scale in this type of semiarid catchment as a result of the loss of water into the bed of ephemeral channels and the decreasing relative size of rainstorm coverage with catchment area for any individual event. Detailed modelling studies were made using KINEROS for three of the catchments, LH-106 (0.34 ha), LH-104 (4.4 ha) and WG-11 (6.31 km^2). The model was calibrated by adjusting three multipliers applied to the distributed patterns of overland flow roughness, mean soil hydraulic conductivity and the coefficient of variation of hydraulic conductivity (which is assumed to have a log normal distribution as in Woolhiser and Goodrich (1988)), to improve the predictions for 10 calibration events. Twenty additional events were used to evaluate the calibrated models. The model was relatively successful in predicting the responses for the two small catchments, but less so for WG-11 where the performance on the validation events was much worse than in calibration. The results, however, confirmed the tendency towards increasing nonlinearity at the larger scale.

Other models that have been used at Walnut Gulch include THALES (Grayson *et al.* 1992a) which includes both infiltration excess and subsurface runoff components. Grayson *et al.* showed that both the goodness of fit of the model and the runoff generation mechanisms simulated were both very sensitive to the parameters of the model, which were difficult to estimate from the information available. By calibrating parameters, catchment outflows could be adequately predicted using both Hortonian infiltration excess and partial area runoff generation mechanisms. They suggest that the difficulties of validating such models are acute (see also the discussion in Grayson *et al.* 1992b).

Finally, a recent study of Houser *et al.* (1998) has applied a variant of the simplified distributed model TOPMODEL (see Chapter 6), called TOPLATS, to Walnut Gulch. This study is included here because it is one of the few studies in rainfall–runoff modelling that has attempted to include measurements of a distributed variable within a *data assimilation* or updating framework. Data assimilation is well established in other distributed modelling fields, such as numerical weather forecasting, but has not been widely used in distributed hydrological modelling. In this study the measured variable was surface soil moisture which was available from the Push Broom Passive Microwave Radiometer (PBMR) carried on an aircraft platform on six days during the MONSOON90 intensive field campaign, including a dry initial condition and the dry down period following a rainstorm of 50 mm. The resulting PBMR soil moisture images showed a very strong correlation with the pattern of rainfall volume interpolated for this storm, resulting in a strong spatial correlation. There are a number of difficulties in trying to make use of such data including the conversion from PBMR brightness temperature to surface soil moisture, the dependence of predicted surface soil moisture on numerous model parameters (the model has some 35 soil and vegetation parameters in all), and the choice of a method for updating the model given the PBMR images over only part of the catchment area. The study compared different methods of varying complexity (and computational burden), suggesting that there is a trade-off between the complexity of the method used and the ability to make use of all the data available. The study confirmed the importance of the rainfall forcing on the hydrologic response, not only for runoff generation but also for the spatial pattern of evapotranspiration and sensible heat fluxes back to the atmosphere.

5.7 Case Study: Modelling the R-5 Catchment at Chickasha, Oklahoma

The R-5 catchment at Chickasha, Oklahoma (0.1 km^2, Figure 5.14) has been the subject of a series of modelling papers using versions of the Quasi-Physically Based Rainfall–Runoff Model (QPBRRM) originally developed by Engman and Rogowski (1974). QPBRRM has many similarities to KINEROS but uses a 1-D analytical infiltration component based on the Philip equation (see Box 5.2); a 1-D kinematic wave overland flow component applied on a constant width hillslope plane discretization of the catchment, and a 1-D kinematic wave channel network routing algorithm. The model takes account of spatially variable runoff generation by an infiltration excess mechanism and the downslope infiltration of runoff as runon onto areas that are not yet saturated. The requirements of the model are therefore the spatially variable infiltration parameters for every grid element in the catchment discretization and the effective parameters of the storage–discharge relationship for the overland and channel flow model components.

Loague and Freeze (1985) applied the QPBRRM model using data on soil moisture of the three main soil types reported by Sharma *et al.* (1980). Loague and Gander

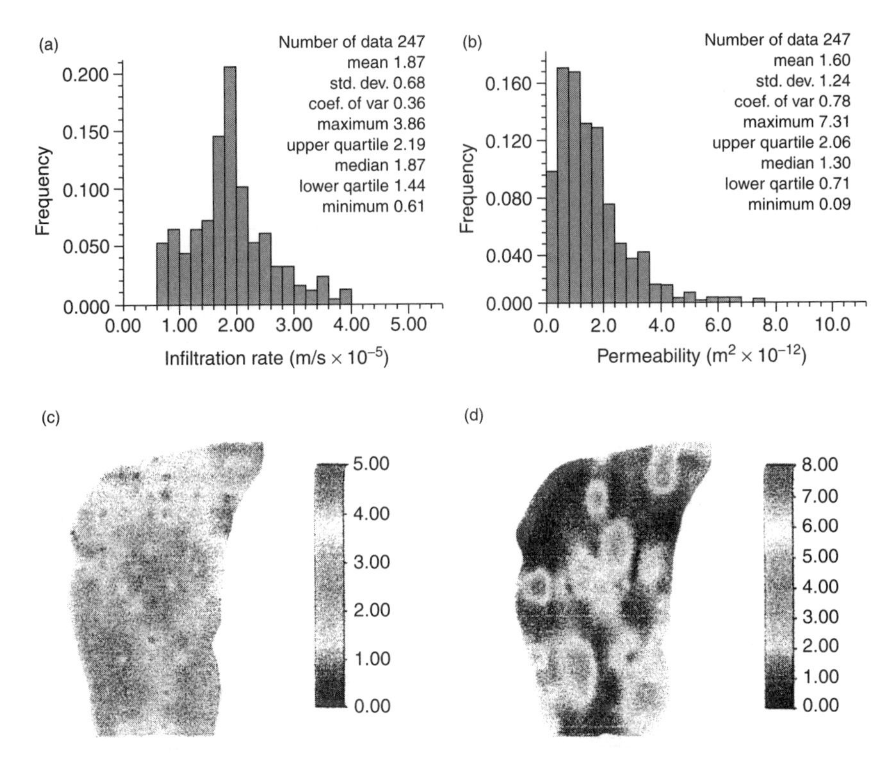

Figure 5.14 *Patterns of infiltration capacity on the R-5 catchment at Chickasha. (a) Distribution of 247 point measurements of infiltration rates. (b) Distribution of derived values of intrinsic permeability with correction to standard temperature. (c) Pattern of saturated hydraulic conductivity derived using a kriging interpolator. (d) Pattern of permeability derived using a kriging interpolator (after Loague and Kyriakidis 1997). Reproduced from Water Resources Research 33: 2883–2896, 1997, copyright by the American Geophysical Union*

(1990) added a further 247 infiltration measurements, based on a grid of 25 m spacing (157 sites) and two transects with measurements at spacings of 2 m and 5 m. In Loague and Kyriakidis (1997), a geostatistical kriging analysis is used to interpolate these data to any 1 m^2 grid point in the catchment, taking account of the effects of temperature on the density and viscosity of the infiltrating water that had been ignored in a similar interpolation by Loague and Gander (1990) and the modelling studies that had made use of their data (e.g. Loague 1990, 1992a,b). Loague and Kyriakidis (1997) have then applied the QPBRRM model to simulate three large rainfall–runoff events (rainfall depths between 33 and 68 mm), with a spatial grid of 2.5 m and a time step of 60 s. The catchment discretization used 959 overland flow plane segments, each with different infiltration parameters. Given all this detailed information, it turned out that applying the model with only field measured parameter values resulted in poorer hydrograph predictions than in the original Loague and Freeze (1985) study (Table 5.2). Adjusting the infiltration parameters to take account of the temperature effects in the original measurements did result in an improvement in the predictions, as did taking the ensemble average of a number of stochastically generated parameter fields relative to the mean values interpolated by kriging. The model significantly underestimates

Table 5.2 *Results of the application of the QPBRRM model (ensemble mean of predictions from 10 stochastic permeability field realizations) to the R-5 catchment without any parameter calibration (after Loague and Kyriakidis 1997).* P_D *is total storm rainfall (in mm),* I_{max} *is peak 2-min rainfall intensity (in mm h,* Q_D *is total stormflow volume (in mm),* Q_{PK} *is hydrograph peak (in* $l\,s^{-1}$*, and* t_{PK} *is time-to-peak in hours*

Storm	Summary Variable	Observed	Model
1	P_D	68	
	I_{max}	86	
	Q_D	27.36	0.75
	Q_{PK}	131.87	16.57
	t_{PK}	5.55	20.56
2	P_D	33	
	I_{max}	214	
	Q_D	22.43	0.49
	Q_{PK}	375.77	13.69
	t_{PK}	2.26	2.51
3	P_D	50	
	I_{max}	72	
	Q_D	10.15	1.65
	Q_{PK}	220.16	44.41
	t_{PK}	2.97	2.77

the total storm runoff for the three storms. Loague and Kyriakidis suggest that the runon process is a major limitation of the model in that even this very detailed representation does not adequately represent the nature of the overland flow pathways. In the model representation, too much overland flow infiltrates on higher conductivity elements downslope. The original model of Loague and Freeze (1985), which used much larger flow planes, had less spatial variability in infiltration rates and produced more runoff. They also suggest that the concepts that underlie the QPBRRM model may not be totally appropriate for this catchment and that subsurface processes and saturation excess runoff generation may play a much larger role in the storm response.

They conclude:

> One would expect that simulating rainfall–runoff events for the data rich R-5 catchment to be somewhat more straightforward and certainly more rewarding than what is reported here ... Why then should we model? It is our opinion that the best use of deterministic physically-based simulation models is in a concept-development mode for the design of future field experiments and optimal data collection strategies (Loague and Freeze 1985, p. 2895).

5.8 Validation or Evaluation of Distributed Models

Validation of distributed models is an issue that has received a great deal of recent attention in the field of groundwater modelling following a number of studies in which predictions of groundwater behaviour were not borne out by subsequent experience (see Konikow and Bredehoeft 1992). Some of this discussion has, in fact, suggested that validation is not an appropriate term to use in this context, since no model approximation can be expected to be a valid representation of a complex reality (e.g. Oreskes *et al.* 1994). Model evaluation has been suggested as a better term. Because distributed models make distributed predictions, there is a lot of potential for evaluating not only the predictions of discharge at a catchment outlet, but also the internal state variables such as water table levels, soil moisture levels, channel flows at different points on the network, etc. (see Bathurst 1986a,b; Refsgaard 1997). It appears that there have still been relatively few studies of distributed models that have attempted such an evaluation. Most models are evaluated only on the basis of predicted discharge, which leaves plenty of scope for the runoff to be simulated by a variety of different mechanisms (see Beven 1989, 1993; Grayson *et al.* 1992a,b).

The lack of evaluation with respect to internal *state variables* is clearly partly due to the expense of collecting widespread measurements of such internal state variables. There are also some difficulties in measuring quantities that can truly be compared with model predictions since, as noted in Section 1.8, the scale of the measurements may be significantly different from the model element scale at which the predictions of the model are made. In addition, even in the very first application of this type of model to a field site, Stephenson and Freeze (1974) pointed out that because of uncertainties in the boundary conditions, initial conditions and parameter values of a distributed model, it is unlikely that a true model validation will ever be possible since the errors in representing the system and specifying the inputs will surely induce unavoidable errors in the simulations, however well a model appears to have been calibrated.

There is a further interesting interaction with the problem of model calibration for the case of distributed models. Suppose that the parameters of a distributed model have initially been calibrated only on the basis of prior information about soil and vegetation type, with some adjustment of values being made to improve the simulation of measured discharges. The model might well do a very good job of simulating the catchment discharge but we would have little idea of how well it was doing in predicting the internal state variables such as water table levels. In fact, because of the lack of information about the internal responses of the catchment, the model user would probably use effective values of model parameters such as hydraulic conductivity over wide regions of the flow domain.

Suppose that after this initial calibration a decision is made to collect more spatially distributed information about the catchment response. Measurements might be made of water table heights and soil moisture storage, and some internal stream gauging sites might be installed. We would expect that the predictions of the calibrated distributed model turn out to be wrong in many places, since the calibration has taken little account of local heterogeneities in the catchment characteristics (other than the broad classification of soil and vegetation types). There is now the potential to use the new internal measurements not to evaluate the model, but to improve the local calibration, a process that will not necessarily improve the prediction of catchment discharge which was the subject of the original calibration (see also the TOPMODEL case study of Section 6.5). It will generally mean a much greater improvement in the prediction of the internal state variables for which measurements have now been made available. But if the new data are being used to improve the local calibration, more data will be needed to make a model evaluation. In fact, in practice, there is generally little model evaluation; rather the model is adapted to take account of the new data, without necessarily having any impact on the variables of greatest interest in prediction (discharge in rainfall–runoff modelling). This is discussed further in Chapter 7.

5.9 Discussion of Distributed Models Based on Process Descriptions

We are at an early stage in the development of physically based distributed hydrological modelling. There is no doubt that the SHE model, in particular, has been important in the development of distributed modelling technology. It represents the greatest collective effort that has ever been put into developing a catchment modelling strategy and has set a standard for others to follow. The application of this type of model is not, however, without limitations, some of which have been discussed in Beven (1989), Grayson *et al.* (1992b) and Beven (1996a,b). The latter paper summarizes some of the problems as follows:

> There is a continuing need for distributed predictions in hydrology but a primary theme of the analysis presented here is that distributed modelling should be approached with some circumspection. It has been shown that the process descriptions used in current models may not be appropriate; that the use of effective grid scale parameter values may not always be acceptable; that the appropriate effective parameter values may vary with grid scale; that techniques for parameter estimation are often at inappropriate scales; and that there is sufficient uncertainty in model structure and spatial discretization in practical applications that these models are very difficult (if not impossible) to validate (Beven 1996a, p. 273).

It is directly followed by a response from Danish Hydraulics Institute SHE modellers (Refsgaard *et al.* 1996). Their response concludes as follows:

> In our view the main justification for the distributed physically-based codes are the demands for prediction of effects of such human intervention as land use change, groundwater abstractions, wetland management, irrigation and drainage and climate change as well as for subsequent simulations of water quality and soil erosion. For these important purposes we see no alternative to further enhancements of the distributed physically-based modelling codes, and we believe that the necessary codes in this respect will be much more comprehensive and complex than the presently existing ones (Refsgaard *et al.* 1996, p. 286).

A review of experience in using distributed models has also been published by Bronstert (1999). He makes the following suggestions:

- There is still a lack of knowledge about how to represent some processes at or near the soil surface, including surface crusting and preferential flows.
- In many practical applications the data available on soil characteristics and boundary conditions, with all their spatial and temporal variability, may not be adequate to support the use of a fully distributed model.
- In some cases experience suggests that the predicted responses are highly sensitive to small changes in parameters, initial and boundary conditions. It may be necessary

to allow for this sensitivity by considering ranges of parameters values and the propagation of the resulting uncertainty through the model. Such uncertainty may mean that processes that account for only a small part of the water balance may be relatively unpredictable.

The justifications that underlie the development of physically based distributed models are not in dispute. The need for prediction of the effects of distributed changes in a catchment continues to increase (see Chapter 9). However, it is somewhat difficult to see how such advances will be made unless new measurement techniques for effective parameter values or grid scale fluxes are developed. There are theoretical problems and practical numerical problems that will need to be overcome in the future development of this type of model but the problem of parameter identification, particularly for the subsurface, will be even greater. We will return to the question of the future of distributed models in Chapter 10.

5.10 Key Points from Chapter 5

- Fully three-dimensional models for surface and subsurface flow processes are just beginning to become computationally feasible but most current physically based distributed rainfall–runoff models discretize the catchment into lower dimension subsystems. Such a discretization will lead to only approximate representations of the processes and may lead to numerical problems in some cases. The effects of heterogeneities of soil properties, preferential flows in the soil and irregularities of surface flows are not generally well represented in the current generation of models.
- The SHE model is an example of a grid-element-based model, using one-dimensional finite-difference solutions for channel reaches and the unsaturated zone in each grid element, and two-dimensional solutions in plan for the saturated zone and overland flow processes.
- The IHDM model is an example of a hillslope element model that uses finite-element unsaturated–saturated flow solutions for a two-dimensional vertical slice through the hillslope, coupled to one-dimensional downslope overland flow and channel flow solutions.
- Both models require effective parameter values to be specified at the scale of the calculation elements which may be different from values measured in the field.
- In some circumstances it may be possible to use simpler solutions based on the kinematic wave equation for both surface and subsurface flows.
- Distributed predictions mean that distributed data can be used in model calibration but evaluation of this type of model may be difficult due to differences in scale of predictions and measurements and the fact that the initial and boundary conditions for the model cannot normally be specified sufficiently accurately.
- There are theoretical problems and practical numerical problems that will need to be overcome in the future development of this type of model but the problem of parameter identification, particularly for the subsurface, will be even greater, and significant progress will undoubtedly depend on the development of improved measurement techniques.

Box 5.1 Descriptive Equations for Subsurface Flows

The generally used description for both saturated and unsaturated subsurface flows is based on Darcy's law (Darcy 1856), which assumes that the discharge per unit area or Darcian velocity can be represented as the product of a gradient of hydraulic potential and a scaling constant called the hydraulic conductivity. Thus

$$v_x = -K\frac{\partial \Phi}{\partial x} \tag{5.15}$$

where v_x [L T^{-1}] is the Darcian velocity in the x direction, K [L T^{-1}] is the hydraulic conductivity, and Φ [L] is the total potential ($\Phi = \psi + z$ where ψ [L] is capillary potential and z is elevation above some datum). In the case of unsaturated flow, the hydraulic conductivity will change in a nonlinear way with moisture content so that

$$K = K(\theta) \tag{5.16}$$

where θ [–] is volumetric moisture content.

Combining Darcy's law with the three-dimensional mass balance equation gives

$$\frac{\partial \rho\theta}{\partial t} = -\frac{\partial \rho v_x}{\partial x} - \frac{\partial \rho v_x}{\partial x} - \frac{\partial \rho v_x}{\partial x} - \rho E_T(x, y, z, t) \tag{5.17}$$

where ρ is the density of water [M L^3] (often assumed constant) and $E_T(x, y, z, t)$ [T^{-1}] is a rate of evapotranspiration loss expressed as a volume of water per unit volume of soil that may vary with position and time. This is the nonlinear partial differential equation now known as the Richards equation after L. A. Richards (1931):

$$\frac{\partial \rho\theta}{\partial t} = \frac{\partial}{\partial x}\left[\rho K(\theta)\frac{\partial \Phi}{\partial x}\right] + \frac{\partial}{\partial y}\left[\rho K(\theta)\frac{\partial \Phi}{\partial y}\right] + \frac{\partial}{\partial z}\left[\rho K(\theta)\frac{\partial \Phi}{\partial z}\right] - \rho E_T(x, y, z, t) \tag{5.18}$$

or, remembering that $\Phi = \psi + z$,

$$\frac{\partial \rho\theta}{\partial t} = \frac{\partial}{\partial x}\left[\rho K(\theta)\frac{\partial \psi}{\partial x}\right] + \frac{\partial}{\partial y}\left[\rho K(\theta)\frac{\partial \psi}{\partial y}\right] + \frac{\partial}{\partial z}\left[\rho K(\theta)\left(\frac{\partial \psi}{\partial z} + 1\right)\right] - \rho E_T(x, y, z, t) \tag{5.19}$$

or, in a more concise form using the differential operator ∇ (see Box 2.2),

$$\frac{\partial \rho\theta}{\partial t} = \nabla\left[\rho K(\theta)\nabla\psi\right] + \frac{\partial \rho K(\theta)}{\partial z} - \rho E_T(x, y, z, t) \tag{5.20}$$

This form of the equation assumes that the hydraulic conductivity at a given moisture content is equal in all flow directions, i.e. that the soil is *isotropic*, but in general the soil or aquifer may be *anisotropic*, in which case the hydraulic conductivity will have the form of a matrix of values. This equation involves two solution variables ψ and θ. It can be modified to have only one solution variable by making additional assumptions about the nature of the relationship between ψ and θ. For example, defining the *specific moisture capacity* of the soil as $C(\psi) = d\theta/d\psi$, and assuming that the density of water is constant, then the Richards equation may be written

$$C(\psi)\frac{\partial \psi}{\partial t} = \nabla\left[\underline{K}(\psi)\nabla\psi\right] + \frac{\partial K_z(\psi)}{\partial z} - E_T(x, y, z, t) \tag{5.21}$$

where $\underline{K}(\psi)$ and $K_z(\psi)$ now indicate that hydraulic conductivity depends on direction and is treated as a function of ψ for the unsaturated case. To solve this flow equation, it is necessary to define the functions $C(\psi)$ and $\underline{K}(\psi)$ (see Box 5.4). To make things more complicated, both $C(\psi)$ and $\underline{K}(\psi)$ may not be simple single-valued functions of ψ or θ but may be subject to *hysteresis*, i.e. varying with the history of wetting and drying at a point. Various models of hysteretic soil characteristics have been proposed (see Jaynes 1990) but are not often used in rainfall–runoff modelling. For saturated soil and in the saturated zone of an aquifer then $\underline{K}(\psi)$ will approach the saturated conductivity $\underline{K}_s$ and $C(\psi)$ will take on a very small value due to the compressibility of the soil or aquifer.

An equivalent form of the Richards equation may be written with soil moisture content θ as the dependent variable as

$$\frac{\partial\theta}{\partial t} = \nabla\left[\underline{K}(\theta)\frac{d\psi}{d\theta}\nabla\theta\right] + \frac{\partial K_z(\theta)}{\partial z} - E_T(x, y, z, t) \quad (5.22)$$

The product $\underline{K}(\theta)(d\psi/d\theta)$ is known as the *diffusivity* of the soil and is often written as $\underline{D}(\theta)$.

The Richards equation applies to both saturated and unsaturated flow through a porous medium. For near-surface flows it is normally assumed that the water is of constant density and that the soil is also incompressible. For deep aquifers, such an assumption may not be valid and it may be necessary to take account of the compressibility of both water and rock. A typical parametrization takes the following form:

$$n(p) = \theta_s(1 + bp) \quad (5.23)$$

where $n(p)$ is the porosity of the soil at pore pressure p $[\mathrm{M\,L^{-1}\,T^{-2}}]$ for $p > 0$, θ_s is the porosity at atmospheric pressure, and b is a compressibility coefficient $[\mathrm{M^{-1}\,L\,T^2}]$ that will vary with the nature of the soil or rock.

The temperature dependence of hydraulic conductivity due may also be important and may be expressed in the form:

$$K(\psi) = k_s k_r(\psi)\frac{\rho g}{\mu} \quad (5.24)$$

where k_s is called the intrinsic permeability $[\mathrm{L^2}]$ of the soil at saturation which should be a characteristic only of the porous medium, $k_r(\psi)$ is a relative conductivity $(0 < k_r(\psi) < 1)$ varying with capillary potential ψ, g is the dynamic gravity acceleration constant $[\mathrm{L\,T^{-2}}]$, and μ is the dynamic viscosity of the soil. $[\mathrm{M\,L^{-1}\,T^{-1}}]$ Both ρ and μ vary with temperature.

The Richards equation does not have an analytical solution for most cases of interest and solutions must be obtained by approximate numerical solutions (see Box 5.3).

Summary of Model Assumptions

A summary of the assumptions made in developing this form of the Richards equation is as follows:

- *A1.* For both saturated and unsaturated flow, flow velocity may be assumed to be a linear function of the gradient of hydraulic potential in accordance with Darcy's law.
- *A2.* Functional relationships can be specified for the soil moisture characteristics curves to relate moisture content, capillary potential and hydraulic conductivity of the soil.
- *A3.* Fluxes of water in vapour form can be neglected.

Additional assumptions that are often made in applying the Richards equation are as follows:

- *A4.* The soil moisture characteristics are non-hysteretic.

- *A5.* The hydraulic conductivity tensor is isotropic.
- *A6.* The porous medium is incompressible.
- *A7.* The water is of constant temperature and density.

These last assumptions reduce the number of parameters that must be specified before the model can be run.

Box 5.2 Estimating Infiltration Rates at the Soil Surface

There have been many hydrological models that have had as their basis the Horton infiltration excess concept of runoff generation. These have included some distributed models, in which the only consideration of subsurface flow processes has been the prediction of infiltration at the soil surface (see Section 5.5). Other models, including some unit hydrograph models, have also used an infiltration excess concept to calculate how much of a rainstorm to route as runoff. Thus, the estimation of infiltration at the soil surface has traditionally been an important part of hydrological theory.

Most infiltration experiments show that at the start of rainfall infiltration, rates are high and then decline gradually over time (there are rare exceptions where, for example, the breakdown of a surface crust or reduction in hydrophobicity may result in increasing infiltration rates over time). The initially high rates are due to the effects of capillary potential drawing water into the dry soil in addition to the effects of gravity. If there is a time before the surface of the soil reaches saturation then initially the infiltration may be limited by the rainfall rate, i.e. all the rainfall infiltrates into the soil. The soil may only start to limit infiltration rates once this *time to ponding* has been reached. After a long period of heavy rainfall, the wetting front will have moved some distance into the soil, the effects of capillary potential will be small, the potential gradient will be dominated by gravity and the infiltration rate will approach the effective hydraulic conductivity of the soil (see Figure B5.2.1).

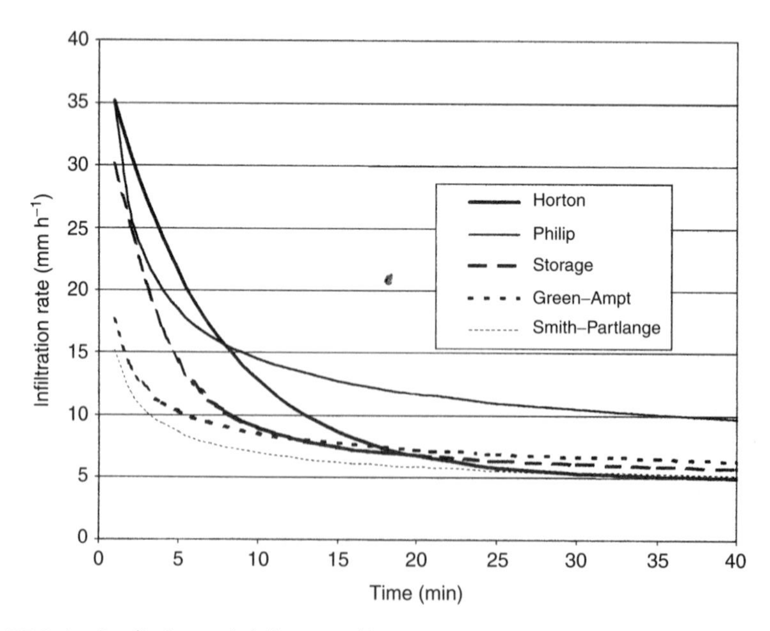

Figure B5.2.1 *Predictions of different infiltration equations under conditions of surface ponding*

This decline of infiltration capacity can be predicted for any arbitrary initial soil profile and pattern of rainfall intensities by a numerical solution of the Richards equation (see Box 5.1). Such solutions can also easily take account of arbitrary layering in the soil, including surface crusting, given some knowledge of the soil moisture characteristics for the different layers. Dual porosity solutions can make some attempt at representing the effects of macropores on infiltration (e.g. those of Jarvis *et al.* 1991; Bronstert and Plate 1997; Mohanty *et al.* 1998).

However, an accurate simulation of the time to ponding and infiltration capacity of the soil may require very small depth and time increments to resolve the rapid changes in hydraulic gradients in time and space that control the change in infiltration capacity. Computer run time therefore becomes an issue, particularly where distributed predictions of infiltration capacity may be required to simulate the generation of surface runoff in a heterogeneous catchment. Thus, there remains a need for simpler analytical solutions of infiltration into the soil surface. There are many such equations reported in the literature. A selection of the most widely used are discussed below.

The Horton Infiltration Equation

Horton (1933, 1940) described this type of curve by an empirical function of the form

$$f(t) = (f_o - f_c)\exp\{-kt\} + f_c \tag{5.25}$$

where f_o is an initial infiltration capacity, [L T^{-1}], f_c is a final infiltration capacity [L T^{-1}] and k is an empirical coefficient [T^{-1}]. The three parameters, f_o, f_c, and k, are a function of soil type but may also depend on the antecedent state of the soil. The final infiltration capacity, f_c, will be close to the hydraulic conductivity of the soil at field saturation. Although this form of equation was based on empirical evidence, Eagleson (1970) has shown that it is an approximate solution of the Richards equation under certain simplifying assumptions.

The Green–Ampt Infiltration Equation

There are other, more direct, ways of deriving an infiltration equation from the Richards equation. Soil physical theory suggests that infiltration can be described (at least in the absence of major macropores) by the Richards equation which is based on the nonlinear form of Darcy's law for partially saturated flow (see Box 5.1). There are no general analytical solutions to the Richards equations but a number of different solutions are available for infiltration at the soil surface based on different simplifying assumptions. Green and Ampt (1911), for example, assumed that the infiltrating wetting front forms a sharp jump from a constant initial moisture content ahead of the front to saturation at the front. This allows a simple form of Darcy's law to be used to represent the infiltration such that infiltration rate f is calculated as

$$f(t) = \tilde{K}_s\left(\frac{h_o + \psi_f}{z_f} + 1\right) \tag{5.26}$$

where $\tilde{K}_s$ is the hydraulic conductivity of the soil at field saturation [L T^{-1}], h_o is the depth of ponded water on the soil surface, [L], ψ_f is a parameter related to the difference in capillary potential across the wetting front [L], and z_f is the depth of penetration of the wetting front [L]. The first term in the brackets in this equation is due to the capillary potential gradient, here estimated from an effective difference in capillary potential across the wetting front averaged over the depth of penetration. The magnitude of this term gets less as the wetting front goes deeper, leading to the decline in infiltration capacity of the soil. The second term (1) is the gravitational term. It stays at unity, regardless of the depth of the wetting front, and when multiplied by the effective saturated hydraulic conductivity gives the final infiltration capacity.

Morel-Seytoux and Khanji (1974) have shown that a parameter called the capillary drive, with units of length and defined as $C_D = \psi_f \delta\tilde{\theta}$, is a relatively constant parameter for a range of initial moisture conditions defined as $\delta\tilde{\theta} = (\tilde{\theta} - \theta_i)$ which is the change in moisture content between the initial state of the soil, θ_i and field saturation, $\tilde{\theta}$. The Green–Ampt equation is then better applied in the form:

$$f(t) = \frac{\tilde{K}_s}{B}\left(\frac{h_o\delta\tilde{\theta} + C_D}{z_f\delta\tilde{\theta}} + 1\right) \tag{5.27}$$

where B is an additional parameter proposed by Morel-Seytoux and Khanji (1974) to allow for air pressure effects called the viscous resistance correction factor ($1 < B < 1.7$).

The original Green and Ampt infiltration equation assumes constant soil characteristics with depth. An analysis of a wide variety of soil moisture characteristics data by Rawls *et al.* (1983; see also Box 5.5) has led to a classification of the Green and Ampt parameters by soil texture (see Figure B5.2.2). These types of relationships should be used with care, however, since they are based on measurements on small samples brought back to the laboratory, not field measurements at the plot scale. Beven (1984) has produced a solution with equivalent assumptions for the case where hydraulic conductivity declines exponentially with depth, often a useful approximation of real soil characteristics.

The Philip Infiltration Equation

Philip (1957) obtained an analytical solution to the Richards equation by assuming a delta function change in diffusivity for the soil across the wetting front. His widely used infiltration equation has the form

$$f(t) = 0.5St^{-0.5} + A \tag{5.28}$$

where S is called the sorptivity of the soil [L T$^{-0.5}$] and is calculated from knowledge of the soil moisture characteristics of the soil, and A is a final infiltration capacity [L T^{-1}] equivalent to the f_c of the Horton equation or $\tilde{K}_s$ of the Green–Ampt equation. The effects of the sorptivity term gradually reduce with increasing time, eventually leaving the final infiltration capacity as a function of the effective saturated hydraulic conductivity of the soil.

The Smith–Parlange Infiltration Equation

Under assumptions of a diffusivity that changes exponentially with moisture content, Smith and Parlange (1978) derived another widely used infiltration equation that takes the following form:

$$f(t) = \tilde{K}_s \frac{\exp F(t)/C_D}{\exp F(t)/C_D - 1} \tag{5.29}$$

where $F(t) = \int_0^t f(t)dt$ is the total infiltrated volume of water per unit area at time t [L], and C_D is the capillary drive [L] as above.

A Storage Capacity Based Infiltration Equation

Treating infiltration capacity as a function of infiltrated volume can also be used to treat the case where overland flow is produced as a result of the topsoil layer becoming saturated due to a limitation on vertical flow at some depth within the soil. This can occur where

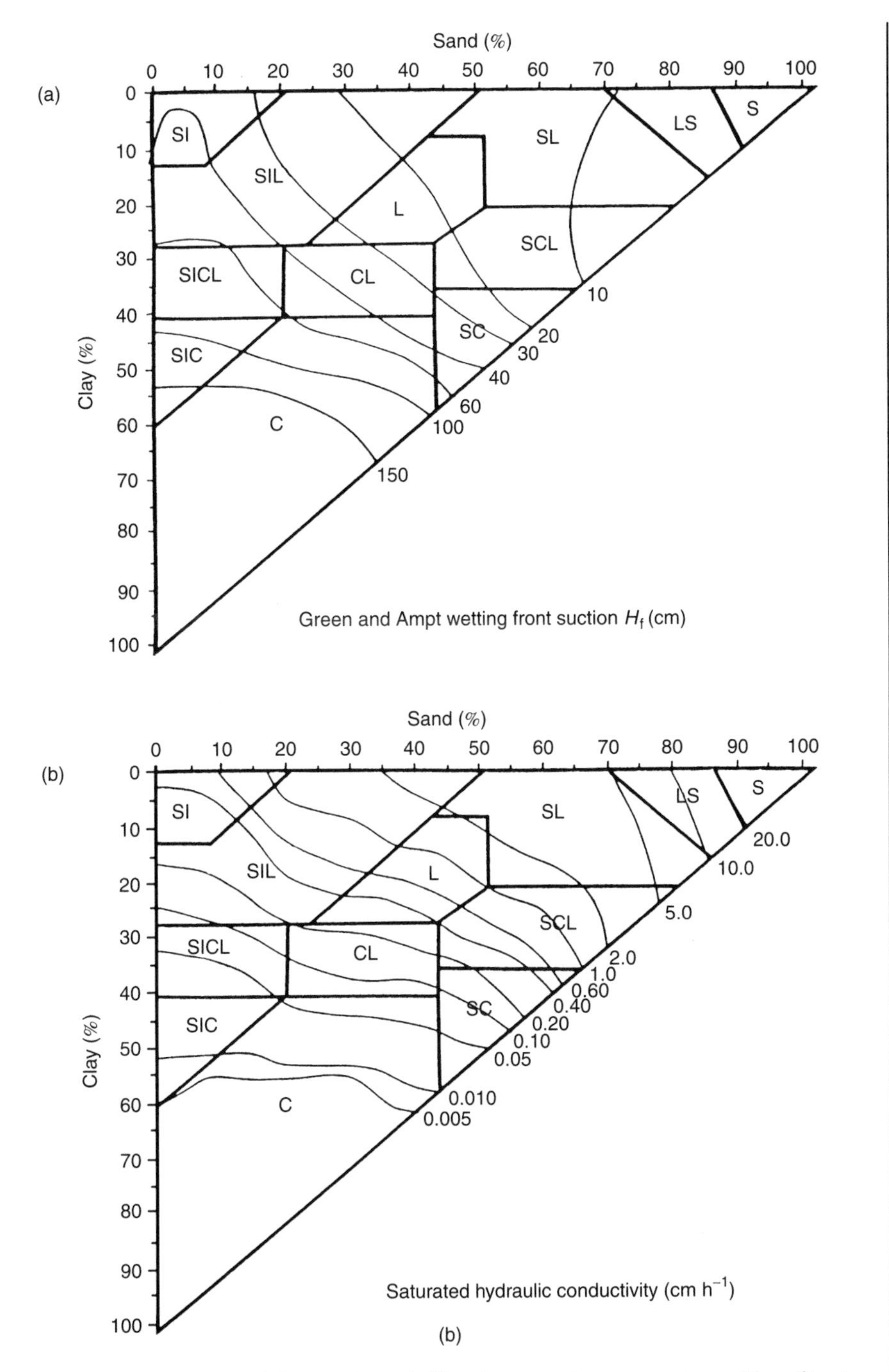

Figure B5.2.2 *Variation of Green–Ampt infiltration equation parameters with soil texture (after Rawls and Brakensiek, 1989). Reproduced with kind permission from Kluwer Academic Publishers*

either a thin soil overlies an impermeable bedrock or where there is some horizon of lower permeability at some depth into the soil profile (e.g. Taha *et al.* 1997). In these circumstances, infiltration rates might be controlled more by a saturation excess than by a surface infiltration excess process. This situation was addressed by Kirkby (1975; see also Scoging and Thornes 1982) using an infiltration equation of the following form:

$$f(t) = B/H + A \tag{5.30}$$

where H is the current depth of storage [L], A is the long-term infiltration rate [L T^{-1}] (which may now be controlled at depth) and B is a constant [L^{-2} T^{-1}]. Updating of H at successive time steps allows application to any irregular sequence of rainfall inputs using:

for $r(t) > f(t)$, $dH/dt = r(t) - f(t)$
for $r(t) < f(t)$, $dH/dt = r(t) - A$

For steady rainfall inputs this results in infiltration capacity being an inverse function of time rather than of the square root of time as in the Philip (1957) equation.

Time to Ponding and the Time Compression Assumption

Whatever type of function is used to describe infiltration, effective rainfalls will be calculated as the excess of rainfall over infiltration following the time to ponding. For rainfall intensities that are irregular in time, this is often most easily done in terms of comparing the cumulative rainfall and cumulative infiltration rates, estimating the time of ponding as the point at which the cumulative infiltration satisfies one of the solutions above (this is called the *time compression assumption*).

Thus, using the Green–Ampt equation as an example, the cumulative infiltration at any time up to the time of ponding is the integral of the sequence of rainfall intensities:

$$F(t) = \int_0^{t_p} r(t)dt \tag{5.31}$$

Under the sharp wetting front assumption this will also be equal to

$$F(t) = z_f(\tilde{\theta} - \theta_i) \tag{5.32}$$

Thus, at each time step, the depth of the wetting front can be calculated and used in (5.26) or (5.27) to calculate the current infiltration capacity, f_t. If this is less than the current rainfall intensity, r_t, then the time of ponding has been reached. Similar arguments can be used for other surface control infiltration equations to determine the time of ponding. For storage-based approaches, a minimum storage before surface saturation can be introduced as an additional parameter if required.

Infiltration in Storms of Varying Rainfall Intensity

In storms of varying rainfall intensity, the input rate may sometimes exceed the infiltration capacity of the soil and sometimes not. When input rates are lower than the infiltration capacity of the soil then there may be the chance for redistribution of water within the soil profile to take place, leading to an increase in the infiltration capacity of the soil. This is handled easily in storage capacity approaches to predicting infiltration, such as that of Kirkby (1975) noted above, but such approaches do not have a strong basis in soil physics. Clapp *et al.* (1983) used an approach based on a Green–Ampt or kinematic approach to the infiltration and redistribution of successive wetting fronts. More recently, Corradini *et al.* (1997) have produced a similar approach based on a more flexible profile shape than the piston-like wetting front of the Green–Ampt approach. It is worth noting that methods based on soil physics may also be limited in their applicability in this respect since they take no account of the effects of macropores in the soil on infiltration and redistribution (though see Beven and Clarke (1986) for one attempt at an infiltration model that takes account of a population of macropores of limited depth).

Estimating Infiltration at Larger Scales

All of the equations above are for the prediction of infiltration at the point scale. They have often been wrongly applied at hillslope or catchment scales, as if the same point scale equation applied for the case of heterogeneous soil properties. However, because of the nonlinearities inherent in any of the infiltration equations, this will not be the case. It is therefore necessary to predict a distribution of point scale infiltration rates before averaging up to larger scales. Similar media theory (see Box 5.4) can be useful in this case, if it can be assumed that the soil can be treated as a distribution of parallel non-interacting heterogeneous columns. Clapp *et al.* (1983), for example, have based a field-scale infiltration model on this type of representation. Philip (1991) has attempted to extend the range of analytical solutions to the case of infiltration over a sloping hillslope, albeit a hillslope of homogeneous soil characteristics.

There have been many hypothetical studies of the effects of random variability of soil properties and initial conditions on infiltration at the hillslope and catchment scales, some taking account of the infiltration of surface flow from upslope, others not (see for example the recent study of Corradini *et al.* 1998). These have mostly been based on purely stochastic heterogeneity but it is worth noting that some variation may be systematic (see for example the dependence of infiltration on depth of overland flow and vegetation cover included in the model of Dunne *et al.* 1991). To apply such models to a practical application will require considerable information on the stochastic variation in soil properties and even then, will not guarantee accurate predictions of runoff generation (see the discussion of the R-5 catchment case study in Section 5.7).

Storage-based approaches need not be considered only as point infiltration models. A function such as that of (5.30) may equally be considered as a conceptual representation of basin-wide infiltration, with the storage as a mean basin storage variable. Other storage approaches have taken a more explicit representation of a distribution of storage capacities in a catchment in predicting storm runoff. These include the Stanford Watershed model and variable infiltration capacity models (see Section 2.4 and Box 2.2), and the probability distributed model (see Section 6.2).

A further widely used method for predicting runoff at larger scales, often interpreted as an infiltration model, is the USDA Soil Conservation Service (SCS) approach. This is discussed in more detail in the next chapter and Box 6.1.

Box 5.3 Solution of Partial Differential Equations: Some Basic Concepts

As noted in the main text, it is nearly always not possible to obtain analytical solutions to the nonlinear differential equations describing hydrological flow processes for cases of real interest in practical applications such as rainfall–runoff modelling. The approximate numerical solution of nonlinear differential equations is, in itself, a specialism in applied mathematics, and writing solution algorithms is something that is definitely best left to the specialist. A major reason for this is that it is quite possible to produce solution algorithms that are inaccurate or are not *consistent* with the original differential equation (i.e. the approximate solution does not converge to the solution of the original equation as the space and time increments used become very small). It is also easy to produce solutions for nonlinear equations that are not *stable*. A stable solution means that any small errors due to the approximate nature of the solution will be damped out. An unstable solution means that those small errors become amplified, often resulting in wild oscillations of the solution variable at adjacent solution nodes or successive time steps. The aim of this box is to make the reader aware of some of the issues involved in approximate numerical solutions and what to look out for in using a model based on one of the numerical algorithms available.

The differential equations of interest to the hydrologist generally involve one or more space dimensions and time (such as the Richards equation of Box 5.1). An approximate solution then requires a discretization of the solution in both space and time to produce a grid of points at which a solution will be sought for the dependent variable in the equation. Figure B5.3.1 shows some different ways of subdividing a cross-section through

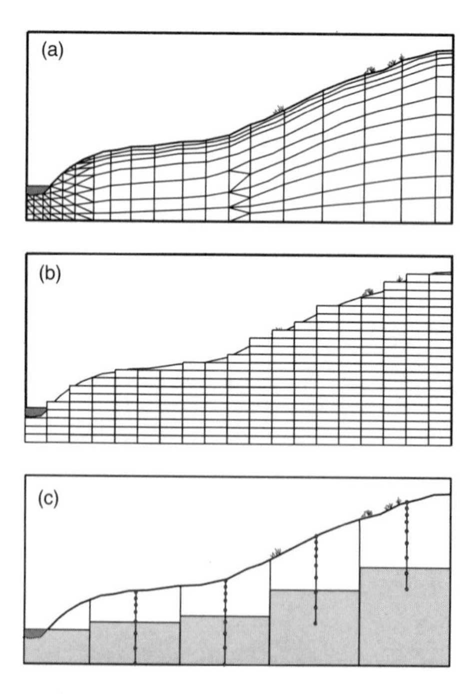

Figure B5.3.1 *Different discretizations of a hillslope for approximate solutions of the subsurface flow equations. (a) Finite-element discretization (as in the IHDM). (b) Rectangular finite-difference discretization (as used by Freeze 1972). (c) Square grid in plan for saturated zone, with one-dimensional vertical finite-difference discretization for the unsaturated zone (as used in the SHE model)*

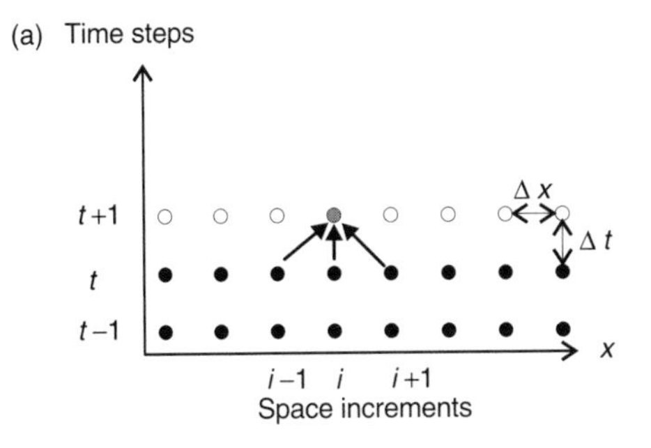

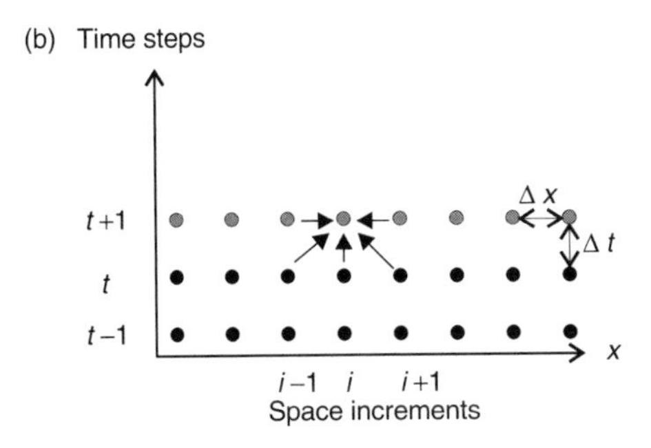

Figure B5.3.2 *Schematic diagram of (a) Explicit and (b) Implicit time stepping strategies in approximate numerical solutions of a partial differential equation, here in one spatial dimension,* x*. Arrows indicate the nodal values contributing to the solution for node (*i*,* t + 1*). Nodal values in black are known at time* t*; nodal values in grey indicate dependencies in the solution for time* t + 1

a hillslope into a grid of solution points or nodes in space, while Figure B5.3.2 shows a regular discretization into both time and space increments, Δt and Δx, for a single spatial dimension.

Starting a numerical solution always requires a complete set of nodal values of the solution variable at time $t = 0$. Note that it is not always easy to specify a complete field of initial values and that it is generally good practice to allow a generous run-in period before the results are evaluated. The algorithm then aims to step the solution through time, using time steps of length Δt, to obtain values of the variable of interest for all the nodes at each time step. The most easily understood numerical approximation method is the finite-difference method. The SHE model discussed in Section 5.2.2 uses a finite-difference solution of the surface and subsurface flow equations on a regular spatial grid. The differentials in the flow equation are replaced directly by differences. For example, to calculate a spatial differential for a variable ψ at node i in the interior of a grid at time step j, one possible difference approximation is

$$\left\{\frac{\partial\psi}{\partial x}\right\}_{i,j} \approx \frac{\psi_{i+1,j} - \psi_{i-1,j}}{2\Delta x} \tag{5.33}$$

This form is called a centred difference approximation. It is a more accurate approximation than the forward difference that is often used for the time differential:

$$\left\{\frac{\partial\psi}{\partial t}\right\}_{i,j} \approx \frac{\psi_{i,j+1} - \psi_{i,j}}{\Delta t} \tag{5.34}$$

The forward difference is convenient to use for any differential in time, since given the nodal values of ψ at time step j, the only unknowns are the values at time step $j+1$. A centred approximation for a second-order spatial differential involving a spatially variable coefficient K may be written as

$$\left\{\frac{\partial}{\partial x}\left(K\frac{\partial\psi}{\partial x}\right)\right\}_{i,j} \begin{array}{l} \approx \dfrac{K_{i,j}}{\Delta x}\left(\dfrac{\psi_{i+1,j}-\psi_{i,j}}{\Delta x}-\dfrac{\psi_{i,j}-\psi_{i-1,j}}{\Delta x}\right) \\ \approx K_{i,j}\left(\dfrac{\psi_{i+1,j}-2\psi_{i,j}-\psi_{i-1,j}}{\Delta x^2}\right) \end{array}$$

When an equation involves both space and time differentials, the simplest (but not necessarily most accurate) solution is to evaluate all the spatial differentials at time step j and use them to solve for the dependent variable at time step $j+1$. In this strategy the calculations for each node can be carried out independently of all the other nodal values at time $j+1$ as shown in Figure B5.3.2(a). This is known as a solution that is *explicit* in time. It results in a solution algorithm that has many attractions in that it is not necessary to assemble large matrices and the solution is easily implemented on parallel computers. The disadvantage is that the solution is only conditionally stable, especially for problems that are strongly nonlinear such as those encountered in hydrology. Thus it is necessary to ensure that the time step is short enough that the solution remains stable. For nonlinear problems this may mean that very small time steps are required. It is possible to calculate a stability criterion and to ensure that the time step is small enough to ensure stability. In general, the smaller the spatial discretization used, the smaller the time step that will be required to ensure stability. Such criteria will take the form $\Delta t \leq c\Delta x$, where c is a local wave speed and Δx is the local space increment of the model discretization.

There are some types of differential equations, known as elliptic equations, for which explicit methods are not suitable since the effective wave speed is theoretically infinite. In hydrology, flow in a saturated soil or aquifer, where the effective specific moisture capacity or storage coefficient is very small, results in a quasi-elliptic equation which is one reason why some physically based models have had significant problems with stability of solutions.

A more robust method arises if some form of averaging of the estimates of the spatial differentials at time steps j and $j+1$ is used, e.g.

$$\left\{\frac{\partial\psi}{\partial x}\right\}_{i,j} \approx \varpi\left(\frac{\psi_{i+1,j+1}-\psi_{i-1,j+1}}{2\Delta x}\right)+(1-\varpi)\left(\frac{\psi_{i+1,j}-\psi_{i-1,j}}{2\Delta x}\right) \qquad (5.35)$$

where ϖ is a weighting coefficient in time. An explicit solution has $\varpi = 0$. Any solution algorithm with $\varpi > 0$ is known as an *implicit* scheme (Figure B5.3.2(b)). A central difference or Crank–Nicholson scheme has $\varpi = 0.5$, while a backward difference or fully implicit scheme has $\varpi = 1$. For linear problems, all implicit algorithms with $\varpi > 0.5$ can be shown analytically to be unconditionally stable, but no such general theory exists for nonlinear equations. The central difference approximation is superior to the backward difference scheme in principle, since the truncation error of the approximation is smaller. However, the backward difference scheme has been found to be useful in highly nonlinear problems. Details of a four-point implicit finite-difference scheme to solve the one-dimensional St Venant channel flow equations are given in Fread (1973, 1985), while a simpler implicit scheme to solve the one-dimensional kinematic wave equation is given in Li *et al.* (1975).

With an implicit scheme, the solution for node i at time step $j+1$ will involve values of the dependent variable at other surrounding nodes, such as at $i+1$ and $i-1$ at time step $j+1$ which are themselves required as part of the solution (as shown in Figure B5.3.2(b)). Thus it is necessary to solve the problem as a system of simultaneous equations at each time step. This is achieved by an iterative method, in which an initial guess of the values at time $j+1$ is used to evaluate the spatial differentials in the implicit scheme. The system of equations is then solved to get new estimates of the values at time $j+1$ which are then used to refine the estimates of the spatial differentials. This iterative procedure is continued until the solution converges towards values that change by less than some specified tolerance threshold between successive iterations. A successful algorithm results in a stable solution with rapid convergence (a small number of iterations) at each time step. At each iteration, the system of equations is assembled as a matrix equation, linearized at each iteration for nonlinear problems in the following form:

$$[A]\{\Psi\} = \{B\} \tag{5.36}$$

where $[A]$ is a two-dimensional matrix of coefficients known at the current iteration, $\{\Psi\}$ is a one-dimensional vector of the unknown variable and $\{B\}$ is a one-dimensional vector of known values. Explicit schemes can also be expressed in this form but will only be solved once at each time step. For a large number of nodes, the $[A]$ matrix may be very large and sparse (i.e. having many zero coefficients) and many algorithms use special techniques, such as indexing of non-zero coefficients, to speed up the solution.

Solution methods may be direct, iterative or a combination of the two. Direct methods carry out a solution once and provide an exact solution to the limit of computer round-off errors. The problem with direct methods is that with a large sparse $[A]$ matrix, the computer storage required may be vast. Examples of direct methods include Gaussian elimination and Cholesky factorization.

Iterative methods attempt to find a solution by making an initial estimate and then refining that estimate over successive iterations until some error criterion is satisfied. For implicit solutions of nonlinear problems, the nonlinear coefficients can be refined at each iteration. Iterative methods may not converge for all nonlinear problems. Steady-state problems, for example, often pose greater problems in obtaining a solution than transient problems and convergence may depend on making a good initial guess at the solution. This is often easier in the transient case where a good initial guess will be available by extrapolating the behaviour of the solution variable from previous time steps. Iterative methods can, however, be fast and generally require much less computer storage than direct methods for large problems. Examples of iterative solution methods include Picard iteration, Newton iteration, Gauss–Seidal iteration, successive over-relaxation and the conjugate gradient method. Paniconi *et al.* (1991) have compared Picard and Newton iteration in solving the nonlinear Richards equation by a finite-element method. The Newton iteration method requires the calculation of a gradient matrix at each iteration, but will generally converge in fewer iterations. They suggested that it was more efficient in certain strongly transient or highly nonlinear cases (but also recommend the non-iterative Lees method for consideration). The pre-conditioned form of the conjugate gradient method is very popular in the solution of groundwater problems and readily implemented on parallel computers (e.g. Binley and Beven 1992).

Although we have discussed the considerations of explicit and implicit schemes, direct and iterative solutions, stability and convergence in the context of finite-difference approximations, they apply to all solution algorithms including finite-element and finite-volume techniques. Note that with all these schemes, it is possible to have an algorithm that is stable and consistent but not accurate, if an injudicious choice of space and time increments is made. It may be very difficult to check accuracy in a practical application except by testing the sensitivity of the solution to reducing the space and time increments. This is something that is often overlooked in the application of numerical methods since checks on the accuracy of the solution may often be expensive to carry out.

Finite-element solutions are also commonly used in hydrological problems, such as in the Institute of Hydrology Distributed Model (IHDM) of Calver and Wood (1995) (see Section 5.4.3). The finite-element method has an important advantage over finite-difference approximations in that flow domains with irregular external and internal boundaries are more realistically represented by elements with straight or curved sides. No-flow and specified flux boundaries are also more easily handled in the finite-element method. The solution nodes in the finite-element method lie (mostly at least) along the boundaries of the elements. Spatial gradients are represented by interpolating the nodal values of the variable of interest within each element using *basis functions*. The simplest form of basis function is simple linear interpolation. Higher-order interpolation can be used, but will require more solution nodes within each element. Ideally, the form of interpolation should be guided by the nature of the problem being solved but this is often difficult for problems involving changes over time and in which different solution variables are nonlinearly related, implying that different basis functions should be used. A higher-order finite-element interpolation for time differentials can also be used.

There is a vast literature on numerical solution algorithms for differential equations. A good detailed introduction to the types of algorithms used in hydrology may be found in Pinder and Gray (1977). It is always important to remember that all these methods are approximations, especially in the case of nonlinear problems, and that approximations will involve inaccuracies even if the solution is stable and convergent. One form of inaccuracy commonly encountered is that of numerical dispersion. It is an important consideration in problems involving strong advection, such as the propagation of a sharp wetting front into the soil, the movement of a steep flood wave down a river or the transport of a contaminant in a flow with a steep concentration gradient. Numerical solutions using a fixed nodal grid will inevitably smear out any rapid changes in moisture content or concentration in such problems. This smearing is called numerical dispersion and will become greater as the grid gets coarser, even if the solution apparently stays stable. Sharp fronts may also lead to oscillations in the solution for many solution schemes, a product of the approximate solution not the process. There have been some techniques developed to try to minimize such problems, especially in one-dimensional advection problems, but the lesson is that these types of solution must be used with care.

Box 5.4 Soil Moisture Characteristic Functions for Use in the Richards Equation

Use of the Richards equation to predict water flow in unsaturated soil (see Box 5.1) requires the specification of the nonlinear functions $C(\psi)$ and $K(\psi)$ if solving for capillary potential ψ, or $D(\theta)$ and $K(\theta)$ with moisture content θ as the dependent variable. These functions are time-consuming to measure directly, even in the laboratory, and are generally complicated by being multi-valued functions dependent on the history of wetting and drying. For modelling purposes, it is often assumed that simpler, single-valued functions can be used. Two sets of widely used functions will be presented here: those suggested by Brooks and Corey (1964) and those of van Genuchten (1980). Both specify forms for $\theta(\psi)$ and $K(\theta)$, from which the specific moisture capacity, $C(\psi)$ and diffusivity, $D(\psi)$, can be derived. Smith *et al.* (1993) recently proposed a form that is an extension of both the Brooks–Corey and van Genuchten forms but this has not yet been widely used. Some recent theoretical developments have involved developing functional forms based on assumptions of a fractal pore space (e.g. Tyler and Wheatcraft 1992; Pachepsky and Timlin 1998), while Jaynes (1990) reviews multi-valued hysteretic forms.

The Brooks–Corey Functions

In the Brooks–Corey functions, moisture content and capillary potential are related as

$$\frac{\theta - \theta_r}{\theta_s - \theta_r} = \left(\frac{\psi_o}{\psi}\right)^{\lambda} \tag{5.37}$$

while hydraulic conductivity and moisture content are related as

$$\frac{K(\theta)}{K_s} = \left(\frac{\theta - \theta_r}{\theta_s - \theta_r}\right)^{3+2\lambda} \tag{5.38}$$

These relationships have five parameters that define the shape of the functions: θ_s is the saturated porosity of the soil [–]; θ_r is a residual moisture content; [–]; K_s is the saturated hydraulic conductivity [L T^{-1}]; ψ_o is called the bubbling potential [L]; and λ is called the pore size index [–].

The van Genuchten Functions

In the van Genuchten functions, moisture content and capillary potential are related as

$$\frac{\theta - \theta_r}{\theta_s - \theta_r} = \left(\frac{1}{1 + \{\psi/\psi_o\}^{\lambda+1}}\right)^{\frac{\lambda}{\lambda+1}} \tag{5.39}$$

while hydraulic conductivity and moisture content are related as

$$\frac{K(\theta)}{K_s} = \left(\frac{\theta - \theta_r}{\theta_s - \theta_r}\right)^{\lambda+1} \left[1 - \left(1 - \left\{\frac{\theta - \theta_r}{\theta_s - \theta_r}\right\}^{\frac{\lambda+1}{\lambda}}\right)^{\frac{\lambda}{\lambda+1}}\right]^2 \tag{5.40}$$

These relationships similarly have five parameters that define the shape of the functions: θ_s is the saturated porosity of the soil; θ_r is a residual moisture content; K_s is the saturated hydraulic conductivity; ψ_o is a bubbling potential; and λ is also a pore size index. Note that both ψ_o and λ are best viewed as fitting rather than physical parameters and may take different numerical values in the Brooks–Corey and van Genuchten equations if both functions are fitted to the same set of data.

The two functions, calculated with identical values of the five parameters, are plotted in Figure B5.4.1. The smoother van Genuchten functions in the area close to saturation are numerically advantageous when used in modelling (and in most cases probably a more realistic representation of the soil wetting and drying characteristics). A variant on the Brooks–Corey relationships has been suggested by Clapp and Hornberger (1978), using a parabolic function to smooth the sharp break at the bubbling potential, ψ_o, while Mohanty *et al.* (1997) have suggested piecewise functions that can take account of a rapid increase in hydraulic conductivity due to preferential flows close to saturation.

Similar Media Scaling in Representing the Soil Moisture Characteristic Functions

An early attempt to deal with the heterogeneity of soil properties that has proven to be quite robust is the local-scale scaling theory of Miller and Miller (1956). They derived relationships for the soil characteristics of different soils that were geometrically similar in terms of the shape and packing of the particles, differing only in terms of a characteristic length ratio (Figure B5.4.2). In this case the water content of the soil, θ, can be related to the capillary potential, ψ, as

$$\theta = \Omega_1\left(\frac{\alpha}{\sigma}\rho g \psi\right) \tag{5.41}$$

where σ is surface tension, [M T^{-2}], ρ is the density of water [M T^{-3}], g is the gravity acceleration constant [L T^{-2}], α is the characteristic length scaling [–], and the function $\Omega_1()$ is the same for all the soils scaled in this way.

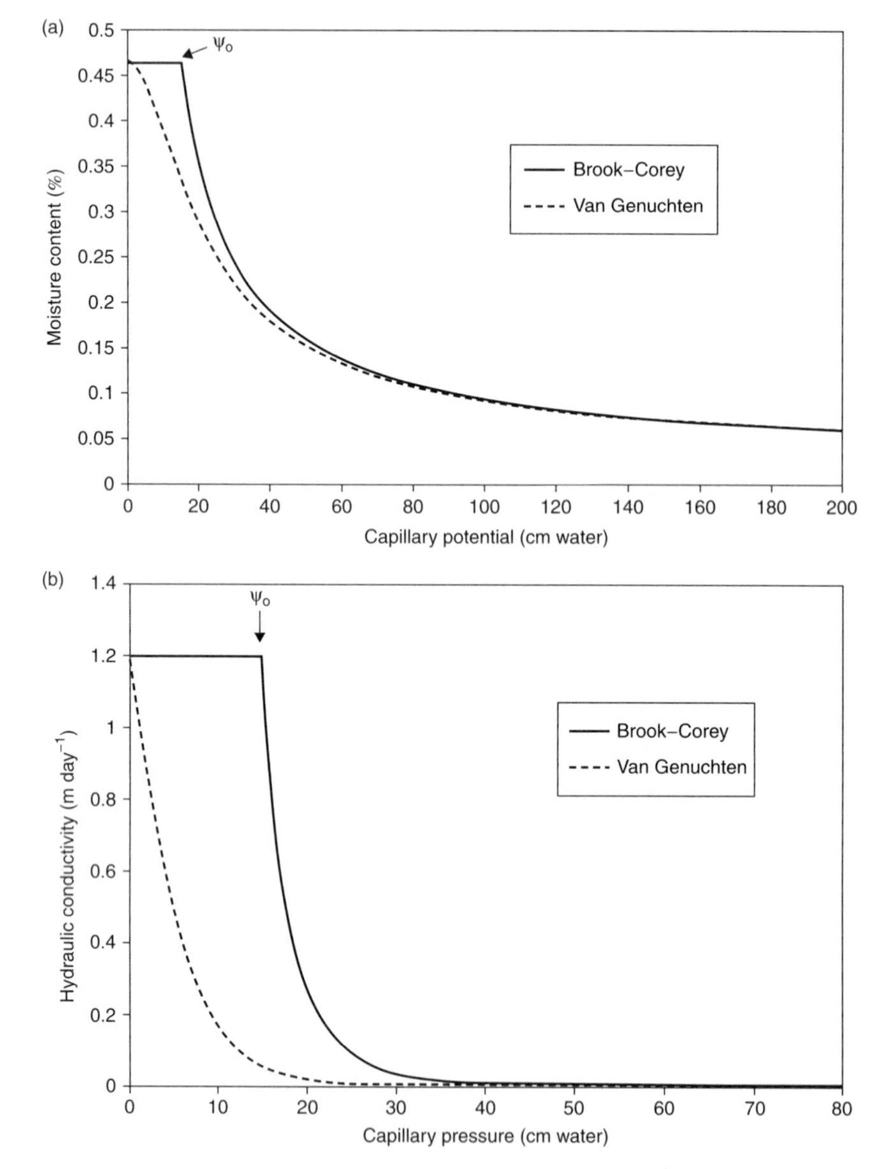

Figure B5.4.1 *Comparison of the Brook–Corey and Van Genuchten soil moisture characteristic functions for different values of capillary potential. (a) Soil moisture content and (b) Hydraulic conductivity*

A similar relationship can be derived for the hydraulic conductivity of similar soils as

$$K = \frac{\alpha^2}{\rho g \mu} \Omega_2(\theta) \tag{5.42}$$

where μ is the dynamic viscosity of water and $[M L^{-1} T^{-1}]$ Ω_2 is again constant for soils that are geometrically similar.

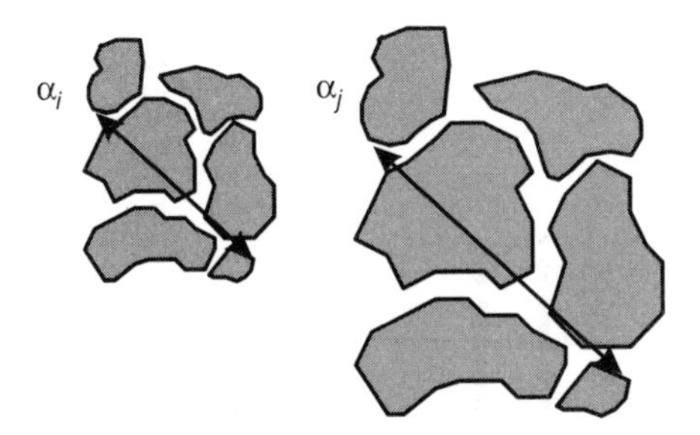

Figure 5.4.2 *Scaling of two similar media with different length scales α*

Thus, the soil moisture characteristics of similar soils can be scaled by representing their variability in terms of the distribution of the characteristic length α. Note that a consequence of the geometric similarity is that the porosity of the soil θ_s is a constant for all similar soils. This type of scaling has been applied in many studies (e.g. Simmons *et al.* 1979; Clausnitzer *et al.* 1992); its relevance here is as a way of incorporating a simple description of soil variability into predictions of infiltration and runoff generation (e.g. Luxmoore and Sharma 1980; Clapp *et al.* 1983; Warrick and Hussen 1993).

Using the Brooks–Corey characteristics as an example, assuming that we have knowledge of the parameters for one particular soil with $\alpha = 1$, then for any other similar soil:

$$\frac{\theta - \theta_r}{\theta_s - \theta_r} = \left(\alpha \frac{\psi_o}{\psi}\right)^{\lambda} \tag{5.43}$$

while saturated hydraulic conductivity is given by

$$K_s(\alpha) = \alpha^2 K_s^1 \tag{5.44}$$

where K_s^1 is the saturated hydraulic conductivity of the reference soil. The expression for relative hydraulic conductivity, (e.g. 5.40), stays the same.

The analysis can be extended to other parameters dependent on the soil moisture characteristics of the soil. Thus, the Philip infiltration equation (see Box 5.2) can be scaled as:

$$f(t) = 0.5\alpha^{1.5} S^1 t^{-0.5} + \alpha^2 A^1 \tag{5.45}$$

where S^1 and A^1 are the sorptivity and final infiltration capacity parameters for a reference soil with $\alpha = 1$. An analysis of multiple infiltration curves is one way of calculating a distribution of α values for a particular soil type (see for example, Shouse and Mohanty 1998). An example of the scaling of infiltration curves in this way is demonstrated in Figure B5.4.3.

The Miller and Miller similar media concept is just one possible scaling theory that could be used to provide a simple representation of heterogeneity of soil properties. A number of other possibilities have been reviewed by Tillotson and Nielsen (1984). Mohanty (1999) has recently proposed a method for scaling the properties of soils with macropores, treating their soil moisture characteristics as continuous curves spanning the matrix/macropore pore sizes. This assumes that such a dual porosity medium will respond as a continuum, which will not always be a good assumption.

Identification of the Soil Moisture Characteristic Function Parameters

Both the Brooks–Corey and the van Genuchten soil moisture characteristic functions require the calibration of a number of different parameter values before they can be used

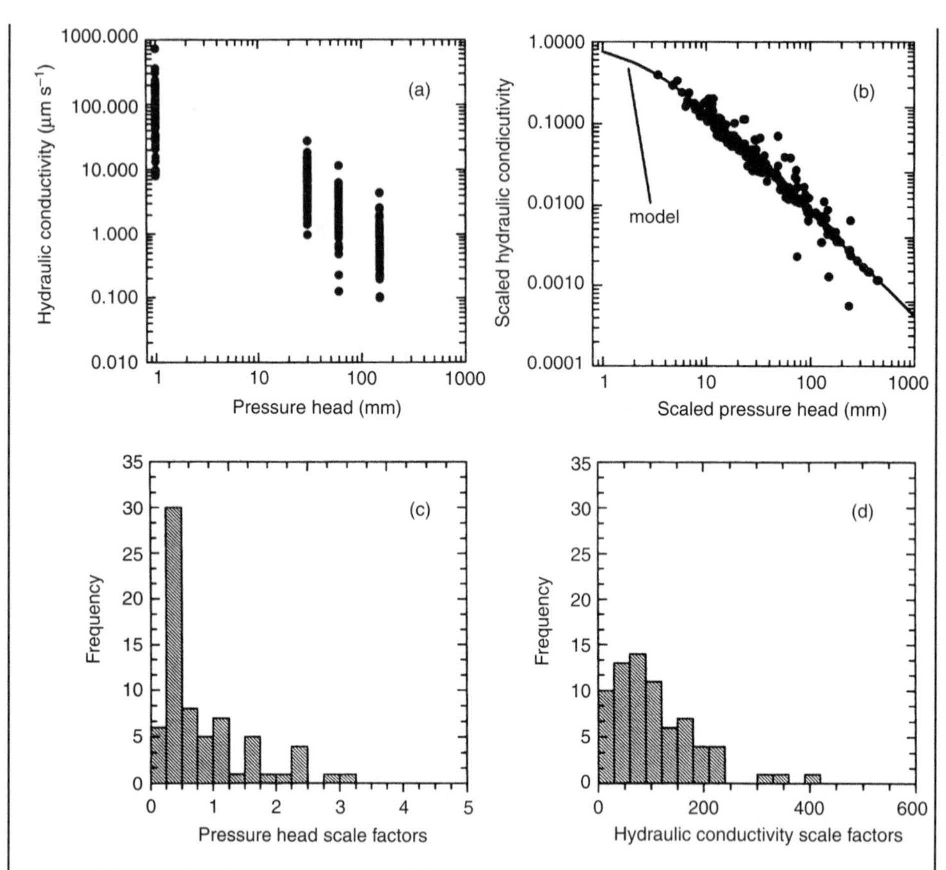

Figure B5.4.3 *Scaling of unsaturated hydraulic conductivity curves derived from field infiltration measurements at 70 sites under corn rows on Nicollet soil, near Boone, Iowa (after Shouse and Mohanty 1998). Reproduced from Water Resources Research 34, 1195–1205, 1998, copyright by the American Geophysical Union*

in a predictive model. Both laboratory and field measurements may be used in calibration, as well as indirect techniques based on *pedotransfer functions* which are discussed in Box 5.5. The resulting parameter values are not independent of the measurement techniques and parameter identification techniques used.

In principle, the soil moisture characteristics and conductivity characteristics at different capillary pressures can be measured directly in the laboratory. Different functions can then be fitted directly to the measured curves using a nonlinear least-squares regression procedure (e.g. Kool *et al.* 1987). However, such measurements are only currently possible on small samples which tend to show significant heterogeneity from sample to sample, especially close to saturation.

Identification is also possible from field-measured moisture profiles (or large undisturbed columns in the laboratory) by an inverse identification using a model based on the Richards equation. The parameter values are adjusted until a good fit is obtained between observed and predicted discharges. Because of interactions between the parameters, and physical effects such as hysteresis and, perhaps, preferential flows, it is often not possible to find a single parameter set that gives a best fit to the data (see Mishra and Parker 1989; Abeliuk and Wheater 1990).

Box 5.5 Pedotransfer Functions

Pedotransfer functions allow the estimation of the soil moisture characteristic curves on the basis of information about more easily measured variables, particularly textural variables. Pedotransfer functions are developed from experimental measurements made on a large number of samples. Two types of pedotransfer functions may be distinguished. In the first, equations are developed for values of moisture content and hydraulic conductivity for specific values of capillary potential (e.g. Gupta and Larson 1979). In the second, equations are developed to estimate the parameters of functional forms of the soil moisture characteristics, such as the Brooks–Corey or van Genuchten curves of Box 5.4, or parameters of application specific functions, such as the infiltration equations of Box 5.2 (e.g. Rawls and Brakensiek 1982; Cosby *et al.* 1984; van Genuchten *et al.* 1989; Vereeken *et al.* 1989, 1990; Schaap and Leij 1998).

Rawls and Brakensiek (1982) provided regression equations for the Brooks–Corey functions as a function of soil properties based on several thousand sets of measurements collected by the US Department of Agriculture. A summary of this regression approach is given by Rawls and Brakensiek (1989). A typical equation for K_s in terms of the variables C (percent clay, $5 < C < 60$), S (percent sand, $5 < S < 70$), and porosity, θ_s, is:

$$\begin{aligned}K_s = \exp[&19.52348\theta_s - 8.96847 - 0.028212C + 0.00018107S^2\\&- 0.0094125C^2 - 8.395215\theta_s^2 + 0.077718S\theta_s - 0.00298S^2\theta_s^2\\&- 0.019492C^2\theta_s^2 + 0.0000173S^2C + 0.02733C^2\theta_s\\&+ 0.001434S^2\theta_s - 0.0000035C^2S]\end{aligned}$$

where K_s is in cm h^{-1}, and the porosity θ_s can be estimated from measured dry bulk density ρ_d as

$$\theta_s = 1 - \frac{\rho_d}{\rho_s} \tag{5.46}$$

where ρ_s is the density of the soil mineral material ($\approx$2650 kg m^{-3}), or from another equation to estimate θ_s in terms of C, S, percent organic matter, and cation exchange capacity of the soil. These are all variables that are often available in soil databases. Rawls and Brakensiek also provide equations for adjusting porosity to allow for entrapped air, to correct for frozen ground and for surface crusting, to account for the effects of management practices, and for the parameters of various infiltration equations including the Green–Ampt and Philip equations of Box 5.2. Some special pedotransfer functions have also been developed, such as those of Brakensiek and Rawls (1994), to take account of the effects of soil stoniness on infiltration parameters.

It is necessary to use all these equations with some care. Equations such as that for K_s above have been developed from data generally collected on small samples (the USDA standard sample for the measurement of hydraulic conductivity was a 'fist-sized fragment' (Holtan *et al.* 1968) which would exclude any effects of larger-scale macroporosity). There is also considerable variability within each textural class. The apparent precision of the coefficients in this equation is therefore to some extent misleading. Each coefficient will be associated with a significant standard error, leading to a high uncertainty for each estimate of K_s. In the original paper of Rawls and Brakensiek (1982), the order of magnitude of these standard errors is given. In some later papers, this is no longer the case. The estimates provided by these equations are then apparently without uncertainty. This gives plenty of potential for being wrong, especially when in the application of a catchment-scale model it is the effective values of parameters at the model grid element scale that are needed. Some evaluations of the predictions of pedotransfer functions relative to field-measured

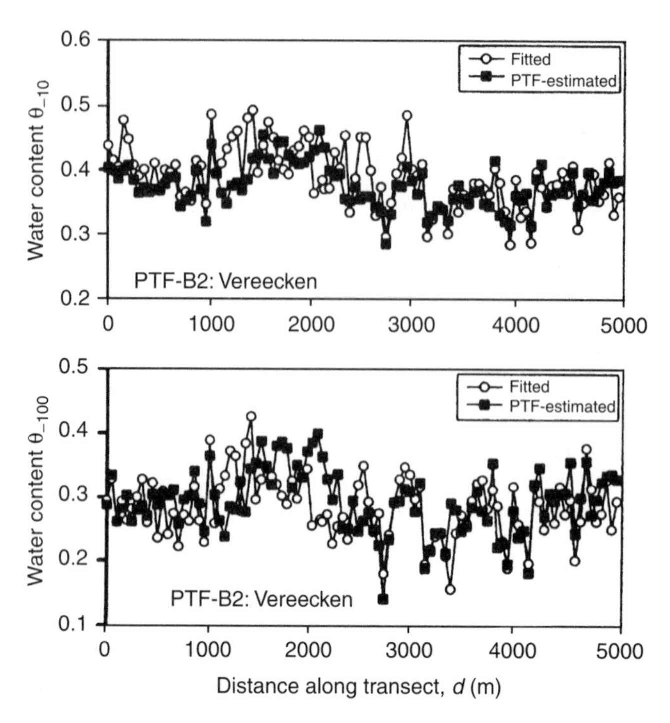

Figure B5.5.1 *A comparison of values of soil moisture at capillary potentials of −10 and −100 cm predicted by curves fitted to measured data and estimated using the pedotransfer functions of Vereeken et al. (1989) for different locations on a transect (after Romano and Santini 1997). Reprinted from Journal of Hydrology 202, 137–157, copyright (1997) with permission from Elsevier Science*

soil characteristics have been provided by Espino *et al.* (1995), Tietje and Tapkenhinrichs (1993), Romano and Santini (1997) and Wagner *et al.* (1998).

A recent development has been the use of neural network methods to derive pedotransfer functions (e.g. Schaap and Bouten 1996; Schaap *et al.* 1998). Uncertainty in the resulting functions cannot then be predicted by standard regression confidence estimates but can be estimated using a bootstrap method (Schaap and Leij 1998). Figure B5.5.1 gives an indication of the significant uncertainty in estimating soil moisture characteristics using pedotransfer functions for samples along a transet.

Pedotransfer functions for different soil texture classes have been included in the US STATSGO soils database (USDA SCS 1992) and the European HYPRES soils database (Wösten 1999).

Box 5.6 Descriptive Equations for Surface Flows

We will consider only a one-dimensional (downslope or downstream) description of surface flows here (for a review of two-dimensional models, see Lane 1998). In one-dimension, it is assumed that the flow can be adequately represented by a flow velocity, v [L T^{-1}], averaged over the local cross-sectional area, A [L^2] such that discharge $Q = vA$ [L^3 T^{-1}]. Thus, since both v and A will vary with discharge, there are two solution variables, but for surface flow it is less usual to assume a simple functional relationship between them (although see the discussion of the kinematic wave approximation below). Thus two equations are required to solve for the two variables. These equations were first formulated by the Barré de St Venant in terms of a mass balance and a balance of momentum. Using the cross-sectional area of the flow to represent storage per unit length of channel, then as shown in Box 2.3, the mass balance is given by

$$\begin{aligned}\frac{\partial A}{\partial t} &= -\frac{\partial Q}{\partial x} + i \\ &= -\frac{\partial vA}{\partial x} + i \\ &= -A\frac{\partial v}{\partial x} - v\frac{\partial A}{\partial x} + i\end{aligned}$$

where x [L] is distance downslope or downstream and i[L^2 T^{-1}] is the net lateral inflow rate per unit length of channel.

A second equation linking v and A can be developed from the momentum balance of the flow. The control volume approach of Box 2.3 can also be used to derive the momentum balance which may be expressed in words as:

spatial change in hydrostatic pressure	+	loss in potential energy	−	friction loss	=	temporal change in local momentum	+	spatial change in momentum flux

or

$$-\frac{\partial Ap}{\partial x} + \rho g A S_o - \tau P = \frac{\partial \rho A v}{\partial t} + \frac{\partial \rho A v^2}{\partial x} \tag{5.47}$$

where g [L T^{-2}] is the acceleration due to gravity, S_o [–] is the channel bed slope, P is the wetted perimeter of the channel [L], τ is the boundary shear stress [M L^{-1} T^{-2}], and p is the local hydrostatic pressure at the bed [M L^{-1} T^{-2}]. We can substitute for p and τ as

$$p = \rho g h \tag{5.48}$$

and

$$\tau = \rho g R_h S_f \tag{5.49}$$

where h [L] is an average depth of flow, S_f [–] is the friction slope which is a function of the roughness of the surface or channel, and R_h [L] is the hydraulic radius of the flow (= A/P).

With these substitutions, and dividing through by ρ under the assumption that the fluid is incompressible, the momentum equation may be rearranged in the following form:

$$\frac{\partial Av}{\partial t} + \frac{\partial Av^2}{\partial x} + \frac{\partial Agh}{\partial x} = gA(S_o - S_f) \tag{5.50}$$

The friction slope is usually calculated by assuming that the rate of loss of energy is approximately the same as it would be under uniform flow conditions at the same water surface slope so that one of the *uniform flow* equations holds locally in space and time. Thus, for the Darcy–Weisbach uniform flow equation

$$v = \left[\frac{2g}{f} S_f R_h\right]^{0.5} \tag{5.51}$$

where f is the Darcy–Weisbach resistance coefficient, so that an alternative form of the momentum equation is

$$\frac{\partial Av}{\partial t} + \frac{\partial Av^2}{\partial x} + \frac{\partial Agh}{\partial x} = gAS_o - gP\frac{f}{2g}v^2 \tag{5.52}$$

Again the St Venant equations are nonlinear partial differential equations that do not have analytical solutions except for some very special cases and approximate numerical solutions are necessary. As noted in the main text, the first attempt at formulating a numerical solution was due to Stoker (1957). This used explicit time stepping which generally requires very short time steps to achieve adequate accuracy. Most solution algorithms used today are based on implicit time stepping (see explanation of Box 5.3), such as the four-point implicit method described by Fread (1973, 1985).

Summary of Model Assumptions

We can summarize the assumptions made in developing this form of the St Venant equations as follows:

- *A1.* The flow can be adequately represented by the average flow velocity and average flow depth at any cross-section.
- *A2.* The amplitude of the flood wave is small relative to its wavelength so that pressure in the water column at any cross-section is approximately hydrostatic (pressure is directly proportional to depth below the water surface).
- *A3.* The water is incompressible and of constant temperature and density.
- *A4.* The friction slope may be estimated approximately using one of the uniform flow equations (such as the Darcy–Weisbach equation used above) with actual flow velocities and depths.

Simplifications of the St Venant Equations

The St Venant equations are based on hydraulic principles but they are clearly an approximation to the fully three-dimensional flow processes in any stream channel. There are various further approximations to the St Venant equations that are produced by assuming that one or more of the terms in the momentum equation, (5.50) can be neglected. The two main approximate solutions are the diffusion wave approximation:

$$\frac{\partial A\rho gh}{\partial x} = \rho gA(S_o - S_f) \tag{5.53}$$

and the kinematic wave approximation:

$$\rho gA(S_o - S_f) = 0 \tag{5.54}$$

so that

$$S_o = S_f \tag{5.55}$$

which reflects the assumption for the kinematic wave equation that the water surface is always parallel to the bed. This is also, of course, the assumption made in equations describing a uniform flow, so assuming again that the Darcy–Weisbach uniform flow equation is a good approximation for the transient flow case,

$$v = \left[\frac{2g}{f} S_o R_h\right]^{0.5} \tag{5.56}$$

For a channel that is wide relative to its depth, or an overland flow on a relatively smooth slope, then $R_h \approx h$ and this equation has the form of a power law relationship

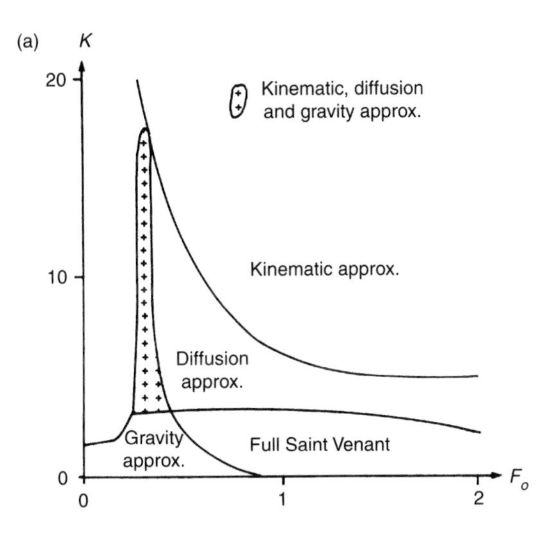

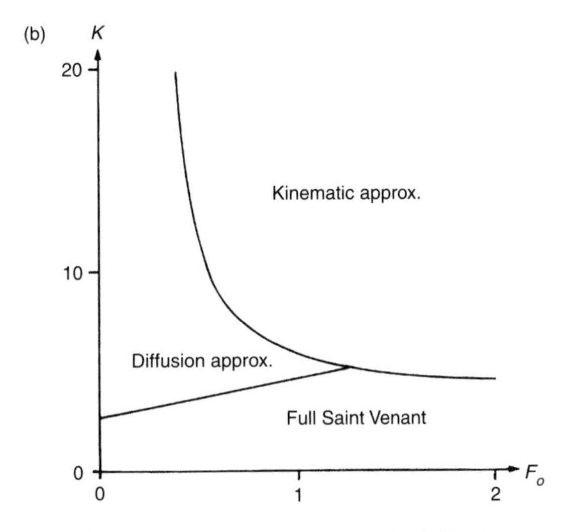

Figure B5.6.1 *Ranges of validity of approximations to the full St Venant equations defined in terms of the dimensionless Froude and kinematic wave numbers (after Daluz Vieira 1983)*

between velocity and storage:

$$v = bh^{0.5} \tag{5.57}$$

or, for discharge per unit width q (= vh):

$$q = bh^{1.5} \tag{5.58}$$

This type of power law has been widely used in kinematic wave descriptions of overland and channel flows (see Box 5.7). Use of a different uniform flow equation, such as the Manning equation (5.7), may result in a different power (1.67 in the case of the Manning equation).

A number of studies have examined the theoretical limits of acceptability of approximations to the full St Venant equations for different surface flows. A study by Daluz Vieira (1983), based on extensive numerical analysis, provided ranges of validity for different approximations to the St Venant equations, including the kinematic wave equation, in terms of two dimensionless numbers (see Figure B5.6.1): a Froude number

$$F_o = C\left(\frac{\tan\beta}{g}\right)^{0.5} \tag{5.59}$$

and a kinematic wave number

$$\kappa = \left(\frac{g^3 L \sin\beta}{C^4 q^2}\right)^{0.333} \tag{5.60}$$

where β is the local slope angle.

These studies suggest where the simplified models give a reasonable theoretical agreement with the full St Venant equations. However, uncertainty in effective parameter values and boundary conditions may mean that the simplified models are useful under a wider range of conditions. The flood routing example of Figure 5.9 in the main text is a good illustration of this. In a real application the shear stress and lateral inflow terms are not usually well known. Both terms may vary in both space and time as the river stage or depth of overland flow changes during an event. For river flow, in particular, there may be important changes in the effective shear stress if the flow exceeds the bankfull stage and starts to spill onto a flood plain (e.g. Knight *et al.* 1994).

Box 5.7 Derivation of the Kinematic Wave Equation

The kinematic wave equation arises from the combination of a mass balance or continuity equation expressed in terms of storage and flows, and a functional relationship between storage and flow that may be nonlinear but is single-valued, i.e. there is a single value of flow at a point corresponding to any value of storage at that point. Consider a one-dimensional downslope overland flow on a slope of constant width. Let x be distance along the slope [L], h the depth of flow (which acts as the storage variable) [L], and q the mean downslope discharge per unit width [L^{-2} T^{-1}] at any x (which is the flow variable). The mass balance equation can then be expressed as the differential equation

$$\frac{\partial h}{\partial t} = -\frac{\partial q}{\partial x} + r \tag{5.61}$$

where r is a rate of addition or loss of water per unit length and width of slope [L T^{-1}] at point x, and t is time [T].

The functional relation between h and q may be of many forms but a common assumption is the power law (see for example the kinematic approximation for surface flows in Box 5.6):

$$q = bh^a \tag{5.62}$$

Thus, assuming a and b are constant and combining these two equations, to yield a single equation in h

$$\frac{\partial h}{\partial t} = -abh^{(a-1)}\frac{\partial h}{\partial x} + r \tag{5.63}$$

or

$$\frac{\partial h}{\partial t} = -c\frac{\partial h}{\partial x} + r \tag{5.64}$$

where $c = dq/dh = abh^{(a-1)}$ is the kinematic wave velocity or *celerity* [$L\,T^{-1}$]. This is a kinematic wave equation. The celerity is the speed with which any disturbance to the system will be propagated downslope. It is worth noting again that kinematic wave equations can only propagate the effects of disturbances in a downslope or downstream direction. They cannot therefore predict any backwater effects in rivers or drawdown effects due to a channel for subsurface hillslope flows. They do have the advantage, however, that for some simple cases, such as the case of constant input rates, analytical solutions exist for both surface and subsurface cases (e.g. Eagleson 1970; Singh 1996). For more complex cases, such as variable width slopes and arbitrary patterns of inputs, it may be necessary to resort to a numerical solution but a finite difference approximation is very easy to formulate and include in a hydrological model. The four-point implicit finite-difference scheme of Li *et al.* (1975) has proven to be robust in a variety of applications. Li *et al.* point out that solving for the flow rate, q, rather than the storage variable, h, has some numerical advantages. The equivalent kinematic wave equation in q is

$$\frac{\partial q}{\partial t} = -c\frac{\partial q}{\partial x} + cr \tag{5.65}$$

The kinematic wave equation is easily extended to the case of a slope or channel for which the width, W_x, is varying downslope so that

$$W_x\frac{\partial q}{\partial t} = -c\frac{\partial W_x q}{\partial x} + cW_x r \tag{5.66}$$

For surface runoff, both overland flow and channel flow, the kinematic wave approach will be a good approximation to the full dynamic equations as the roughness of the surface or channel gets greater and as the bedslope gets steeper (see Box 5.6).

The kinematic wave approach can also be adapted for the case of saturated downslope subsurface flow, in which h represents a depth of saturation above a water table and account must be taken of the effective storage deficit in the unsaturated zone above the water table which will affect the rise and fall of the water table (Beven 1981). In the subsurface flow case, downslope flow rate per unit width of slope can often be approximated by a function such as

$$q = K_s h \sin\beta \tag{5.67}$$

for a saturated hydraulic conductivity, K_s, that is a constant with depth of saturation, or for a soil in which hydraulic conductivity falls with increasing distance below the soil surface, by a function of the following form:

$$q = K_o \exp(-f\{D - h\}) \sin\beta \tag{5.68}$$

where K_o is the saturated hydraulic conductivity of the soil at the soil surface [L T^{-1}], D is the depth of the soil to the impermeable layer [L], β is the local slope angle, and f is a coefficient that controls how rapidly hydraulic conductivity declines with depth [L^{-1}]. These functions can be interpreted as a form of Darcy's law in which the effective downslope hydraulic gradient is assumed to be equal to the local slope, $\sin\beta$. The kinematic wave equation may then be written as

$$\varepsilon W_x \frac{\partial h}{\partial t} = -\frac{\partial W_x q}{\partial x} + W_x r \tag{5.69}$$

where ε is an effective storage coefficient [–] (here assumed constant). Substituting for h gives the same equation as for surface flow on a variable width slope:

$$W_x \frac{\partial q}{\partial t} = -c\frac{\partial W_x q}{\partial x} + cW_x r \tag{5.70}$$

but here $c = (K_s \sin\beta)/\varepsilon$ for the constant conductivity case and

$$c = \frac{K_o \sin\beta \exp(f\{D - h\})}{f}$$

for the exponentially declining conductivity case.

For saturated subsurface runoff on a hillslope, Beven (1981) showed that the kinematic wave description was a good approximation to a more complete Dupuit–Forchheimer equation description, if the value of a non-dimensional parameter defined by

$$\chi = \frac{4K_s \sin\beta}{i} \tag{5.71}$$

where i the effective rate of storm recharge to the slope, was greater than about 0.75. When this condition is met, any drawdown of the water table at the lower end of the slope due to an incised channel, is unlikely to have a great effect on the predicted discharges.

Summary of Assumptions of Kinematic Wave Models

There is only one primary assumption underlying the derivation of the kinematic wave equation:

- *A1.* A functional relationship between storage and discharge can be specified for the particular flow process being studied.

Several examples of such relationships for both surface and subsurface flow have been demonstrated above. The limitations of the kinematic wave approach must be appreciated, but a major advantage is that it is not restrictive in its assumptions about the nature of the flow processes, only that discharge should be a function of storage. Analytical solutions of the kinematic wave equation require that this functional relationship should be univalued (and generally of simple form). Numerical solutions do not have this restriction and it is possible to envisage a kinematic wave solution that would have a hysteretic storage–discharge relationship that would more closely mimic the solution of the full surface or subsurface flow equations (in the same way that hysteretic soil moisture characteristics are sometimes used in unsaturated zone models; see Jaynes 1990). It seems that no-one has tried to implement such a model in hydrology.

6

Hydrological Similarity and Distribution Function Rainfall–Runoff Models

Faced with this situation, it has been usual to incorporate what knowledge we have about the operation of the processes into some conceptual model of the system. It is common that some parts of a complex conceptual model may be more rigorously based in physical theory than others. Even the most physically-based models, however, cannot reflect the true complexity and heterogeneity of the processes occurring in the field. Catchment hydrology is still very much an empirical science.

George Hornberger *et al.* 1985

6.1 Hydrological Similarity and Hydrological Response Units

In any catchment the hydrologist is faced with a wide variety of geology, soils, vegetation and land use, and topographic characteristics that will affect the relationship between rainfall and runoff. One way of taking these characteristics of any individual catchment into account is the type of fully distributed model that was discussed in the last chapter but, as was shown there, such models are difficult to apply because of their demands of both input data, much of which is not directly measureable, and computational resources. However, in any catchment, there may be many points that act in a hydrologically similar way with a similar water balance and similar runoff generation characteristics, whether by surface or subsurface flows. If it were possible to classify points in the catchment in terms of their hydrological similarity, then a simpler form of model could be used based on a distribution of functional hydrological responses in the catchment without the need to consider every individual point separately.

There are three main approaches that have been used in attempting to use such a distribution of different responses in the catchment to model rainfall–runoff processes. The first is a statistical approach, based on the idea that the range of responses in a catchment area can be represented as a probability distribution of conceptual stores without any explicit consideration of the physical characteristics that control the distribution of responses. This approach therefore has much in common with the transfer function models of Chapter 4, and the example we will outline in the next section, the probability distributed model of Moore and Clarke (1981), uses a similar parallel transfer function for the routing of the generated runoff (Section 6.2).

This purely statistical approach does not require any formal definition of similarity for different points in the catchment. Another type of distribution function model that attempts to define similarity more explicitly is that based on the idea of *hydrological response units* or HRUs. These are parcels of the landscape differentiated by overlaying maps of different characteristics, such as soils, slope, aspect, vegetation type, etc. This type of classification of the landscape has been very much easier to achieve now that maps of such characteristics can be held on the databases of geographical information systems, so that producing overlay maps of joint characteristics is a matter of a few simple mouse clicks on a personal computer or workstation. An example of the resulting landscape classification has already been seen in Figure 2.7. Models of this type differ in the type of conceptualization used for each HRU (see Section 6.3).

The third approach to be described is based on an attempt to define the hydrological similarity of different points in a catchment based on simple theory using topography and soils information. For catchments with moderate to steep slopes and relatively shallow soils overlying an impermeable bedrock, topography does have an important effect on runoff generation, at least under wet conditions, arising from the effects of downslope flows. This was the basis for the index of hydrological similarity introduced by Kirkby (1975) that was developed into a full catchment rainfall–runoff model, TOPMODEL, by Beven and Kirkby (1979) (see Section 6.4). The basic assumption of TOPMODEL is that all points in a catchment with the same value of the topographic index (or one of its variants described below) will respond in a hydrologically similar way. It is then not necessary to carry out calculations for all points in the catchment, but only for representative points with different values of the index. The distribution function of the index will then allow the calculation of the responses at the catchment scale.

These distribution function models are easy to implement and should require much less computer time than fully distributed models. They are clearly an approximation to a fully realistic distributed representation of runoff generation processes, but then so are the current generation of 'physically based' distributed models discussed in Chapter 5. It is not yet clear whether the latter are distinctly advantageous in practical applications. Certainly, there are limits to the accuracy of the distribution function models but the use of this type of model has led to some important insights.

6.2 The Probability Distributed Moisture Model (PDM)

In many ways, the PDM is a simple extension of some of the lumped storage models developed in the 1960s (and later) to the case of multiple storages representing a spatial distribution of different storage capacities in a catchment. It is a logical extension in that we would expect that a distribution of storages might be a better representation of the variability in the catchment than simple lumped storage elements. However, in the original form outlined by Moore and Clarke (1981) the model makes no real attempt to relate the distribution of storages to any physical characteristics of the catchment. In fact one of their main reasons for introducing a distribution of storages was to make the calibration problem easier since they found that they obtained smoother response surfaces for their new model formulation in comparison with models based on ESMA-type storage elements lumped at the catchment scale. A smoother response surface will, in general, make it easier for an automatic parameter optimization routine to find the best fitting set of parameter values (but see the discussion of parameter calibration in Chapter 7).

The basic idea of the PDM model is illustrated in Figure 6.1. The multiple storage elements are allowed to fill and drain during rainstorm and inter-storm periods respectively. If any storage is full then any additional rainfall is assumed to reach the channel quickly as storm runoff. A slow drainage component is allowed to deplete the storages between storms, contributing to the recession discharge in the channel and setting up the initial storages prior to the next storm. Evapotranspiration is also taken from each store during the inter-storm periods.

In any storm, clearly those stores with the smallest storage capacity will be filled first and will start to produce rapid runoff first. Each storage capacity is assumed to be representing a certain proportion of the catchment so that as the stores fill, the proportion of the area producing fast runoff can also be calculated. This area will expand during rainstorms and contract between rainstorms so that, in essence, the distribution of stores is representing a dynamic contributing area for runoff generation. In the published descriptions of the PDM models, the authors distinguish between surface runoff and baseflow components. This is not, however, a necessary interpretation and as with the transfer function models of Chapter 4, it is sufficient to recognize these as fast and slow runoff. The distribution of storages used in the model is only a distribution of conceptual storages and the question arises as to what form of distribution might be appropriate for a given catchment.

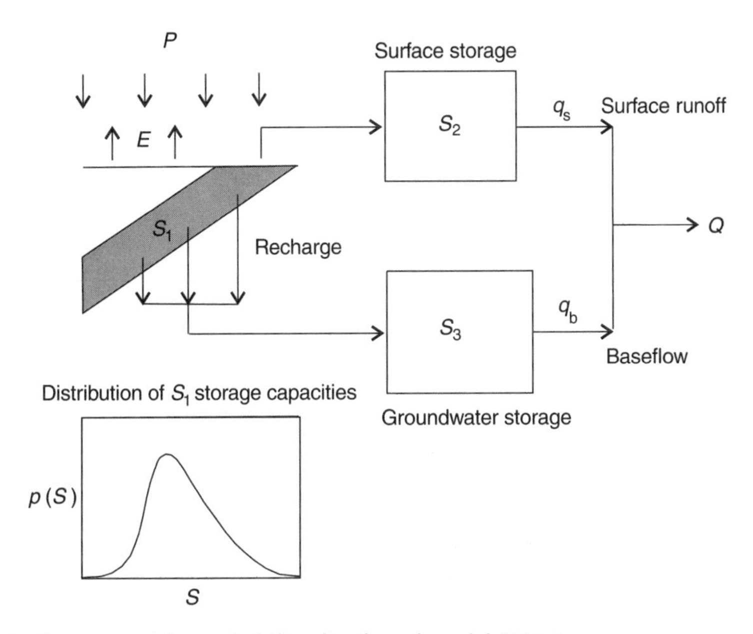

Figure 6.1 *Structure of the probability distributed model (PDM)*

Moore and Clarke (1981) show that a variety of distributions can be easily incorporated into this type of model structure and they derive analytical equations for the responses of different distributions. Their work was extended by Hosking and Clarke (1990) who show how the model can be used to derive a relationship between the frequencies of storm rainfall and flow peak magnitudes in an analytical form. Moore (1985) examines the case where the stores lose water to deep drainage and evapotranspiration, while Moore and Clarke (1983) link the model to predicting sediment production as well as discharges. A recent review of PDM concepts and equations has been provided by Clarke (1998). The model continues to be used and developed. Recent work at the UK Institute of Hydrology has seen the model used for long runs to derive flood frequencies (Lamb 1999), and also in a more distributed application with radar rainfall and snowmelt inputs for flood forecasting (Moore *et al.* 1994; Bell and Moore 1998; Moore *et al.* 1999). In the latter application, a separate PDM is used for each radar rainfall pixel (an element being 2 km by 2 km for UK radar rainfalls) so that any effect of the spatial distribution of rainfalls is preserved. Some attempt has also been made to reflect the different soil and topographic characteristics of these landscape units by varying the parameters of the distribution of stores in each element according to the soil type and average slope angle. The PDM has also recently been applied coupled to a distributed snowmelt model (Moore *et al.* 1999), a form of the model has been used as a macroscale hydrological model (Arnell 1999), and an alternative method of redistributing storage between the storage elements has been suggested (Senbeta *et al.* 1999).

The advantages of the PDM model are its analytical and computational simplicity. It has been shown to provide good simulations of observed discharges in many applications so that the distribution of conceptual storages can be interpreted as a reasonably realistic representation of the functioning of the catchment in terms of runoff generation. However, no further interpretation in terms of the pattern of responses is possible, since there is no way of assigning particular locations to the storage elements. In this sense, the PDM remains a lumped representation at the catchment (or subcatchment element in the distributed version) scale.

In fact, an analogy can be drawn between the structure of the PDM model and some lumped catchment models such as the VIC model, which use a functional relationship between catchment storage and the area producing rapid runoff (see Figure B2.2.1 in Box 2.2). The form of this relationship is controlled by parameters that are calibrated for a particular catchment area but will then imply a certain distribution of storage capacities in the catchment in a similar way to the PDM. Both models also use parallel transfer function routing for fast and slow runoff (surface runoff and baseflow in Figure 6.1), similar to the transfer function models discussed in Chapter 4.

6.3 Hydrological Response Unit Models

It would be useful to be able to relate storm runoff generation more directly to units of the landscape, but how then to take more account of the distribution of physical characteristics of a catchment without resorting to the fully distributed models of the last chapter? One method has arisen naturally out of the use of geographical information systems (GIS) in hydrological modelling. A GIS is commonly used to store data derived from soil maps, geological maps, a digital elevation map and a vegetation classification. As we have discussed earlier, these different maps cannot provide information of direct use in hydrological modelling, but they provide information that is certainly relevant to hydrological modelling. By overlaying the different types of information, a classification of different elements of the landscape into hydrological response units (HRUs) can be obtained (e.g. Figure 2.6). This is a relatively easy task with a modern GIS, or at least relatively easy once all the different sources of information have been stored and properly spatially registered in the GIS database (which can be very time consuming). The HRUs defined in this way may be irregular in shape where overlays of vector data are used, or based on regular elements where a raster (grid or pixel) database is used. Similar HRUs within the catchment will often be grouped together into a single unit for calculation purposes, as in the grouped response units of the SLURP model of Kite (1995) (Figure 6.2). It is these groupings, or the individual units, which then allow the prediction of the distribution of responses within the catchment.

The difficult bit in this type of modelling is how to represent the hydrological response of each HRU, which varies significantly between different models of this type. In some models a conceptual storage model is used to represent each HRU element (e.g

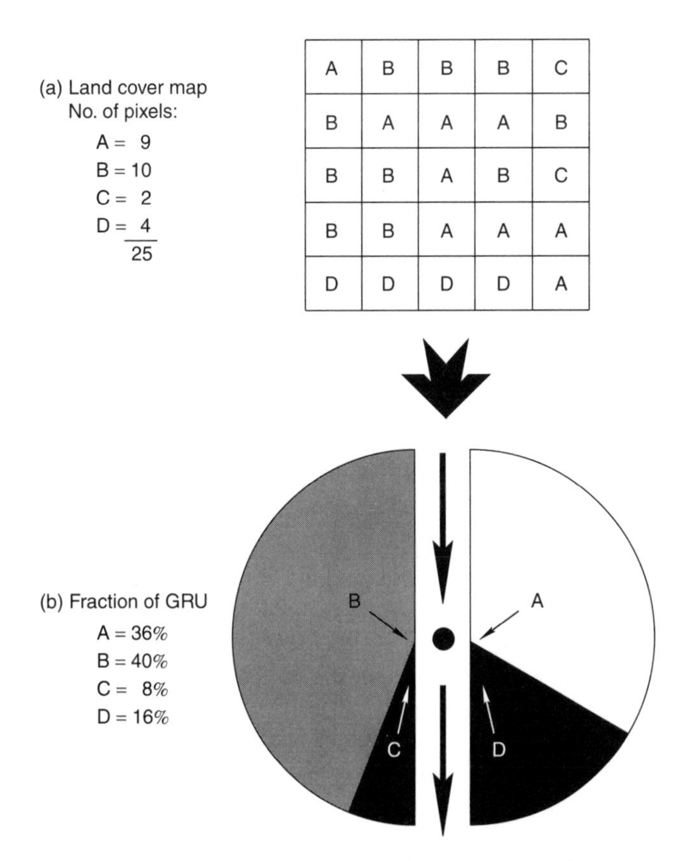

Figure 6.2 *Grouped response units as used in each grid square of the SLURP model (after Kite 1995)*

the SLURP model; the HBV96 model of Lindstrom *et al.* (1997); Modèle Couplé of Girard *et al.* (1981; also Ambroise *et al.* 1995); and the ARC/EGMO model of Becker and Braun (1999)). In others, a loss function is used to calculate a rainfall excess which is then routed to the catchment outlet, in some cases assuming a distribution of storage capacities within each HRU (e.g. Schumann and Funke 1996). As the HRU element scale becomes finer, and the hydrological description becomes more physically based, then this type of model will approach the fully distributed physically based models of the last chapter; the distinction we will draw here by including HRU models in this chapter about distribution function models, is that they do not explicitly aim to solve descriptive flow equations for surface and subsurface flow but allow the grouping of elements to reduce the number of calculations required.

Within this broad definition a wide variety of GIS-based models can be included. The distinction is not, in fact, a sharp one. For example, there are a number of raster GIS models where runoff generation calculations are made for every pixel, and flow is routed from pixel to pixel. However, not all such models use process representations that have a firm basis in physical theory but rather use conceptual functions, e.g. in the USGS PRMS system of Leavesley and Stannard (1995) and Flüghl (1995) in which fast runoff is generated by a simple variable contributing area function and the USDA SWAT model of Arnold *et al.* (1998) which is based on the USDA SCS (Soil Conservation Service) curve number method.

In fact, there are a number of examples of this type of model that use the SCS curve number method for predicting runoff generation (see Box 6.1). This is a method that has an interesting history, and will continue to be used because of the way in which databases of the SCS curve number can be related to distributed soil and vegetation information stored within a GIS. The SCS method has its origins in empirical analyses of rainfall–runoff data on small catchments and hillslope plots. It is commonly regarded as a purely empirical method for predicting runoff generation with no basis in hydrological theory. It is also commonly presented in hydrology texts as an infiltration equation or a way of predicting Hortonian infiltration excess runoff (e.g. Bras 1990), and a recent study by Yu (1998) has attempted to give it a basis in physical theory by showing that partial area infiltration excess runoff generation on a statistical distribution of soil infiltration characteristics gives similar runoff generation characteristics to the SCS method.

This is of some interest in itself, but the method becomes even more interesting if we return to the origins of the method as a summary of small catchment rainfall–runoff measurements by Mockus (1949). Mockus related storm runoff to rainfalls and showed that the ratio of cumulative discharge to cumulative storm rainfall shows a characteristic form (see Figure 6.3). In the past, the storm runoff may have been widely interpreted as infiltration excess runoff, but this is not a necessary interpretation. At the small catchment scale, the measured runoff in some of the original experiments will almost certainly have included some subsurface-derived water due to displacement, preferential flows or subsurface contributions from close to the channel. Certainly the method has since been applied to catchments and hydrological response units that are not dominated by infiltration excess runoff generation. Steenhuis *et al.* (1995) have already interpreted the SCS method in terms of a variable saturated contributing area, excluding, in their analysis, some data from high-intensity events that might have

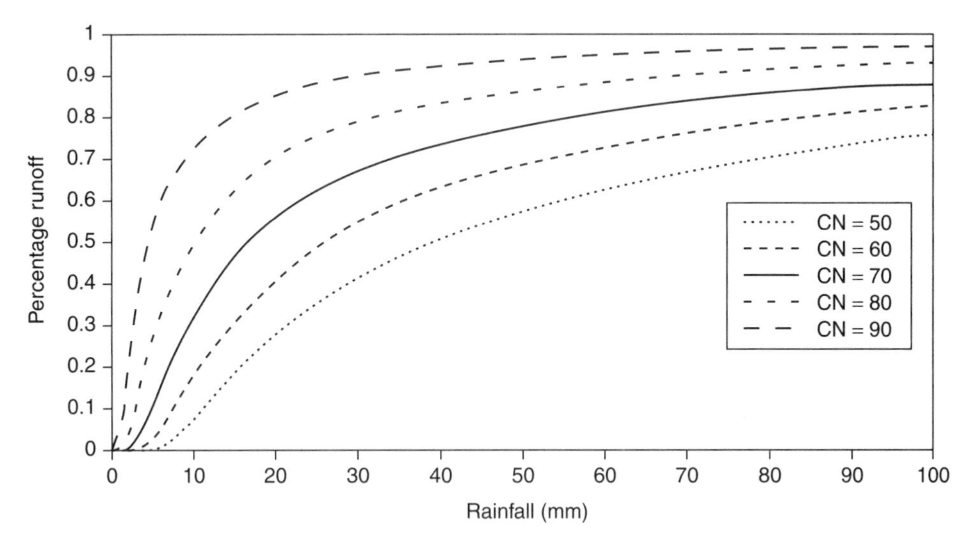

Figure 6.3 *Relationship between storm rainfall and percentage runoff predicted by the USDA SCS method for different curve numbers*

produced an infiltration excess runoff. A sufficient view of the method is that it incorporates some empirical knowledge of fast runoff generation, by whatever method, at the small catchment scale, into a simple functional form. It may be necessary to check whether that form is appropriate in any particular application but it may be an appropriate method to use at the HRU scale since it encapsulates knowledge gained at similar scales. In considering the scale dependency of an HRU model, therefore, it might be more appropriate than, say, any of the point scale infiltration equations described in Box 5.2, even though these would normally be considered more 'physically based'. Some further background (and limitations) to the SCS method is given in Box 6.1.

Scale is an issue in HRU modelling. The type of HRU representation used to predict runoff generation might be expected to vary both with the hydrological environment and with the spatial scale at which the HRU elements are defined, and at least one modelling system (the modular modelling system, MMS, under development by the USGS to replace the PMRS system) allows the description to be chosen interactively by the user. Each HRU is generally considered as homogeneous in its parameter values and response so that, for example, if surface runoff is calculated to occur, it will do so over the whole HRU. The HRUs are also often treated independently with no explicit routing of downslope surface or subsurface flows between HRU elements, only routing of runoff to the nearest channel. This assumption of independence of position in the catchment will, in fact, be necessary if HRUs with similar characteristics are to be grouped together. A typical example of a HRU-based model structure that does include some routing between elements is shown in Figure 6.4.

One advantage of the HRU approach is that the calculated responses can be mapped back into space using visualization routines in a GIS so that this can, in principal at least, provide information for a spatial evaluation of the predictions. The major

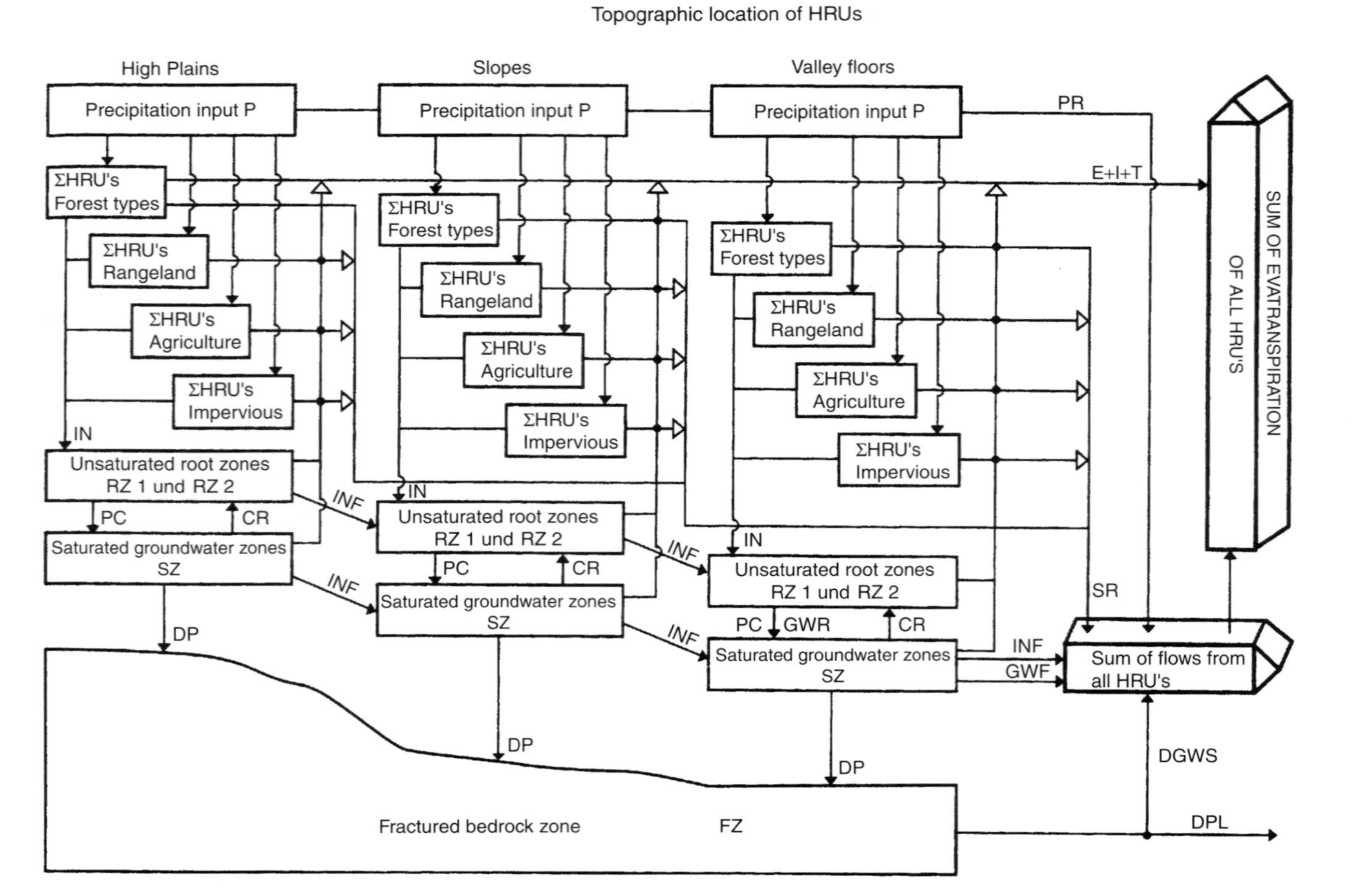

Figure 6.4 *A model structure for a representative catchment hillslope based on hydrological response units (after Flügel 1995). Reproduced*

disadvantage is the way in which each HRU is considered to be spatially homogeneous – an assumption that must become poorer as the HRU gets larger. At very large scales, new methods may be needed, as in the macroscale hydrological models that are discussed in Chapter 9. At the current time, we have no theory for predicting the appropriate model structure or parameter values at one scale, given information at another scale. Indeed, it has been argued that we will never have such a theory, and that we will need to resort to models that are essentially scale-dependent (Beven 1995, 2000).

In applying HRU models, the scale dependence of the parameter values used in representing each unit should be considered. Because there are many HRUs and several (or many) parameters will be needed for each HRU, it will not be easy to calibrate parameters by an optimization process. In this, such models face similar problems to fully distributed physically based models. The GIS may store soil type and vegetation type but the information about model parameters for each classification may be highly uncertain and may not be independent (e.g. rooting depth of a certain vegetation type may depend on soil type or the hydraulic characteristics of a soil type might depend on the type of land use). The real hydrological response of an HRU may depend on the heterogeneity within the element, that might not be well represented by homogeneous 'effective' parameter values. This has to be an important limitation on this type of model structure but, as discussed in Chapter 5, it is essentially a limitation of all types of model given the limitations of our knowledge of how to represent the detailed variability of hydrological systems. Again, it suggests that the predictions of such models should be associated with some estimate of uncertainty, but to my knowledge there are no reported cases where this has been done for an HRU model.

6.4 TOPMODEL

A simpler approach to predicting spatial patterns of responses in a catchment is represented by TOPMODEL (see Beven *et al.* 1995; Beven 1997). TOPMODEL may be seen as a product of two objectives. One is the development of a pragmatic and practical forecasting and continuous simulation model. The other is the development of a

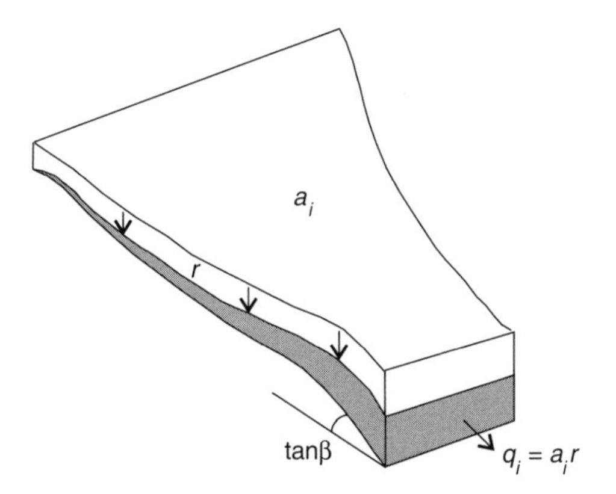

Figure 6.5 *Definition of the upslope area draining through a point within a catchment*

theoretical framework within which perceived hydrological processes, issues of scale and realism and model procedures may be researched. Parameters are intended to be physically interpretable and their number is kept to a minimum to ensure that values determined by a calibration exercise should be more easily identifiable, while still allowing a mapping of the predictions back into the catchment based on the pattern of a topographic index derived from an analysis of flow paths in the catchment. The model, in practice, represents an attempt to combine the computational and parametric efficiency of a distribution function approach with the link to physical theory and possibilities for more rigorous evaluation offered by a fully distributed model.

6.4.1 TOPMODEL – The Background Theory

TOPMODEL can be considered to be a further approximation to the kinematic wave description of the subsurface flow system of Section 5.4. This link is made explicit by Kirkby (1997) and Wigmosta and Lettenmaier (1999). It is premised upon two basic assumptions: that the dynamics of the saturated zone can be approximated by successive steady-state representations of the saturated zone on an area a draining to a point on a hillslope (Figure 6.5); and that the hydraulic gradient of the saturated zone can be approximated by the local surface topographic slope measured with respect to plan distance, $\tan\beta$ (Box 6.2).

These assumptions lead to simple relationships between catchment storage (or storage deficit below saturation) in which the main factor is the Kirkby topographic index ($a/\tan\beta$) (Kirkby 1975). The Kirkby index represents the propensity of any point in the catchment to develop saturated conditions. High values will be caused by either long slopes or upslope contour convergence, and low slope angles. Points with the same value of the index will be predicted as having the same hydrological responses. The topographic index approach was developed into a complete rainfall–runoff model by Beven and Kirkby (1979) and has been generalized since to allow for differences in soil characteristics within the catchment (see below and Box 6.1). The assumptions are similar to those used in the development of the 'wetness' index developed independently by O'Loughlin (1981, 1986) and used in the model of Moore *et al.* (1986).

TOPMODEL in its original form takes advantage of the mathematical simplifications allowed by a third assumption: that the distribution of downslope transmissivity with depth is an exponential function of storage deficit or depth to the water table:

$$T = T_o e^{-D/m} \tag{6.1}$$

where T_o is the lateral (horizontal) transmissivity when the soil is just saturated [$L^2\ T^{-1}$], D is a local storage deficit below saturation expressed as a water depth [L], and m is a model parameter controlling the rate of decline of transmissivity in the soil profile, also with dimensions of length [L]. A physical interpretation of the decay parameter m is that it controls the effective depth or active storage of the catchment soil profile. A larger value of m effectively increases the active storage of the soil profile. A small value generates a shallow effective soil, with a pronounced transmissivity decay.

Given this exponential transmissivity assumption, it can be shown that the appropriate index of similarity is $\ln(a/\tan\beta)$ or if the value of T_o is allowed to vary in space

$\ln(a/T_o \tan\beta)$ such that given a mean storage deficit over a catchment area $\overline{D}$, a local deficit at any point can be calculated as (see Box 6.2):

$$D_i = \overline{D} + m[\gamma - \ln(a/T_o \tan\beta)] \tag{6.2}$$

where γ is the mean value of the index over the catchment area. Thus, every point having the same soil/topographic index value $(a/T_o \tan\beta)$ behaves functionally in an identical manner. The $(a/T_o \tan\beta)$ variable is therefore an index of hydrological similarity. Other forms of transmissivity profile assumption lead to different forms for the index and local deficit calculation (see Box 6.2). Of particular interest are points in the catchment for which the local deficit is predicted as being zero at any time step. These points, or fraction of the catchment, will represent the saturated contributing area that expands and contracts with the change in $\overline{D}$ as the catchment wets and dries (Figure 6.6). The equations can also be derived in terms of water table depth rather than storage deficit but this introduces at least one additional effective storage parameter (Beven *et al.* 1995). In each case, there will be a relationship between the transmissivity profile assumed and the form of the recession curve at the catchment scale produced by soil drainage. For the exponential transmissivity assumption the derived recession curve function is given by

$$Q_b = Q_o e^{-\overline{D}/m} \tag{6.3}$$

where $Q_o = Ae^{-\gamma}$ for a catchment area of A. This equation (and equivalent forms for different transmissivity assumptions) is derived under the assumption that the effective hydraulic gradients for the subsurface flow do not change with time, as would be predicted by a more complete analysis.

The calculation of the index for every point in the catchment requires knowledge of the local slope angle, the area draining through that point and the transmissivity at saturation. The spatial distribution of $(a/\tan\beta)$ (see Figure 6.6) may be derived from analysis of a digital terrain model (DTM) or digital elevation map (DEM) of the catchment (see Section 6.4.2). Specifying a spatial distribution for T_o is generally much more problematic, since there are no good measurement techniques for obtaining this parameter. In most applications it has been assumed to be spatially homogeneous, in which case the similarity index reduces to the form $(a/\tan\beta)$.

To calculate the surface (or subsurface) contributing area, the catchment topographic index is expressed in distribution function form (Figure 6.7). Discretization of the $(a/\tan\beta)$ distribution function brings computational advantages. Given that all points having the same value of $(a/\tan\beta)$ are assumed to behave in a hydrologically similar fashion, then the computation required to generate a spatially distributed local water table pattern reduces to one calculation for each $(a/\tan\beta)$ class; calculations are not required for each individual location in space. This approach should be computationally more efficient than a solution scheme that must make calculations at each of a large number of spatial grid nodes – a potentially significant advantage when parameter sensitivity and uncertainty estimation procedures are carried out.

In a time step with rainfall, the model predicts that any rainfall falling upon the saturated source area will reach the stream by a surface or subsurface route as storm runoff, along with rainfall in excess of that required to fill areas where the local deficit is small. The calculated local deficits may also be used to predict the pattern

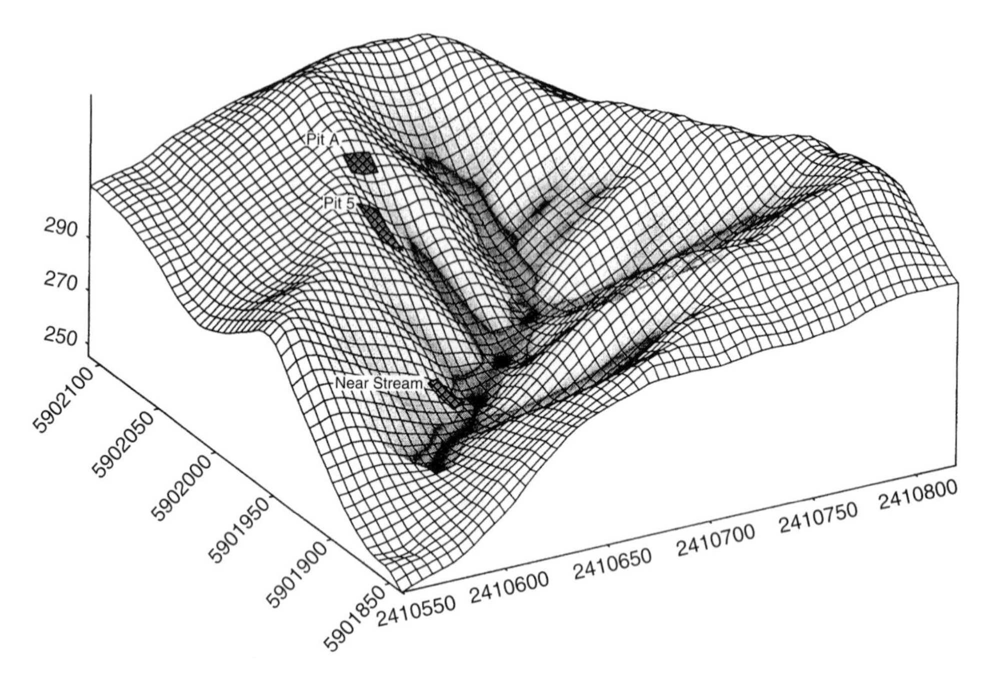

Figure 6.6 *(a) The ln(a/tanβ) topographic index in the small Maimai M8 catchment (3.8 ha), New Zealand, calculated using a multiple flow direction downslope flow algorithm. High values of topographic index in the valley bottoms and hillslope hollows indicate that these areas will be predicted as saturating first (after Freer 1998)*

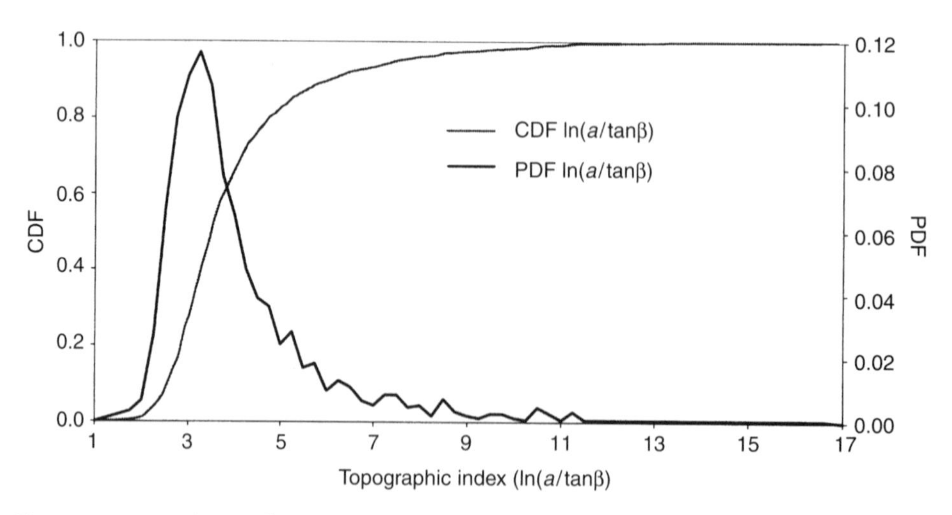

Figure 6.7 *Distribution function and cumulative distribution function of topographic index values in the Maimai M8 catchment (3.8 ha), New Zealand, as derived from the pattern of Figure 6.6*

of subsurface stormflow contributing areas, or flow through different soil horizons (Robson *et al.* 1992) if these can be defined by some threshold value of deficit (or water table depth).

The model is completed by a representation of the unsaturated zone and a flow routing component. Both have been kept deliberately simple to facilitate parameter estimation. It is particularly difficult to account explicitly for the effects of local soil heterogeneity and macroporosity. No adequate mathematical description of unsaturated flow in structured soils with parameters that can be identified at a practical prediction scale is currently available (but see Bronstert and Plate (1997) for an interesting attempt to do so) and if parameter values are to be determined by calibration then a minimal parametrization is advantageous. Current versions of TOPMODEL use two stores to represent the unsaturated zone, one representing interception and root zone storages for which additional deficits due to evapotranspiration are calculated, and a drainage store which controls recharge to the saturated zone. Both introduce one additional parameter.

There is no reason why this part of the model should not be made more complex if sufficient information is available to justify that complexity. Indeed, more complex representations of the soil and vegetation have been linked to TOPMODEL concepts in the TOPLATS formulation of Famiglietti *et al.* (1992), in RHESSys (Band *et al.* 1991, 1993; Fagre *et al.* 1997; Hartman *et al.* 1999) and in the MACAQUE model of Watson *et al.* (1999). TOPLATS can also be used as fully distributed models with calculations made for every pixel within a catchment (e.g. Houser *et al.* 1998). This extension introduces further vegetation and soil parameters. Representations of infiltration excess runoff can also be included and in some other applications of TOPMODEL, more explicitly physically based approaches to infiltration have been adopted (Beven 1986a,b, 1987; Sivapalan *et al.* 1990; Wood *et al.* 1990). However, these approaches also introduce extra parameters to the model which may be difficult to calibrate, especially for the case of spatially heterogeneous soils.

6.4.2 Deriving the Topographic Index

An analysis of catchment topography is required in order to derive the $(a/\tan\beta)$ distribution function (Figure 6.6). In order to obtain discrete values of $(a/\tan\beta)$, some sampling of topography is implied. Early development of TOPMODEL relied upon the manual analysis (based on map and air photo information) of local slope angles, upslope contributing areas and cumulative areas. Beven and Kirkby (1979) outlined a computerized technique used to derive the topographic index distribution function (and the overland flow delay histogram) based on the division of the catchment into sub-basin units. Each unit was then discretized into small 'local' slope elements on the basis of dominant flow paths (inferred from lines of greatest slope in a way similar to the TAPES-C flow path analysis software of Grayson *et al.* 1995). Calculation of $(a/\tan\beta)$ was carried out for the downslope edge of each element. Although an approximation, this method was felt to be justified by its relative efficiency and because field observations of flow paths could be used in defining the slope elements to be analysed. In particular, the effects of field drains and roads in controlling effective upslope contributing areas could be taken into account. Such human modifications of the natural hydrological flow pathways are often important to the hydrological response but are not normally included in the DTMs of catchment areas commonly used in topographic analysis today.

However, given a DTM, more computerized methods are now available. Quinn *et al.* (1995a) demonstrate the use of digital terrain analysis (DTA) programs, based on raster elevation data, in application to catchment modelling studies based on TOPMODEL. There are subjective choices to be made in any digital terrain analysis. Techniques of determining flow pathways from raster, contour and triangular irregular network DTMs have been discussed in Section 3.7. Different DTA techniques will result in different flowpath definitions and therefore different calculations of 'upslope contributing area' for each point in the catchment. The resolution of the DTM data will also have an effect. The DTM must have a fine enough resolution to reflect the effect of topography on surface and subsurface flow pathways adequately. Coarse resolution DTM data may, for example, fail to represent some convergent slope features. However, too fine a resolution may introduce perturbations to flow directions and slope angles that may not be reflected in subsurface flow pathways which, in any case, will not always follow the directions suggested by the surface topography. Freer *et al.* (1997), for example, suggest that bedrock topography may be a more important control in some catchments. The appropriate resolution will depend on the scale of the hillslope features, but 50 m or better data is normally suggested. Anything much larger and in most catchments it will not be possible to represent the form of the hillslopes in the calculated distribution of the $(a/\tan\beta)$ index.

Experience suggests that the scale of the DTM used and the way in which river grid squares are treated in the DTA, do affect the derived topographic index distribution, in particular inducing a shift in the mean value of $(a/\tan\beta)$. There will then be a consequent effect on the calibrated values of parameters (especially the transmissivity parameter) in particular applications. This is one very explicit example of how the form of a model definition may interact with the parameter values required to reproduce the hydrology of a catchment. In this case, two different topographic analyses of a catchment may require different effective transmissivity values for hydrograph simulation. However, in one recent study, Saulnier *et al.* (1997a) have suggested that excluding river grid squares from the distribution results in calibrated transmissivity parameter values that are much more stable with respect to changing DTM resolution. The simplicity of the TOPMODEL structure has allowed this problem to be studied in some detail but similar considerations of an interaction between grid scale and model parameters must apply to even the most physically based, fully distributed models (Beven 1989; Refsgaard 1997; Kuo *et al.* 1999).

It is worth noting that a parametrization of the $(a/\tan\beta)$ distribution may sometimes be useful. Sivapalan *et al.* (1990) introduced the use of a gamma distribution in their scaled version of TOPMODEL. Wolock (1993) also gives details of a gamma distribution version for continuous simulation. An analogy with the statistical PDM model of Section 6.2 becomes apparent in this form. The advantage of using an analysis of topography to define the index distribution beforehand is that there are then no additional parameters to estimate. This will only be an advantage, however, where the topographic analysis allows a realistic representation of the similarity of hydrological responses in the catchment, which clearly depends on the validity of the simplifying assumptions that underlie the index.

6.4.3 Applications of TOPMODEL

Simulation of humid catchment responses

TOPMODEL was originally developed to simulate small catchments in the UK (Beven and Kirkby 1979; Beven *et al.* 1984). These studies showed that it was possible to get reasonable results with a minimum of calibration of parameter values. A summary table of applications is given in Beven (1997). More recent applications include Franks *et al.* (1998) and Saulnier *et al.* (1998) in France; Lamb *et al.* (1998a,b) in Norway (see the case study in Section 6.5); Quinn *et al.* (1998) in Alaska; Cameron *et al.* (1999) in Wales; Dietterick *et al.* (1999) in the US; Donnelly-Makowecki and Moore (1999) in Canada; and Güntner *et al.* (1999) in Germany (Figure 6.8). In most of these cases it has been found that, after calibration of the parameters, TOPMODEL provides good simulations of stream discharges, and broadly believable simulations of variable contributing areas.

Catchments with deeper groundwater systems or locally perched saturated zones may be much more difficult to model. Such catchments tend to go through a wetting up sequence at the end of the summer period in which the controls on recharge to any saturated zone and the connectivity of local saturated zones may change with time. An example is the Slapton Wood catchment in southern England, modelled by Fisher and Beven (1995).

Simulation of drier catchment responses

A model that purports to predict fast catchment responses on the basis of the dynamics of saturated contributed areas may not seem to be a likely contender to simulate the responses of catchments that are often dry, such as in Mediterranean or savannah climates. However, Durand *et al.* (1992) have shown that TOPMODEL can successfully simulate discharges in such catchments at Mont-Lozère in the Cevennes, southern France, at least after the calibration of some parameters.

Experience in modelling the Booro–Borotou catchment in the Côte d'Ivoire (Quinn *et al.* 1991), Australia (Barling *et al.* 1994) and catchments in the Prades mountains of Catalonia, Spain (Piñol *et al.* 1997), suggests that TOPMODEL will only provide satisfactory simulations once the catchment has wetted up. In many low-precipitation catchments, of course, the soil may never reach a 'wetted' state, and the response may be controlled by the connectivity of any saturated downslope flows. TOPMODEL assumes that there is connected downslope saturation everywhere on the hillslope; before such connectivity is established a dynamic index would be required. Such catchments also tend to receive precipitation in short, high-intensity storms. Such rainfalls may lead, at least locally, to the production of infiltration excess overland flow which is not usually included in TOPMODEL (but see Beven (1986a,b) and Sivapalan *et al.* (1990) for example applications including infiltration excess calculations). The underlying assumptions of the TOPMODEL concepts must always be borne in mind relative to the pertinent perceptual model for a particular catchment.

6.4.4 Testing the Hydrological Similarity Concept in TOPMODEL

TOPMODEL may be expected to perform best when tested against catchments where its assumptions are met, in particular those of an exponential saturated zone store,

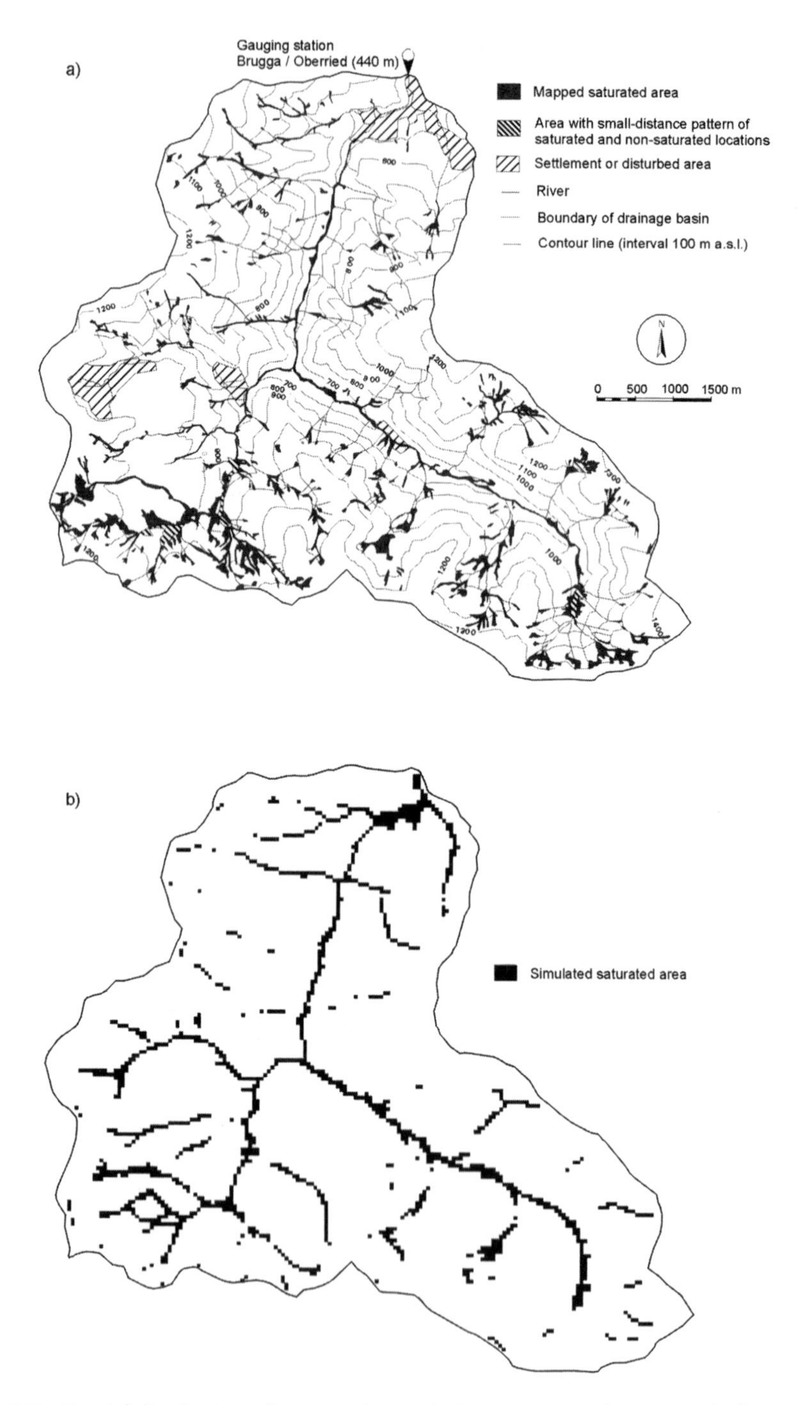

Figure 6.8 *Spatial distribution of saturated areas in the Brugga catchment (40 km²), Germany. (a) Mapped saturated areas (6.2 percent of catchment area). (b) Topographic index predicted pattern at same fractional catchment area assuming a homogeneous soil (after Güntner et al. 1999). Reproduced with permission of John Wiley & Sons Limited*

a quasi-parallel water table and a topographic control on water table depth. A full critique of the TOPMODEL concepts can be found in Beven (1997). There are certainly limitations on both the geographical and seasonal validity of the TOPMODEL concepts, but they do provide a basis for thinking about the distributed nature of catchment responses. It has always been stressed that TOPMODEL is not a fixed model structure but rather a set of concepts that should be modified if it is perceived that a catchment does not conform to the basic assumptions. Ways of relaxing the basic assumptions are discussed in Box 6.2.

The main limitation imposed by the model assumptions is that of the quasi-steady-state water table shape. This has been criticized by Wigmosta and Lettenmaier (1999) who compare the results of TOPMODEL with a dynamic simulation based on a subsurface kinematic wave solution. They show that although TOPMODEL can usually be calibrated to produce reasonable simulation of discharge hydrographs produced by the kinematic wave model, the resulting effective transmissivity values tended to be high and the steady-state assumption did not produce reasonable predictions of the dynamic changes in the water table. In this case, the model simulations assumed recharge to the saturated zone all over the hillslope but clearly this will not always be the case, as noted in the last section. Barling *et al.* (1994) have shown that a better relationship could be found between a saturated area and a topographic index, if the index was calculated using only an effective upslope contributing area rather than the full upslope area all the way to the divide normally calculated in digital terrain map analysis. This effective upslope area would be expected to be small when the catchment was dry and increase as the catchment wets up. In fact, Western *et al.* (1999) show that patterns of near-surface soil moisture only show the effects of a topographic control on downslope flows under relatively wet conditions in the Tarrawarra catchment in Australia. This is, in fact, another reason why calibrated values of transmissivity in TOPMODEL might be high. Since, in the soils topographic index ($a/T_o \tan\beta$), a and T_o appear in ratio, a high value of T_o can compensate for an overestimation of the effective upslope area a.

It is the steady-state assumption that allows TOPMODEL to make use of similarity in greatly increasing computational efficiency. This is useful for a number of purposes, not least of which is the exploration of predictive uncertainty considered next in Chapter 7. There is a possibility that the approach could be modified to allow more dynamic calculations while retaining the concept of the index. This is currently the subject of active research and some new developments are summarized in Box 6.2 (see also the TOPKAPI variant in Section 6.6 below).

6.4.5 The TOPMODEL Software

There are two programs associated with the TOPMODEL demonstration software: one for the initial analysis of a catchment DTM (DTMAnalysis), and the second (TOMODEL99) to simulate hydrographs and contributing areas, and to carry out model sensitivity analysis. Both programs may be downloaded over the Internet. The options available in each program are described in Appendix A.

The DTMAnalysis package requires a raster elevation data file. It is used to create the topographic index distribution and map for use in TOPMODEL99. An example data file for the small Slapton Wood catchment in Devon, UK is provided.

The TOPMODEL99 package requires the topographic index information that is the output of DTMAnalysis together with files of observed rainfalls, discharges and potential evapotranspiration. Example files of hourly data for the Slapton Wood catchment are provided. If a map of topographic index values for the catchment is available, then an animation of the simulated contributing areas in the catchment may be displayed. The user may change parameter values and rerun the simulation to try and improve the fit to the observed discharges or carry out a simple sensitivity analysis by varying one or more parameters across a range chosen by the user. A further option also allows random sets of parameter values to be chosen and the model run many times to create an output file that can be used in the GLUE uncertainty estimation package that is described in Chapter 7.

6.5 Case Study: Application of TOPMODEL to the Saeternbekken Catchment, Norway

In most rainfall–runoff modelling studies, there are generally few internal state measurements with which to check any distributed model predictions. The potential to make such checks with distributed models raises some interesting questions about model calibration and validation. One study where distributed predictions have been checked is in the application of TOPMODEL to the Saeternbekken MINIFELT catchment in Norway (Lamb *et al.* 1997, 1998a,b). This small subcatchment of only 0.75 ha has a network of 105 piezometers and four recording boreholes (Figure 6.9; Myrabo 1997; Erichsen and Myrabo 1990). The distribution of the $\ln(a/\tan\beta)$ topographic index is shown in Figure 6.9(a). Figure 6.9(b) shows the discharge predictions of two different variants of TOPMODEL. In both cases, the models have been calibrated on rainfall–runoff events in 1987, and the results shown are for a separate evaluation period in 1989.

The EXP model in Figure 6.9(b) is essentially the original exponential transmissivity function version of TOPMODEL that is described in detail in Box 6.2. The COMP model uses a technique due to Lamb *et al.* (1997, 1998a,b) in which a recession curve analysis is used to define an arbitrary discharge/storage deficit relation that can replace the exponential transmissivity assumption of the original model. In the case of the Saeternbekken MINIFELT catchment, a composite curve with exponential and linear segments was found to be suitable. Other functions can also be used (see Box 6.2). There was little to choose between the models in discharge prediction.

TOPMODEL, however, also allows the mapping of the predictions of storage deficit or water table level back into the space of the catchment. While, given the approximate assumptions of the TOPMODEL approach, it would not be expected that the results will be accurate everywhere in the catchment, we would want to reject the model if it was shown to give very poor predictions of such internal measurements. Results for the recording boreholes are shown in Figure 6.10 and for the piezometer levels for five different discharges in Figure 6.11. For these internal data the composite model appears to give slightly better results, but clearly both models are limited in their accuracy (notably in predicting the significant depths of ponding recorded in some of the piezometers since neither model has any direct representation of depression storage in the hummocky terrain). In both cases these are predictions using the global parameter

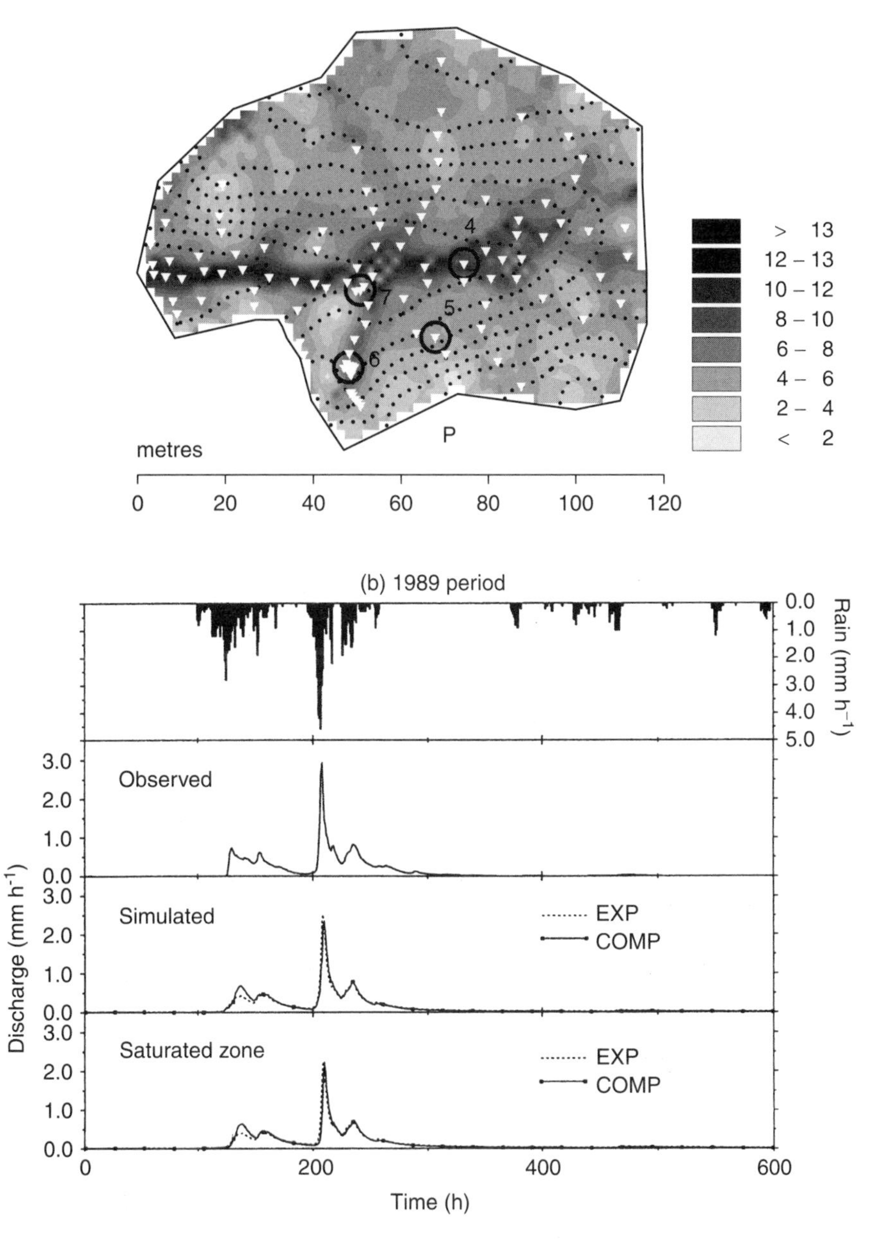

Figure 6.9 *Application of TOPMODEL to the Saeternbekken MINIFELT catchment, Norway (0.75 ha). (a) Topography, network of instrumentation and pattern of the ln(a/tanβ) topographic index indicates positions of piezometers; those circled (4, 5, 6, 7) are continuously recording piezometers. (b) Prediction of stream discharges using both exponential and generalized transmissivity functions (after Lamb et al. 1997)*

values calibrated for the 1987 period. This effectively assumes that the transmissivity function in both models is uniform throughout the catchment. The recording borehole data were used only in the estimation of an effective storage coefficient, for which a catchment median calibrated value was 0.06. The piezometer data were not used in calibration at this stage.

These results are reasonable but could perhaps be improved in a number of ways. The need for some surface depression storage has already been noted. The steady-state

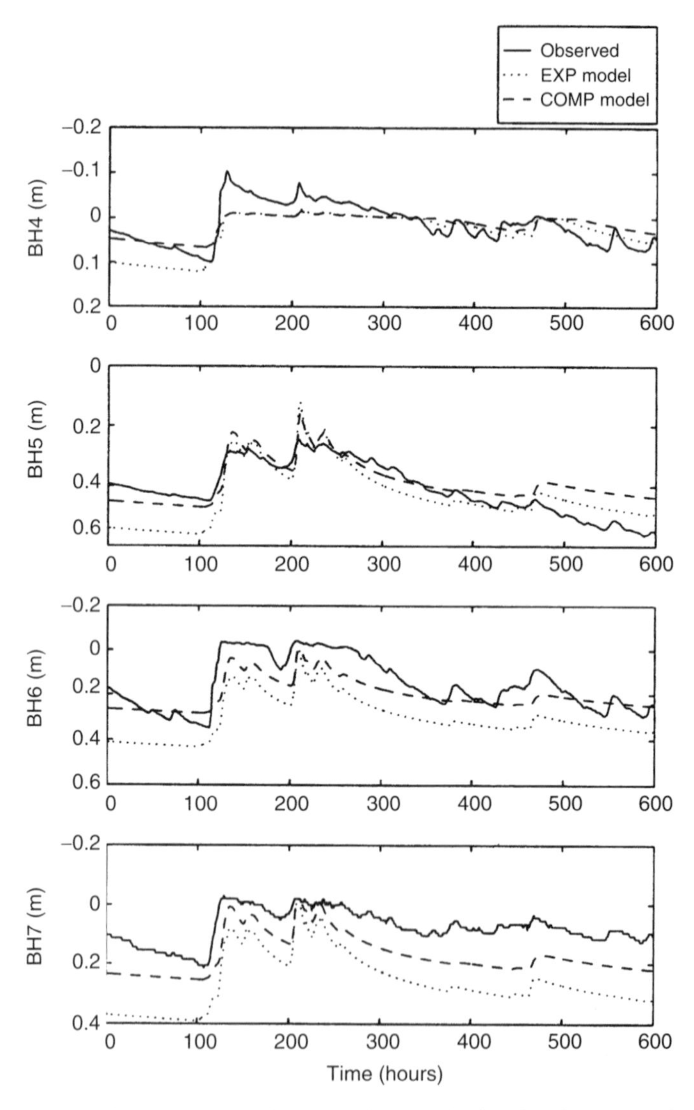

Figure 6.10 Predicted time series of water table levels for the four recording boreholes in the Saeternbekken MINIFELT catchment, Norway, using global parameters calibrated on catchment discharge and recording borehole data from an earlier snow-free period in October–November 1987 (after Lamb et al. 1997)

assumption of the TOPMODEL index approach may not be appropriate. It is also known that soil heterogeneity can be an important source of variability in hydrological response. In the case of these water table predictions, it might be possible to improve the simulations by allowing for some local variability in the transmissivity function. This possibility was investigated for this catchment by Lamb *et al.* (1997).

Figure 6.11 shows the results of local calibration of saturated transmissivity, T_o, and effective porosity, $\delta\theta$, parameters for two of the recording boreholes for the 1987 calibration period. The results show a significant improvement. For the piezometers,

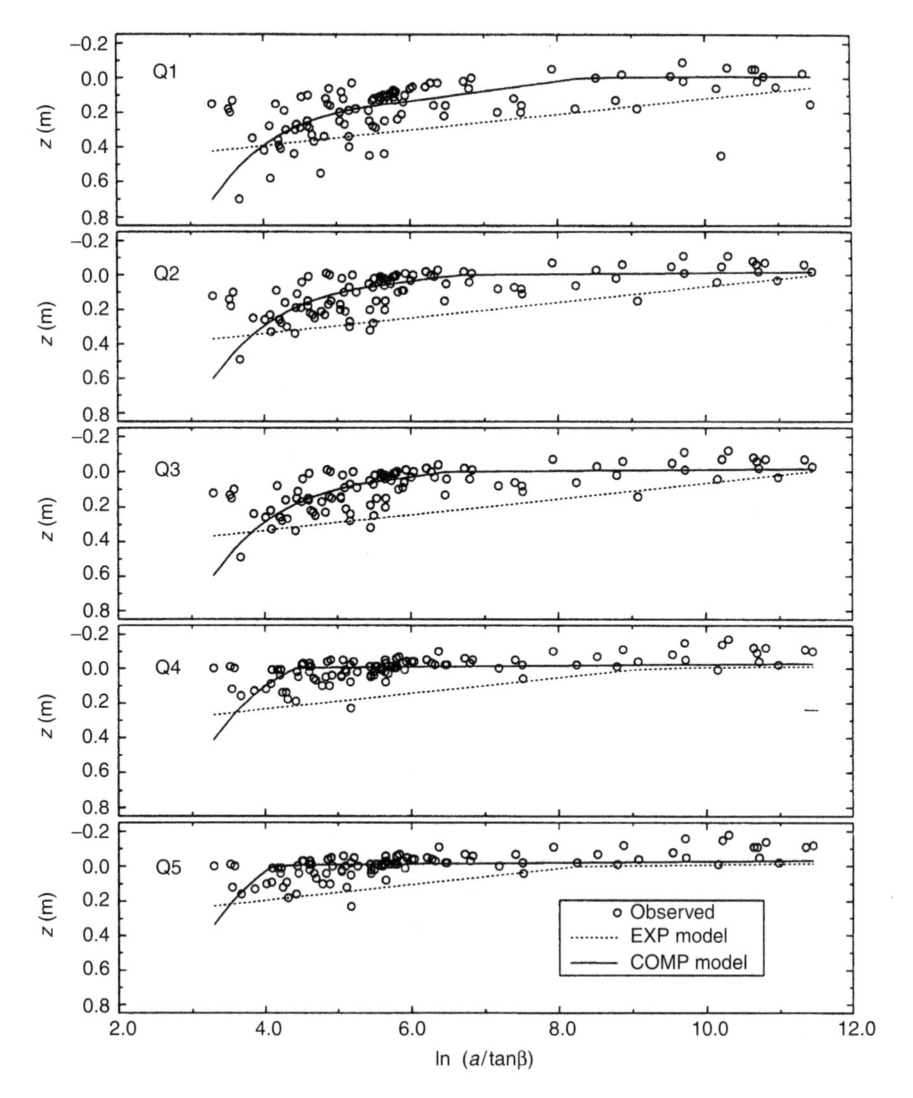

Figure 6.11 *Predicted local water table levels for five different discharges (0.1 to 6.8* mm h^{-1}*) in the Saeternbekken MINIFELT catchment, Norway, using global parameters calibrated on catchment discharge and recording borehole data from October–November 1987 (after Lamb et al. 1997)*

on the basis that the variation of T_o might be much greater than that of the effective porosity, an apparent T_o value was calculated for each piezometer to match the observed and predicted water table elevations at each point for each of the five available data sets (points showing significant ponding were excluded). A similar approach to adjusting local transmissivities has also been used by Jordan (1994) and Seibert *et al.* (1997). In this case it was found that the adjustment necessary was positively correlated with the topographic index, implying higher apparent transmissivities at higher index values. The correlations were stronger for the composite model, with coefficients varying from 0.53 to 0.92 for the five different discharges. The form of this relationship then suggested a modified index of similarity as a power function of $a/\tan\beta$. For this catchment, this resulted in a much more rapid increase in predicted saturated area with discharge than in the unmodified model.

This study raises a number of interesting issues. First, although we do expect from many different field studies that the soil characteristics should be heterogeneous in space, use of local measurements to calibrate local parameter values restricts the value of the internal data in evaluating the model. In addition, such local calibrations of, for example, a transmissivity value should be expected to be both model structure and model grid scale dependent (e.g. Saulnier *et al.* 1997a). Finally, local calibrations can only be made for points at which measurements are available, and the Saeternbekken MINIFELT catchment is very unusual, if not unique, in having so many internal measurement points. Even in this very small catchment there will then be the problem of extrapolating to other points in the catchment. In the Lamb *et al.* study this could be achieved because a suitable correlation with the $a/\tan\beta$ index was found, but it cannot be concluded that this will generally be the case. In fact, such a correlation might be an indication that there is a structural deficiency in the model formulation. The positive correlation between apparent transmissivity and topographic index, for example, might be an indication that the topographic analysis was overestimating the effective upslope contributing areas to each point.

The issues are not specific to this TOPMODEL study but will be generic to any application of any distributed model, including the most process-based models, for which some internal data are available for evaluation of the model predictions. One approach to model evaluation might then be a type of split sample test in which only part of the internal data are used to test whether local parameter calibration might be necessary and the remainder are held back to test the resulting, improved, predictions. Those held back, however, will also have their local characteristics, implying a limit to how far such distributed predictions can be validated. Some uncertainty in such predictions will then be inevitable and should, if possible, be quantified. Uncertainty in the predictions for the Saeternbekken study has been evaluated by Lamb *et al.* (1998b) (see Section 7.8).

6.6 TOPKAPI

An interesting variant on the TOPMODEL approach to distribution function model is TOPKAPI (TOPographic Kinematic Approximation and Integration; Todini 1995;

Ciarapica 1998). TOPKAPI attempts to account for two additional features of hillslope flow processes, relative to the TOPMODEL approach. The first is that downslope flows in the unsaturated zone might also contribute to the storage at any point in the catchment. By including such fluxes, the possibility that there is no downslope saturated zone flow and a very small downslope unsaturated zone effectively allows a more dynamic formulation of the upslope contributing area to a point. Secondly, the assumption of instantaneous redistribution of soil water storage on the hillslope, assumed for the saturated zone in TOPMODEL, is relaxed. In TOPKAPI, an approximate relationship between downslope flux and the integrated profile of soil moisture is assumed as

$$q = \tan\beta K_s L\tilde{\theta}^{3+2\lambda} = C\eta^{3+2\lambda} \tag{6.4}$$

where K_s is the saturated hydraulic conductivity of the soil (assumed constant with depth), L is the depth of the soil layer, $\tilde{\theta}$ is the profile integrated average relative soil moisture content $(=(1/L)\int_0^L(\theta - \theta r)/(\theta_s - \theta_r))$, λ is a coefficient in the Brooks–Corey relationship between relative hydraulic conductivity and moisture content (see Box 5.4) and the total profile moisture content is $\eta = (\theta_s - \theta_r)L\tilde{\theta}$. Letting $\alpha = 3 + 2\lambda$, the coefficient

$$C = \frac{LK_s \tan\beta}{(\theta_s - \theta_r)^{\alpha} L^{\alpha}} \tag{6.5}$$

At a point, the continuity equation may then be written as

$$\frac{\partial\eta}{\partial t} = -\frac{\partial q}{\partial x} + r = -C\frac{\partial\eta}{\partial x} + r \tag{6.6}$$

where x is plan distance and r is the rainfall rate. This is a kinematic wave equation for the change in total profile soil moisture with time. The solution to (6.6) is greatly simplified if it is assumed that the rate of change of η with time may vary over time but is everywhere constant in space. This allows an integration to the basin scale to derive an expression for the rate of change of the total storage volume with time as

$$\frac{\mathrm{d}V}{\mathrm{d}t} = -\overline{C}\left[\frac{\alpha+1}{\alpha N x}\right]^{\alpha} V^{\alpha} + Nxr \tag{6.7}$$

where

$$\frac{1}{\overline{C}} = \left[\sum_{i=1}^{N} \frac{\left(\dfrac{k}{N}\right)^{\frac{\alpha+1}{\alpha}} - \left(\dfrac{k-1}{N}\right)^{\frac{\alpha+1}{\alpha}}}{C_i^{1/\alpha}}\right]^{\alpha} \tag{6.8}$$

where the summation is taken over all the N pixels in the catchment, k represents the total number of pixels contributing to point i, and V is the total storage in the catchment. Equation (6.6) is easily solved for the storage at successive times, and the theory then allows the calculation of local storage given the total storage on the catchment and local values of the $1/\overline{C}$ index in a similar way to the original TOPMODEL formulation. Ciarapica (1998) has applied the TOPKAPI model to the Montano del Reno basin in Italy and the Can Vila basin in Spain with comparisons to the ARNO and SHE models.

6.7 Key Points from Chapter 6

- In considering the variability of hydrological responses within a catchment area, it may be difficult, and may not be necessary, to simulate those responses in a fully distributed way if a simpler way can be found of representing the distribution of responses in the catchment. This implies finding a way to define whether different points in a catchment act in hydrologically similar ways.
- The probability distributed model (PDM) does this in terms of a purely statistical representation of conceptual storage elements. It is an attempt to represent the variability in the catchment but does not allow any mapping of the predictions back into the catchment space for comparison with perceived hydrological processes in the catchment, so can be considered as an extension of a lumped storage model. It has an advantage over lumped storage models in that the parameters may be easier to estimate in an optimization exercise.
- Models based on the definition of hydrological response units (HRUs) derived by overlaying different soil, geology, topography and vegetation characteristics in a geographical information system are becoming increasingly popular. These models differ widely in the representation for each of the HRUs defined in this way, and in the routing of flows to the catchment outlet. Such models do allow the mapping of the distribution of responses back into the catchment space using the GIS. With fine discretizations of the catchment and routing of flow between HRUs, such models may be considered as simplified fully distributed models. Where similar HRU elements are grouped together for calculation purposes, they may be considered as distribution function models.
- The rainfall–runoff model TOPMODEL makes use of an index of hydrological similarity based on topography and soils information that allows the model predictions to be mapped back into space. Calculations are made based on the distribution of the index, which greatly reduces the computer resources required. The TOPMODEL concepts are not, however, applicable everywhere, particularly in catchments subject to strong seasonal drying when the basic assumptions underlying the index break down.
- The simplicity of the TOPMODEL calculations have allowed the interaction between grid resolution of the topographic analysis and calibrated parameter values to be studied in a number of applications. A similar interaction between scale of discretization and effective parameter values should hold for more complex models, including physically based, fully distributed models, but may not be so readily apparent.
- Computer packages for the analysis of digital terrain data and to run TOPMODEL simulations are provided (see Appendix A).
- A number of recent attempts to improve the theory of TOPMODEL while retaining its advantages of simplicity are outlined.

Box 6.1 The SCS Curve Number Model Revisited

The Origins of the Curve Number Method

The USDA Soil Conservation Service curve number (SCS-CN) method has its origins in the unit hydrograph approach to rainfall–runoff modelling (see Chapter 4). The unit hydrograph approach always requires a method for predicting how much of the rainfall contributes to the 'storm runoff'. The SCS-CN method arose out of the empirical analysis of runoff from small catchments and hillslope plots monitored by the USDA. Mockus (1949) proposed that such data could be represented by an equation of the following form:

$$\frac{Q}{P - I_a} = \left[1 - (10)^{-b(P - I_a)}\right] \tag{6.9}$$

or

$$\frac{Q}{P - I_a} = [1 - \exp\{-B(P - I_a)\}] \tag{6.10}$$

where Q is the volume of storm runoff; P is the volume of precipitation, I_a is an initial retention of rainfall in the soil; and b and B are coefficients. Mockus (1949) suggested the coefficient b was related to antecedent rainfall, a soil and cover management index, a seasonal index, and storm duration.

Mishra and Singh (1999) show how this equation can be derived from the water balance equation under an assumption that the rate of change of retention with effective precipitation is a linear function of retention and with the constraint that $B(P - I_a) < 1$. Approximating the right-hand side of equation (6.10) as a series expansion results in an equation equivalent to the standard SCS-CN formulation

$$\frac{Q}{P - I_a} = \frac{P - I_a}{S_{max} + P - I_a} \tag{6.11}$$

where $S_{max} = 1/B$ is some maximum volume of retention. Mishra and Singh (1999) propose a further generalization resulting from a more accurate series representation of equation (6.11) (and giving better fits to data from five catchments) as

$$\frac{Q}{P - I_a} = \frac{P - I_a}{S_{max} + a(P - I_a)} \tag{6.12}$$

This is equivalent to assuming that the cumulative volume of retention $F(t)$ can be predicted as

$$\frac{F(t)}{S_{max}} = \frac{Q}{P - I_a} \tag{6.13}$$

$F(t)$ is often interpreted as a cumulative volume of infiltration, but it is not necessary to assume that the predicted stormflow is all overland flow, since it may not have been in the original small catchment data on which the method is based (application of the method to one of the permeable, forested, Coweeta catchments in Hjelmfelt *et al.* (1982) is a good example of this).

A further assumption is usually made in the SCS-CN method that $I_a = \lambda S_{max}$, with λ commonly assumed to be ≈ 0.2. Thus, with this assumption, the volume of storm runoff may be predicted from a general form of the SCS-CN equation:

$$Q = \frac{(P - \lambda S_{max})^2}{P + (1 - \lambda) S_{max}} \tag{6.14}$$

The Curve Number

The SCS-CN method can be applied by specifying a single parameter called the curve number, CN. The popularity of the method arises from the tabulation of CN values by the USDA for a wide variety of soil types and conditions (see Table B6.1.1). In the standard SCS-CN method, the value of S_{max} (in millimetres) for a given soil is related to the curve number as

$$\frac{S_{max}}{2.54} = \left(\frac{1000}{CN} - 10\right) \tag{6.15}$$

where the constant 2.54 mm converts from the original units of inches to millimetres. Ponce (1989) and others suggest that this is better written as

$$\frac{S_{max}}{C} = \left(\frac{100}{CN} - 1\right) \tag{6.16}$$

where S_{max} now varies between 0 and C. The original equation implies a value of C of 10 inches (254 mm). It is easily seen from equation (6.16) that the range 0 to C is covered by curve numbers in the range 50 to 100. Although values less than 50 are sometimes quoted in the literature, this implies that they are not physically meaningful for the original C of 254 mm.

Curve number range tables are provided for the estimation of the curve number for different circumstances of soil, vegetation and antecedent conditions (e.g. Table B6.1.1). For areas of complex land use, it was normally suggested in the past that a linear combination of the component curve numbers, weighted by the area to which they apply, should be used to determine an effective curve number for the area (e.g. USDA SCS

Table B6.1.1 *Some example curve numbers (after USDA SCS 1985)*

Land use	Hydrologic condition	Hydrologic soil group[a]			
		A	B	C	D
Fallow		77	86	91	94
Row crops: contoured	Poor	70	79	84	88
	Good	65	75	82	86
Small grain: contoured	Poor	63	74	82	85
	Good	61	73	81	84
Rotation meadow: contoured	Poor	64	75	83	85
	Good	55	69	78	83
Pasture or range	Poor	68	79	86	89
	Fair	49	69	79	84
	Good	39	61	74	80
Meadow	Good	30	58	71	78
Woods	Good	25	55	70	77
Farmsteads	–	59	74	82	86
Roads: dirt	–	72	82	87	89
Roads: surfaced	–	74	84	90	92

[a]Soils of Group A have high infiltration rates even if thoroughly wetted. Soils of Group B have moderate infiltration rates and are moderately well drained. Soils of Group C have low infiltration rates with an impeding layer for downward drainage. Soils of Group D have low infiltration rates when wetted, including swelling clays and soils with a permanently high water table.

1985). However, because of the nonlinearity of the runoff prediction equation, this will not give the same results as applying the equation to each individual area and weighting the resulting runoff by the component fractional areas. Grove *et al.* (1998) show how using a distributed curve number calculation will increase the volume of predicted runoff, by as much as 100 percent in their examples, relative to using the composite method. Now that computer limitations are much less restrictive, the latter procedure should be adopted and, in fact, arises naturally when the SCS-CN method is applied to an HRU classification of a catchment based on GIS overlays.

Curve Numbers, Antecedent Moisture Condition, and Contributing Areas

The curve number approach, in its simplest form, takes no explicit account of the effect of the antecedent moisture condition of the catchment in calculating the storm runoff. Yet it is known that there can be a highly nonlinear dependence of runoff on antecedent state. The SCS-CN model allows for this by suggesting that the curve number be modified according to whether the catchment is wet or dry (e.g. Table B6.1.2), noting that the curve number is inversely related to the effective S_{max} value.

The background to the SCS-CN method is purely empirical. That is primarily its strength, but also a limitation in that it does not allow any direct interpretation in terms of process, except in so far as Mockus originally suggested that the value of S_{max} would be the volume of infiltration or available storage, whichever was the smaller. Two recent studies have attempted to extend this interpretation. Steenhuis *et al.* (1995) have interpreted the method as equivalent to assuming a variable contributing area of runoff generation over which the storm rainfall volume is sufficient to exceed the available storage prior to an event. The implied proportion of the catchment contributing to runoff for a given effective rainfall value is then directly related to the value of S_{max} (Figure B6.1.1). They show reasonable fits to observations for several Australian and American catchment areas (ranging from 16.5 to 7000 ha) with permeable soils. Adjusting the value of S_{max} for each catchment resulted in a range of values from 80 to 400 mm (see Figure B6.1.2).

Yu (1998) suggests that a model of the form of the SCS-CN method can be derived on the basis of assumptions about the spatial variability of (time constant) infiltration capacities and the temporal variability in rainfall intensities. Under these assumptions, runoff will be produced anywhere on a catchment where the time-varying rainfall rate exceeds the

Table B6.1.2 *Adjusting curve numbers for antecedent moisture condition (after USDA SCS 1985)*[a]

Curve number (condition 2)	Condition 1 (dry)	Condition 3 (wet)
100	100	100
90	78	96
80	63	91
70	51	85
60	40	78
50	31	70
40	22	60
30	15	50

[a]Condition 2 represents the normal curve number condition prior to the annual maximum flood. Condition 1 represents dry conditions. Condition 3 represents soil that is nearly saturated.

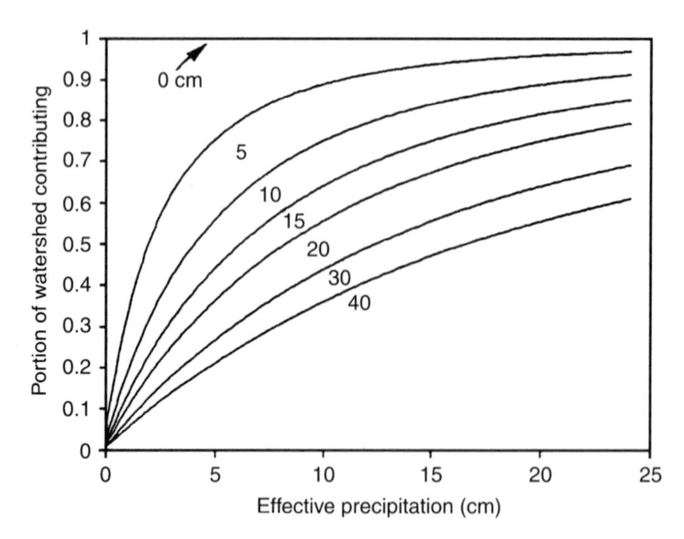

Figure B6.1.1 *Variation in effective contributing area with effective rainfall for different values of S_{max} (after Steenhuis et al. 1995). Effective rainfall is here defined as the volume of rainfall after the start of runoff,* $P - I_a$

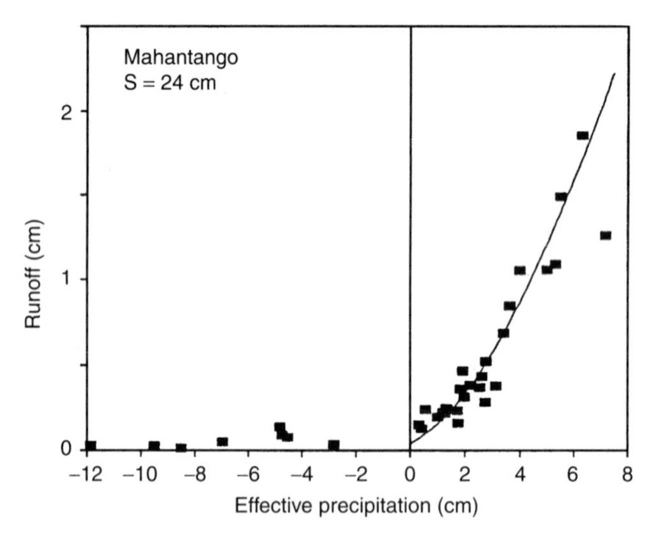

Figure B6.1.2 *Application of the SCS method to data from the Mahatango Creek catchment (55 ha), Pennsylvania (after Steenhuis et al. 1995). Effective rainfall is here the volume of rainfall after the start of runoff*

spatially variable but time-constant infiltration capacity (making no allowances for any run-on process). For an exponential distribution of infiltration capacity in space, and an exponential distribution of rainfall intensity in time, Yu shows that the runoff generated Q is given by the SCS-CN equation.

Applications of the Curve Number Method

There have been a wide variety of models that have been based on the curve number method. It has been used widely in procedures recommended by the USDA, notably in the TR20 and TR55 methods for estimating peak runoff and hydrographs (USDA SCS 1986). A detailed summary of the method, within a traditional interpretation, is given by McCuen (1982). It has also provided a runoff component for a succession of water quality and erosion models, including the Areal Nonpoint Source Watershed Environment Response System, ANSWERS (Beasley *et al.* 1980); the Chemicals, Runoff and Erosion from Agricultural Management Systems model, CREAMS, and its companion model for estimating pesticide loadings to groundwater, GLEAMS (Knisel and Williams 1995); the Simulator for Water Resources in Rural Basins, WRRB (Arnold and Williams 1995); Erosion Productivity Impact Calculator, EPIC (Williams 1995); The Pesticide Root Zone model, PRZM (Carsel *et al.* 1985); the Agricultural Nonpoint Source model, AGNPS (Young *et al.* 1995); the Water Erosion Prediction Project, WEPP (Laflen *et al.* 1991) and, most recently, the distributed Soil Water Assessment Tool, SWAT (Arnold *et al.* 1998).

The SWAT model is intended as a long-term yield model rather than a hydrograph prediction model (Arnold *et al.* 1998; Peterson and Hamlett 1998). The hydrology component uses a daily time step and has components for daily runoff production (using the SCS-CN method), percolation (including a crack flow model), lateral subsurface flow (using a kinematic storage model of Sloan and Moore 1984), groundwater flow (using a linear reservoir), three different evapotranspiration methods (of which the Penman–Monteith equation is the most complex), a snowmelt component (based on the degree-day method), and channel and reservoir routing with a component for transmission losses in ephemeral channels. Additional components are included to predict sediment yield, crop growth, and water quality variables including nitrogen and phosphorus nutrients and pesticides. It is a good example of a comprehensive modelling system based on many conceptual component systems, and requiring many parameters to be specified (Srinavasan *et al.* 1998). A GIS database can help in preparing the information required (Manguerra and Engel 1998).

Limitations of the Curve Number Method

It is worth remembering what is being predicted by the SCS-CN approach. It is the volume of 'storm runoff' in a given storm, after some initial retention before runoff begins, that is then to be routed by the unit hydrograph or some other routing method to predict a storm runoff hydrograph. It is therefore subject to all the problems and limitations associated with separating storm runoff from the total discharge hydrograph, both in the analyses that underlay the original model formulation, and in the calculation of curve numbers for particular situations.

The curve number approach to predicting runoff generation has been the subject of a number of critical reviews (e.g. Hjelmfelt *et al.* 1982; Bales and Betson 1982). Further work is required to clarify under what conditions the method gives good predictions. Mishra and Singh (1999) show that their generalized version of the method (equation (6.14) above) gives better results than the original formulation (with $\lambda = 0.2$ and $a = 0$), as it should, since it has two additional fitting parameters. Hjelmfelt *et al.* (1982) suggest that the curve number, rather than being considered as a characteristic for a given soil–land cover association, might better be considered as a stochastic variable. Their analysis, of the annual maximum storms for two small catchments in Iowa, suggested that the storage capacity parameter, S_{max}, derived for individual storms was approximately log normally distributed with a coefficient of variation of the order of 20 percent. The 10 and 90 percent quantiles of the distributions corresponded well to the modified curve numbers for dry and wet antecedent conditions, following the standard SCS procedures based on the preceeding five-day rainfalls. However, they found no strong correlation between curve number and antecedent condition for the individual storms, suggesting that interactions with individual storm characteristics, tillage, plan growth and temperature were sufficient to mask the effect of antecedent rainfall alone.

Despite its limitations, the SCS-CN method has recently been used quite widely since the tabulated curve number values provide a relatively easy way of moving from a GIS data set on soils and vegetation to a rainfall–runoff model (see for example Berod *et al.* 1999). It does have the important advantage in this respect that it is not formulated on the basis of point scale measurements, but directly from small catchment measurements. It is likely that this type of more empirical method will be revisited in the future as the difficulties of applying models based on point scale process descriptions become more widely appreciated.

Box 6.2 The Theory Underlying TOPMODEL

Fundamental Assumptions of TOPMODEL

The development of the TOPMODEL theory presented here is based on the three assumptions outlined in the main text:

- *A1.* There is a saturated zone in equilibrium with a steady recharge rate over an upslope contributing area a.
- *A2.* The water table is almost parallel to the surface such that the effective hydraulic gradient is equal to the local surface slope, $\tan\beta$.
- *A3.* The transmissivity profile may be described by an exponential function of storage deficit, with a value of T_o when the soil is just saturated to the surface (zero deficit).

Steady Flow in the Saturated Zone and the Topographic Index

Under these assumptions, at any point i on a hillslope the downslope saturated subsurface flow rate, q_i, per unit contour length [$L^2\ T^{-1}$] may be described by the following equation:

$$q_i = T_o \tan\beta \exp(-D_i/m) \tag{6.17}$$

where D_i is local storage deficit per unit plan area, [L], m is a parameter controlling the rate of decline of transmissivity with increasing storage deficit [L], and $T_o[L^2\ T^{-1}]$ and $\tan\beta$ [–] are local values at point i. Note that $\tan\beta$ is used to represent the hydraulic gradient on the basis that the slope is calculated as elevation change per unit distance in plan (rather than along the hillslope), while T_o is also defined with respect to horizontal fluxes (Beven and Freer, 2000).

Then under the assumption that, at any time step, quasi-steady-state flow exists throughout the soil, assuming a spatially homogeneous recharge rate r [$L\ T^{-1}$] entering the water table, the subsurface downslope flow per unit contour length q_i may also be given by

$$q_i = ra \tag{6.18}$$

where a is the area of the hillslope per unit contour length [L^2] that drains through point i.

By combining (6.17) and (6.18) it is possible to derive a formula for any point relating local water table depth to the topographic index $\ln(a/\tan\beta)$ at that point, the parameter

m, the local saturated transmissivity, T_o, and the effective recharge rate, r:

$$D_i = -m \ln\left(\frac{ra}{T_o \tan\beta}\right) \tag{6.19}$$

Note that when the soil is saturated the local deficit will be zero, and that as the soil dries and the water table falls numerical values of storage deficit get larger. An expression for the catchment lumped, or mean, storage deficit ($\overline{D}$) may be obtained by integrating (6.19) over the entire area of the catchment (A) that contributes to the water table. In what follows we will express this areal averaging in terms of a summation over all points (or pixels) within the catchment:

$$\overline{D} = \frac{1}{A}\sum_i A_i\left[-m \ln\frac{ra}{T_o \tan\beta}\right] \tag{6.20}$$

where A_i is the area associated with the i point (or group of points with the same characteristics). In spatially integrating the whole catchment, it is also implicitly required that (6.20) holds even at such locations where water is ponded on the surface ($D_i < 0$). Beven (1991a) justifies this assumption on the basis that the relationship expressed by (6.17) is exponential and that, for many catchments, surface flow is likely to be relatively slow due to vegetation cover. Recently, Datin (1998) has presented a TOPMODEL formulation that avoids this assumption (see below).

By using (6.19) in (6.20), if it is assumed that r is spatially constant, $\ln r$ may be eliminated and a relationship found between mean water table depth, local water table depth, the topographic variables and saturated transmissivity. This has the following form:

$$D_i = \overline{D} + m\left[\gamma - \ln\frac{a}{T_o \tan\beta}\right] \tag{6.21}$$

where $\ln(a/T_o \tan\beta)$ is the soil–topographic index of Beven (1986b), and

$$\gamma = \frac{1}{A}\sum_i A_i \ln\frac{a}{T_o \tan\beta}$$

A separate areal average value of transmissivity may be defined, thus

$$\ln T_e = \frac{1}{A}\sum_i A_i \ln T_o$$

Equation (6.21) may now be rearranged to give

$$\frac{(\overline{D} - D_i)}{m} = -\left[\lambda - \ln\frac{a}{\tan\beta}\right] + [\ln T_o - \ln T_e] \tag{6.22}$$

where $\lambda = (1/A)\sum_i A_i \ln(a/\tan\beta)$ is a topographic constant for the catchment.

Equation (6.22) expresses the deviation between the catchment average water table depth (or deficit) and the local water table depth (or deficit) at any point in terms of the deviation of the local topographic index from its areal mean, and the deviation of the logarithm of local transmissivity from its areal integral value. The relationship is scaled by the parameter m.

Similar relationships can be derived for other transmissivity profile assumptions, relaxing assumption A3 above. Ambroise *et al.* (1996b), for example, analyse the case of linear and parabolic transmissivity functions, while Iorgulescu and Musy (1997) and Duan and Miller

(1997) have generalized the analysis to all power law transmissivity functions. Different assumptions about the transmissivity result in different topographic index functions. For the power law function,

$$q_i = T_o \tan\beta(1 - D_i/M)^n \tag{6.23}$$

where M is a maximum gravity drainage storage in the soil profile expressed as a volume per unit area [L]. The equivalent soil–topographic index is

$$(a/T_o \tan\beta)^{1/n} \tag{6.24}$$

and the equation relating mean storage deficits to local deficits is

$$\frac{(1 - D_i/M)}{(1 - \bar{D}/M)} = \left[\frac{a}{T_o \tan\beta}\right]^{1/n} \Big/ \left[\frac{1}{A}\sum_i A_i \left(\frac{a}{T_o \tan\beta}\right)^{1/n}\right] \tag{6.25}$$

The Topographic Index as an Index of Hydrological Similarity

The implication of (6.21) is that each point in the catchment with the same value of the topographic index, $\ln(a/T_o \tan\beta)$, will be predicted as responding in a hydrologically similar way. Thus it is not necessary to make calculations for every point in space but only for different values of the topographic index. In most applications of TOPMODEL, the distribution function of the topographic index (see, for example Figure 6.5) is discretized into a number of increments representing appropriate proportions of the catchment. At every time step, those increments with high values of the topographic index will be predicted as being saturated or having low storage deficits. This is shown schematically in Figure B6.2.1. The calculations for each increment are completed by an unsaturated zone component.

Moisture Accounting: Unsaturated Zone Fluxes

The basic soil structure illustrated in Figure B6.2.1 may be used to accommodate a variety of unsaturated zone process descriptions as defined by the modeller. One formulation that has been adopted in past TOPMODEL applications assumes that the root zone store for each topographic index value is depleted only by evapotranspiration, and that water is added to the unsaturated zone drainage store only once the root zone reaches field capacity. The drainage is assumed to be essentially vertical and a drainage flux per unit area q_v [L T^{-1}] is calculated for each topographic index class.

Expressed in terms of storage deficit, Beven and Wood (1983) suggested that a suitable functional form for the vertical flux q_v at any point i is

$$q_v = \frac{S_{uz}}{D_i t_d} \tag{6.26}$$

where S_{uz} [L] is storage in the unsaturated (gravity drainage) zone, D_i is the local saturated zone deficit due to gravity drainage, and dependent on the depth of the local water table [L]. Parameter t_d is a time constant, expressed as a mean residence time for vertical flow per unit of deficit [T L^{-1}]. Equation (6.26) is the equation of a linear store but with a time constant $\{D_i t_d\}$ that increases with increasing depth to the water table. Note that there is no physical justification for this functional form, but it has the advantages that it allows for longer residence times and slower drainage rates for lower values of the index where the water table is predicted as being deeper below the surface and yet it only introduces one parameter value. It has generally been found that modelling results are not very sensitive to this parameter.

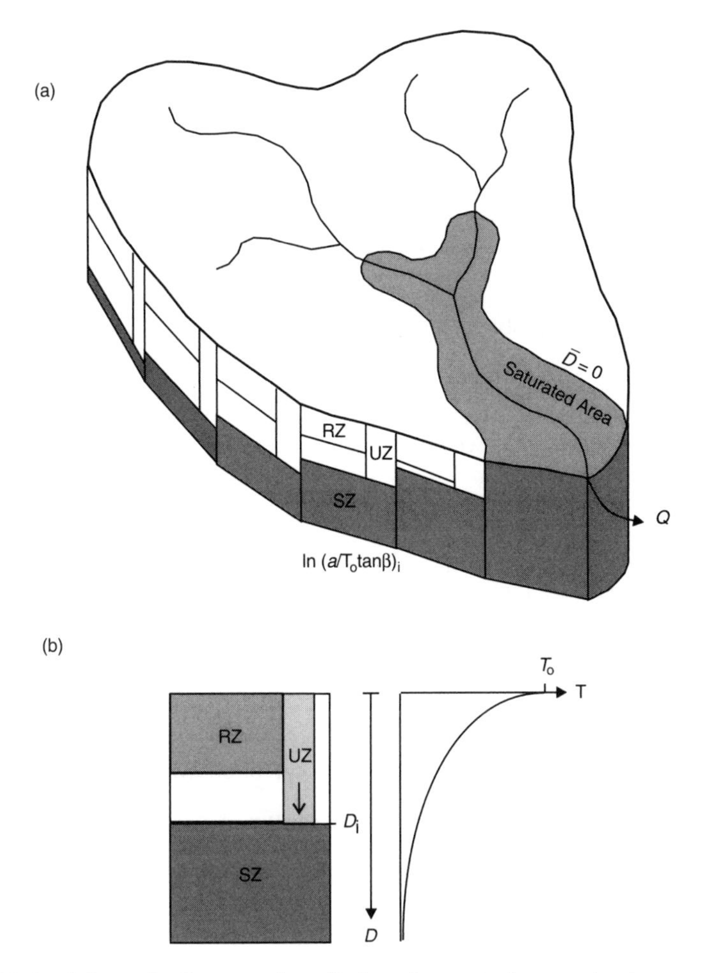

Figure B6.2.1 *Schematic diagram of prediction of saturated area using increments of the topographic index distribution in TOPMODEL*

Accounting for evapotranspiration with a minimal number of parameters poses a problem of similar complexity to that of the unsaturated zone drainage. TOPMODEL follows the widely adopted practice of calculating actual evapotranspiration, E_a, as a function of potential evaporation, E_p, and root zone moisture storage for cases where E_a cannot be specified directly. In the TOPMODEL description of Beven (1991a), evaporation is allowed at the full potential rate for water draining freely in the unsaturated zone and for predicted areas of surface saturation. When the gravity drainage zone is exhausted, evapotranspiration may continue to deplete the root zone store at the rate E_a, given by

$$E_a = E_p \frac{S_{rz}}{S_{r\max}} \tag{6.27}$$

where the variables S_{rz} and $S_{r\max}$ are, respectively, root zone storage [L] and maximum available root zone storage [L]. If some effective root zone depth z_{rz} [L] can be asumed,

$S_{r\max}$ can be estimated approximately from

$$S_{r\max} = z_{rz}(\theta_{fc} - \theta_{wp}) \tag{6.28}$$

where θ_{fc} [−] is moisture content at *field capacity* and θ_{wp} [−] is moisture content at wilting point. For calibration it is only necessary to specify a value for the single parameter $S_{r\max}$. An effective value for $S_{r\max}$ might be greater than that suggested by (6.28) due to capillary rise of water into the root zone under dry conditions.

The flux of water entering the water table locally at any time is q_v. This drainage is also a component of the overall recharge of the lumped saturated zone. To account for the catchment average water balance, all the local recharges must be summed. If Q_v is the total recharge to the water table in any time step, then

$$Q_v = \sum_i q_{v,i} A_i \tag{6.29}$$

where A_i is the area associated with topographic index class i [L^2].

Moisture Accounting: Saturated Zone Fluxes

Output from the saturated zone is given by the baseflow term, Q_b. This may be calculated in a distributed sense by the summation of subsurface flows along each of M stream channel reaches of length l. Recalling (6.17), we may write

$$Q_b = \sum_{j=1}^{M} l_j (T_o \tan\beta) e^{-D_j/m} \tag{6.30}$$

Substituting for S_j using (6.21) and rearranging, it can be shown that

$$Q_b = \sum_j l_j a_j e^{-\gamma - \overline{D}/m} \tag{6.31}$$

Since a_j represents contributing area per unit contour length, then

$$\sum_{j=1}^{m} l_j a_j = A \tag{6.32}$$

Therefore

$$Q_b = A e^{-\gamma} e^{-\overline{D}/m} \tag{6.33}$$

where A is the total catchment area [L^2]. It is therefore possible to calculate baseflow in terms of the average catchment storage deficit ($\overline{D}$):

$$Q_b = Q_o e^{-\overline{D}/m} \tag{6.34}$$

where $Q_o = Ae^{-\gamma}$ is the discharge when $\overline{D}$ equals zero [$L^3\ T^{-1}$]. This is the same form as that originally assumed by Beven and Kirkby (1979). Solution of (6.34) for a pure recession in which recharge is assumed to be negligible shows that discharge has an inverse or first-order hyperbolic relationship to time as

$$\frac{1}{Q_b} = \frac{1}{Q_o} + \frac{t}{m} \tag{6.35}$$

Thus, if (6.34) is an appropriate relationship to represent the subsurface drainage of a given catchment, a plot of $1/Q_b$ against time should plot as a straight line (e.g. Figure B6.2.2) with slope $1/m$. Then, given at least some recession curves that are not greatly influenced

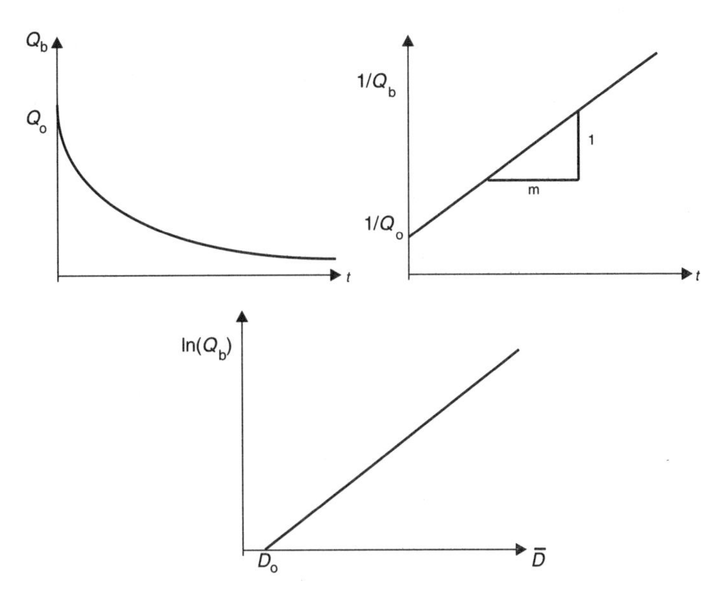

Figure B6.2.2 *Derivation of an estimate for the TOPMODEL* m *parameter using recession curve analysis under the assumption of an exponential transmissivity profile and negligible recharge*

by evapotranspiration or snowmelt processes, it should be possible to set the value of m that will then need minimal calibration.

The catchment average storage deficit before each time step is updated by subtracting the unsaturated zone recharge and adding the baseflow calculated for the previous time step, thus

$$\overline{D}_t = \overline{D}_{t-1} + [Q_{b_{t-1}} - Q_{v_{t-1}}/A] \tag{6.36}$$

Equation (6.34) can be used to initialize the saturated zone of the model at the start of a run. If an initial discharge, $Q_{t=0}$, is known and assumed to be only the result of drainage from the saturated zone, (6.34) can be inverted to give a value for $\overline{D}$ at time $t = 0$ as

$$\overline{D} = -m \ln\left(\frac{Q_{t=0}}{Q_o}\right) \tag{6.37}$$

Once $\overline{D}$ is known, local values of initial storage deficit can be calculated from (6.21). Other forms of transmissivity function can also be used to derive different forms of index and recession curves. Lamb *et al.* (1997) show how an arbitrary recession curve can be used in a generalized TOPMODEL.

Runoff Routing and Subcatchment Structure

For many catchments, especially large ones, it may be inappropriate to assume that all runoff reaches the catchment outlet within a single time step. In the original Beven and Kirkby (1979) TOPMODEL formulation, an overland flow delay function was calculated as a distance-related delay. The time taken to reach the basin outlet from any point was assumed to be given by

$$t_j = \sum_{i=1}^{N} \frac{x_i}{v^* \tan \beta_i} \tag{6.38}$$

where t_j is the time delay [T] for point j, x_i is the plan flowpath length [L], and $\tan\beta$ is the slope of the ith segment of a flow path comprising N segments between point j and the catchment outlet. If the velocity parameter v^* [L T^{-1}] is assumed constant then this equation allows a unique time delay histogram to be derived on the basis of basin topography for any runoff contributing area extent. This is, in effect, a variation of the time–area routing method of Ross (1921) and Clark (1945) (see Figure 2.2), but developed so as to relate the runoff time delay histogram dynamically to the size of the source area.

Channel routing effects were considered by Beven and Kirkby (1979) using an approach based on an average flood wave velocity for the channel network, this being related nonlinearly to total outflow. The approach was an explicit approximation to a kinematic wave channel routing algorithm and is not recommended, since it is not always stable. Most applications have been based on a simple constant wave speed routing algorithm, equivalent to the channel network width function based algorithms used by Surkan (1969), Kirkby (1976) and Beven (1979; Beven and Wood 1993), which has the advantage that it introduces only a single wave speed parameter. In a single event version of TOPMODEL, Saulnier *et al.* (1997a) adopted a routing method based on a unit hydrograph derived by the DPFT-ERUHDIT method of Duband *et al.* (1993). Since the unit hydrograph can be expected to reflect the time response of subsurface as well as surface flow processes, they show that this choice of routing algorithm has an effect on the other parameters of the model. Some other linear routing algorithms are discussed in Section 4.6.

Recent Variations on TOPMODEL

The simplicity of the TOPMODEL formulation as a way of reflecting the topographic controls on runoff generation has proven attractive now that digital terrain models for catchments are more widely available. The simplicity of the ideas has also encouraged both an assessment of the assumptions in relation to perceptions of the processes controlling the hydrological responses in different catchments and attempts to reformulate and improve the theory.

In the first category of variations on the TOPMODEL concept is the idea of the reference level for deeper water tables proposed by Quinn *et al.* (1991) in which the hydraulic gradient is based on a characteristic water table surface rather than the soil surface. This idea was used with an exponential transmissivity profile, which may not be appropriate for a deep water table, but could be extended to more realistic transmissivity profiles. Lamb *et al.* (1997, 1998a) showed how a generalized transmissivity function could be developed on the basis of a recession curve analysis (maintaining the simplifying assumption that the pattern of saturated zone hydraulic gradient in the catchment stays constant); while Lamb *et al.* (1997, 1998b) have shown how data on the spatial distribution of water table depth in a catchment can be used to modify the index distribution to reflect heterogeneity of effective transmissivity values.

Recent attempts to reformulate the index theory include the modified index of Datin (1998), which avoids the assumption that a steady recharge rate and transmissivity function also apply to water ponded on the surface by averaging the local soil moisture deficits only over the non-saturated part of the catchment. Thus equation (6.21) becomes

$$\overline{D}' - D_i = -m\left[\gamma_t' - \ln\frac{a}{T_o\tan\beta}\right] \tag{6.39}$$

where $\overline{D}'$ is now the average deficit over the non-saturated area at time t,

$$\gamma_t' = \frac{1}{A - A_t^c}\sum_i A_i \ln\frac{a}{T_o\tan\beta},$$

A_t^c is the contributing area at time t, A_i is the area of a pixel (or group of similar pixels) as before, and the summation is now taken only over unsaturated pixels.

Since, by definition, the local deficit on the saturated area is everywhere zero, then

$$\overline{D}_t = \frac{A - A_t^c}{A}\overline{D}_t' \qquad (6.40)$$

and combining (6.39) and (6.40),

$$\left(\frac{A}{A - A_t^c}\right)\overline{D} - D_i = -m\left[\gamma_t' - \ln\frac{a}{T_o \tan\beta}\right] \qquad (6.41)$$

with $D_i = 0$ at the critical index value,

$$\left(\ln\frac{a}{T_o \tan\beta}\right)_c = \gamma_t' + \left(\frac{A}{A - A_t^c}\right)\frac{\overline{D}}{m} \qquad (6.42)$$

This can be rearranged to define a function $G(A_t^c)$ that can be calculated from the cumulative distribution function of the soil–topographic index such that

$$G(A_t^c) = \frac{\overline{D}}{m} \qquad (6.43)$$

where

$$G(A_t^c) = \left[\left(\ln\frac{a}{T_o \tan\beta}\right)_c - \gamma_t'\right]\left[\frac{A}{A - A_t^c}\right] \qquad (6.44)$$

An example of the function $G(A^c)$ is shown in Figure B6.2.3. This function will be specific to a catchment but may be calculated during the topographic analysis prior to a simulation. At any time step, knowledge of $\overline{D}$ can be used to calculate the function $G(A^c)$ and thence the contributing area and the average deficit over the nonsaturated area $\overline{D}'$. Thus, this turns out to be quite a simple modification and results in a more satisfactory representation of the time variation in the average (non-saturated) deficit and pattern of deficits in the catchment.

Another alternative index formulation is provided by the TOPKAPI model of Section 6.6. This attempts to relax the assumption of instantaneous redistribution of the saturated

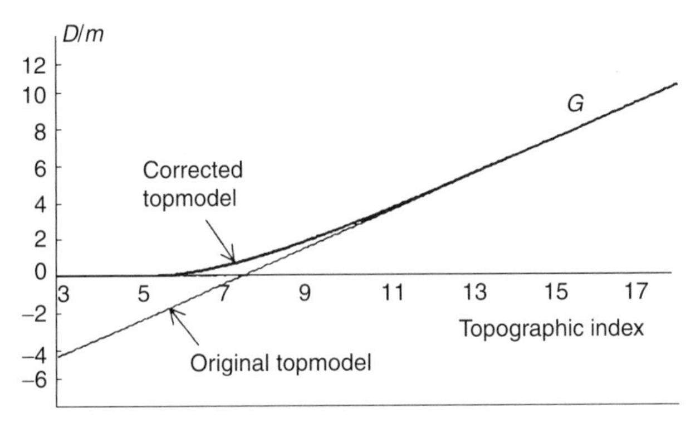

Figure B6.2.3 *Use of the function G(A^c) to determine the critical value of the topographic index at the edge of the contributing area given $\overline{D}$/m, assuming a homogeneous transmissivity (after Datin 1998). Reproduced with permission*

zone at each time step in formulating the new index. The MACAQUE model of Watson *et al.* (1999) addresses the same problem in a different way by introducing a 'lateral redistribution factor' that limits the redistribution towards the steady-state water table configuration allowed at each time step.

One of the limitations of the topographic index approach is the assumption that there is always downslope flow from a upslope contributing area that is constant for any point in the catchment. Improved predictions might be possible if this area was allowed to vary dynamically. Barling *et al.* (1994) showed that an index based on travel times could improve prediction of saturated areas for a single time step prediction, but did not suggest how this might be extended to a continuous time model. A dynamic TOPMODEL can also be derived by an explicit redistribution of downslope fluxes from one group of hydrologically similar points to another, where the definition of hydrologically similar can be based on more flexible criteria than the original topographic index. In the extreme case of every pixel in a catchment being considered separately, this approach would be similar to the distributed kinematic wave model of Wigmosta *et al.* (1994). Grouping of similar pixels results in computational efficiency that might be advantageous in applications to large catchments or where large numbers of model runs are required to assess predictive uncertainty. This is the basis for a new, more dynamic, version of TOPMODEL (Beven and Freer 2000).

7

Parameter Estimation and Predictive Uncertainty

Far better an approximate answer to the right question, which is often vague, than an exact answer to the wrong question, which can always be made precise.

John W. Tukey, 1962

7.1 Parameter Estimation and Predictive Uncertainty

It should be clear from the preceding chapters that limitations of both model structures and the data available on parameter values, initial conditions and boundary conditions, will generally make it difficult to apply a hydrological model (of whatever type) without some form of calibration. In very few cases reported in the literature have models been applied using only parameter values measured or estimated a priori (e.g. Beven *et al.* 1984; Parkin *et al.* 1996; Refsgaard and Knudsen 1996; Loague and Kyriakidis 1997); in the vast majority of cases the parameter values are adjusted to get a better fit to some observed data. This is the model calibration problem discussed in Section 1.8. The question of how to assess whether one model or set of parameter values is better than another is open to a variety of approaches, from a visual inspection of plots of observed and predicted variables, to a number of different quantitative measures of goodness of fit, known variously as objective functions, performance measures, fitness (or misfit) measures, likelihood measures or possibility measures. Some examples of such measures that have been used in rainfall–runoff modelling are discussed in Section 7.3.

All model calibrations and subsequent predictions will be subject to uncertainty. This uncertainty arises in that no rainfall–runoff model is a true reflection of the processes involved, that it is impossible to specify the initial and boundary conditions required by the model with complete accuracy, and that the observational data available for model calibration are not error-free. A good discussion of these sources of uncertainty may be found in Melching (1995). There is a rapidly growing literature on model calibration and the estimation of predictive uncertainty for hydrological models. This chapter can give only a summary of the major themes being explored, and for the purposes of this discussion we will differentiate three major themes as follows:

- Methods of model calibration that assume an optimum parameter set and that ignore the estimation of predictive uncertainty can be found. These methods range from simple trial and error, with parameter values adjusted by the user, to the variety of *automatic optimization* methods discussed in Section 7.4.
- Methods of model calibration that assume an optimum parameter set, but which make certain assumptions about the *response surface* (see Section 7.2) around that optimum to estimate the predictive uncertainty, can be found. These methods will be grouped under the name *reliability analysis* and will be discussed in Section 7.5.
- Methods of model calibration that reject the idea that there is an optimum parameter set in favour of the idea of *equifinality* of models, as discussed in Section 1.8, can be found. Equifinality is the basis of the GLUE methodology discussed in Section 7.6. In this context it is perhaps more appropriate to use model conditioning rather than model calibration since this approach attempts to take account of the many model parameter sets that give acceptable simulations. As a result, the predictions will be necessarily associated with some uncertainty.

In approaching the problem of model calibration or conditioning, there are a number of very basic points to keep in mind. These may be summarized as follows:

- It is most unlikely that there will be one right answer. Many different models and parameter sets may give good fits to the data and it may be very difficult to decide whether one is better than another. In particular, having chosen a model structure, the optimum parameter set for one period of observations may not be the optimum set for another period.
- Calibrated parameter values may only be valid inside the particular model structure used. It may not be appropriate to use those values on different models (even though the parameters may have the same name) or in different catchments.
- The model results will be much more sensitive to changes in the values of some parameters than to changes in others. A basic sensitivity analysis should be carried out early on in a study (Section 7.2).
- Different performance measures will usually give different results in terms of both the 'optimum' values of parameters and the relative sensitivity of different parameters.
- Sensitivity may also depend on the period of data used, and especially whether a particular component of the model is being 'exercised' in a particular period. If it is not (e.g. if an infiltration excess runoff production component only gets to be used under extreme rainfalls), then the parameters associated with these components will generally appear insensitive.
- Model calibration has many of the features of a simple regression analysis in that an optimum parameter set will be one that, in some sense, minimizes the overall error or residuals. There are still residuals, however, and this implies uncertainty in the predictions of a calibrated model. As in regression, these uncertainties will normally get larger as the model predicts the responses for more and more extreme conditions relative to the data used in calibration.

7.2 Parameter Response Surfaces and Sensitivity Analysis

Consider, for simplicity, a model with only two parameters. Some initial values of the parameters are chosen and the model is run with a calibration data set. The resulting predictions are compared with some observed variables and a measure of goodness of fit is calculated and scaled so that if the model was a perfect fit the goodness of fit would have a value of 1.0, and if the fit was very poor it would have a value of 0 (specific performance measures will be discussed in the next section). Assume that the first run resulted in a goodness of fit of 0.72, i.e. we would hope that the model could do better (get closer to a value of 1). It is a relatively simple matter to set up the model to change the values of the parameters, make another run, and recalculate the goodness of fit. This is one of the options provided in the TOPMODEL software (see Appendix A). However, how do we decide which parameter values to change in order to improve the fit?

One way is by simple trial and error, plotting the results on screen, thinking about the role of each parameter in the model, and changing the values to make the hydrograph peaks higher, or the recessions longer, or whatever is needed. This can be very instructive, but as the number of parameters gets larger it becomes more and more difficult to sort out all the different interactions of different parameters in the model and decide what to change next (try it with the demonstration TOPMODEL software where up to five parameters may be changed).

Another way is to make enough model runs to evaluate the model performance in the whole of the *parameter space*. In the simple two-parameter example, we could decide on a range of values for each parameter, use 10 discrete increments on each parameter range, and run the model for every combination of parameter values. The ranges of the parameters define the parameter space. Plotting the resulting values of goodness of fit defines a parameter *response surface* such as that shown as contours in Figure 7.1 (see also the three-dimensional representation of Figure 1.7). In this example, 10 discrete increments would require $10^2 = 100$ runs of the model. For simple models this should not take too long. For example, 100 runs of TOPMODEL with 1000 time steps on a Pentium PC will take about 2 minutes, although complex fully distributed models will take much longer. The same strategy for three parameters is a bit more demanding: 10^3 runs would be required. For six parameters, 10^6 or a million runs (about two weeks of

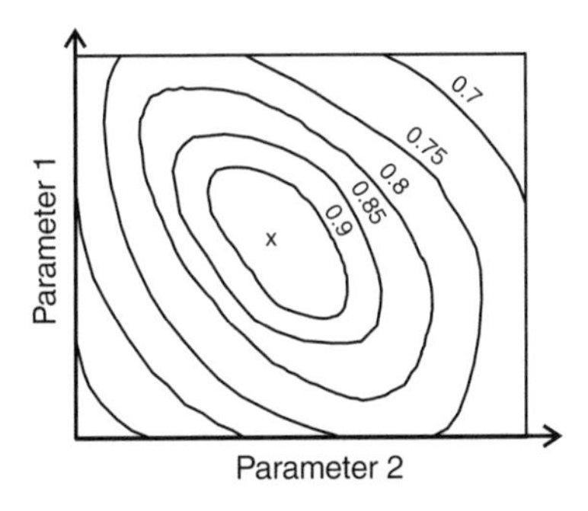

Figure 7.1 *Response surface for two parameter dimensions with goodness of fit represented as contours*

computing for TOPMODEL on a PC, and very much more for more complex models) would be required, and 10 increments per parameter is not a very fine discretization of the parameter space. Not all those runs, of course, would result in models giving good fits to the data. A lot of computer time could therefore be saved by avoiding model runs that give poor fits. This is a major reason why there has been so much research into automatic optimization techniques, which aim to minimize the number of runs necessary to find an optimum parameter set.

The form of the response surface may also become more and more complex as the number of parameters increases, and it is also more and more difficult to visualize the response surface in three or more parameter dimensions. Some of the problems likely to be encountered, however, can be illustrated with our simple two-parameter example. The form of the response surface is not always the type of simple hill shown in Figure 1.7. If it was, then finding an optimum parameter set would not be difficult; any of the so-called hill-climbing automatic optimization techniques of Section 7.4 should do a good job in finding the way from any arbitrary starting point to the optimum.

One of the problems commonly encountered is parameter insensitivity. This will occur if a parameter has very little effect on the model result in part of the range. This may result from the component of the model associated with that parameter not being activated during a run (perhaps the parameter is the maximum capacity of a store in the model and the store never gets filled). In this case part of the parameter response space will be 'flat' with respect to changes in one or more parameters (e.g. Parameter 1 in (Figure 7.2(a)). Changes in that parameter in that area have very little effect on the results. Hill-climbing techniques may find it difficult to find a way off the plateau and towards higher goodness of fit functions if they get onto such a plateau in the response surface. Different starting points may then lead to different final sets of parameter values.

Another problem is parameter interactions. This can lead to multiple optima (Figure 7.2(b)) or 'ridges' in the response surface (Figure 7.2(c)), with different pairs of parameter values giving a very similar goodness of fit. In these latter cases a hill-climbing technique may find the ridge very easily but may find it difficult to converge on a single set of values giving the best fit. Again, different starting values may give different final sets of parameter values.

The problem of multiple *local optima* can make hill-climbing optimization particularly difficult. One of these local peaks will be the *global optimum*, but there may be a number of local optima that give a similar goodness of fit. The response surface may also be very irregular or jagged (see Blackie and Eeles (1985) for a good two-parameter example and also the discussion in Sorooshian and Gupta (1995)). Again, different starting points for a hill-climbing algorithm might lead to very different final values. Most such algorithms will find the nearest local optimum, which may not be the global optimum.

This is not just an example of mathematical complexity; there may be good physical reasons why this might be so. If a model has components for infiltration excess runoff production, saturation excess runoff production or subsurface stormflow (we might expect more than two parameters in this case), then there will likely be sets of parameters that give a good fit to the hydrograph using the infiltration excess mechanism; sets giving good fits using a saturation excess mechanism; sets giving good

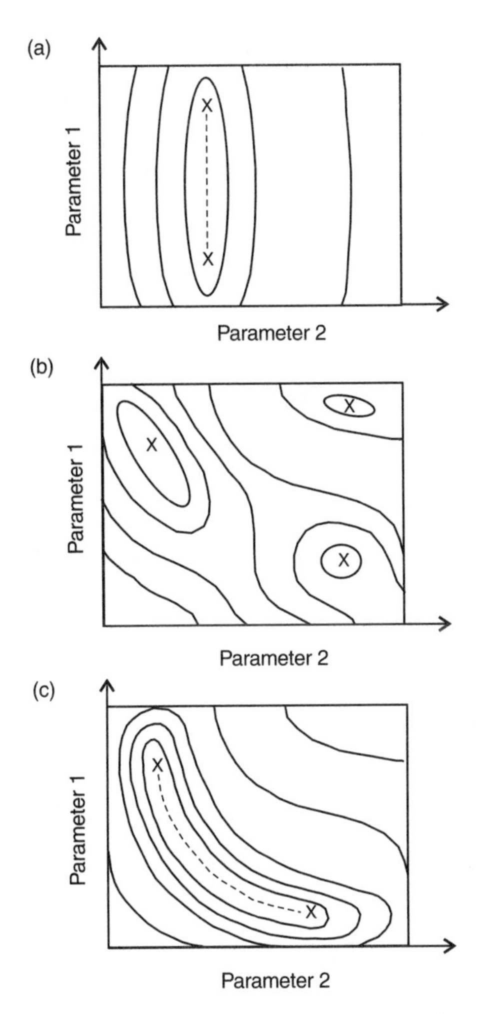

Figure 7.2 *More complex response surfaces in two parameter dimensions. (a) Flat areas of the response surface revealing insensitivity of fit to variations in parameter values. (b) Multiple peaks in the response surface indicating multiple local optima. (c) Ridges in the response surface revealing parameter interactions*

fits by a subsurface stormflow mechanism; and even more sets giving good fits by a mixture of all three processes (see Beven and Kirkby (1979) for an example using the original TOPMODEL). The different local optima may then be in very different parts of the parameter space.

The types of behaviour shown in Figure 7.2 can make finding the global optimum difficult, to say the least. Most parameter optimization problems involve more than two parameters. To get an impression of the difficulties faced, try to imagine what a number of local optima would look like on a three-parameter response surface; then on a four-parameter response surface, and so on. Some advances have been made in computer visualization of higher dimensional response surfaces but trying to picture such a surface soon becomes rather taxing for bears of very little brain (or even expert hydrological modellers). The modern 'hill-climbing' algorithms described in Section 7.4 are designed to be robust with respect to such complexities of the response surface.

There is, however, another way of approaching the problem, i.e. by designing hydrological models to avoid such calibration problems. A model could be structured, for example, to avoid the type of threshold maximum storage capacity parameter that gets activated only for a small number of time steps. Early work on this type of approach in rainfall–runoff modelling was carried out by Richard Ibbitt (see Ibbitt and O'Donnell 1971, 1974) using conceptual ESMA-type models, while, as noted in Section 6.2, the PDM model was originally formulated by Moore and Clarke (1981) with this in mind. Normally, of course, hydrological models are not designed in this way. The hydrological concepts are given priority rather than the problems of parameter calibration, particularly in physics-based models. However, for any model that is subject to calibration in this way, these considerations will be relevant.

There are particular problems in assessing the response surface and sensitivity of parameters in distributed models, not least because of the very large number of parameter values involved and the possibilities for parameter interaction in specifying distributed fields of parameters. This will remain a difficulty for the foreseeable future and the only sensible strategy in calibrating distributed models would appear to be to insist that most, if not all, of the parameters are either fixed (perhaps within some feasible range, as in Parkin *et al.* (1996)) or calibrated with respect to some distributed observations and not catchment discharge alone (such as in Franks *et al.* (1998) and Lamb *et al.* (1998b)). The special problems of calibrating distributed models were discussed in earlier Sections 5.1.1 and 5.7.

7.2.1 Assessing Parameter Sensitivity

The efficiency of parameter calibration would clearly be enhanced if it was possible to concentrate the effort on those parameters to which the model simulation results are most sensitive. This requires an approach to assessing parameter sensitivity within a complex model structure. Sensitivity can be assessed with respect to both predicted variables (such as peak discharges, discharge volume, water table levels, snowmelt rates, etc.) or with respect to some performance measure (see next section). Both can be thought of in terms of their respective response surfaces in the parameter space. One definition of the sensitivity of the model simulation results to a particular parameter is the local gradient of the response surface in the direction of the chosen parameter axis. This can be used to define a normalized sensitivity index of the following form:

$$S_i = \frac{\mathrm{d}Z/\mathrm{d}x_i}{x_i} \tag{7.1}$$

where S_i is the sensitivity index with respect to parameter i with value x_i, and Z is the value of the variable or performance measure at that point in the parameter space (see McCuen 1973). The gradient will be evaluated locally, given values of the other parameters, either analytically for simple models, or numerically by a finite difference, i.e. by evaluating the change in Z as x_i is changed by a small amount (say 1 percent). Thus, since the simulation results depend on all the parameters, the sensitivity S_i for any particular parameter i will tend to vary through the parameter space (as illustrated by the changing gradients for the simple cases in Figure 7.2). Because of this, sensitivities are normally evaluated in the immediate region of a best estimate parameter set or an identified optimum parameter set after a model calibration exercise.

This is, however, a very local estimate of sensitivity in the parameter space. A more global estimate might give a more generally useful estimate of the importance of a parameter within the model structure. There are a number of global sensitivity analysis techniques available, but one that makes minimal assumptions about the shapes of the response surface is variously known as generalized sensitivity analysis (GSA), regionalized sensitivity analysis (RSA) or the Hornberger–Spear–Young (HSY) method (see Hornberger and Spear 1981; Young 1983; Beck 1987), which was a precursor of the GLUE methodology described in Section 7.6. The HSY method is based on Monte Carlo simulation. Monte Carlo simulation makes use of many different runs of a model, with each run using a randomly chosen parameter set. In the HSY method the parameter values are chosen from uniform distributions spanning specified ranges of each parameter. The ranges should reflect the feasible parameter values in a particular application. The idea is to obtain a sample of model simulations from throughout the feasible parameter space. Those simulations are classified in some way into those that are considered *behavioural* and those that are considered *non-behavioural* in respect of the system being studied. Behavioural simulations might be those with a high value of a certain variable or performance measure; non-behavioural simulations might be those with a low value.

HSY sensitivity analysis then looks for differences between the behavioural and non-behavioural sets for each parameter. It does so by comparing the cumulative distribution of that parameter in each set (e.g. Figure 7.3). Where there is a strong difference between the two distributions for a parameter, it may be concluded that the simulations are sensitive to that parameter (Figure 7.3(b)). Where the two distributions are very similar, it may be concluded that the simulations are not very sensitive to that parameter (Figure 7.3(c)). A quantitative measure of the difference between the distributions can be calculated using the nonparametric Kolmagorov–Smirnov d statistic, although for large numbers of simulations this test is not robust and will suggest that small differences are statistically significant. The d statistic can, however, be used as an index of relative difference. This approach may be extended, given enough Monte Carlo simulation samples, to more than two sets of parameters (the GLUE software, for example, uses 10 different classes in assessing sensitivity). Other examples of the use of the HSY approach in rainfall–runoff modelling include Hornberger *et al.* (1985) using TOPMODEL, and Harlin and Kung (1992) using the HBV model. The HSY approach is essentially a *nonparametric method* of sensitivity analysis in that it makes no prior assumptions about the variation or covariation of different parameter values, but only evaluates sets of parameter values in terms of their performance.

7.3 Performance Measures and Likelihood Measures

The definition of a parameter response surface as outlined above and shown in Figures 7.1 and 7.2 requires a quantitative measure of performance or goodness of fit. It is not too difficult to define the requirements of a rainfall–runoff model in

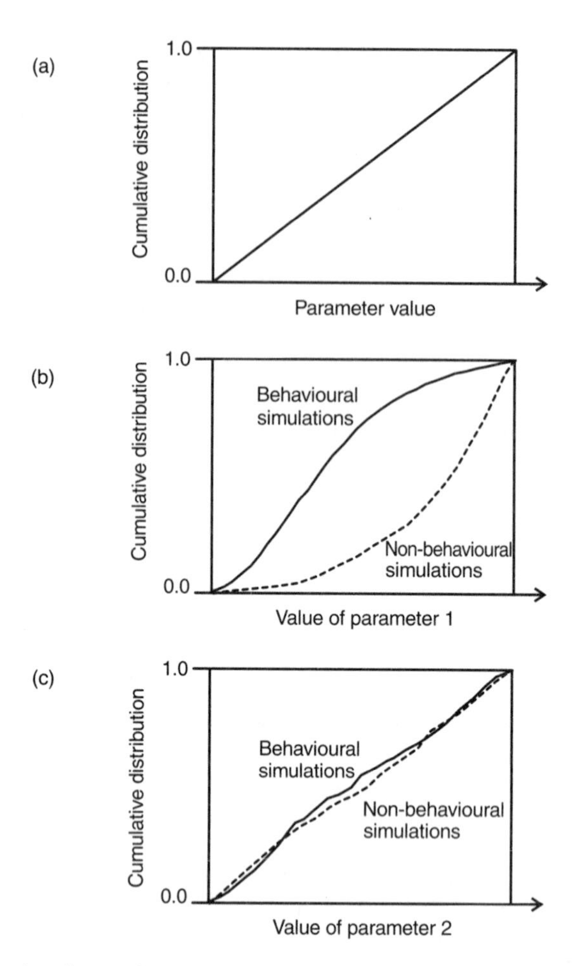

Figure 7.3 *Generalized (Hornberger–Spear–Young) sensitivity analysis. (a) Initial cumulative distributions of parameter values for uniform sampling of prior parameter values across a specified range. (b) Cumulative distributions of parameter values for behavioural and non-behavioural simulations for a sensitive parameter. (c) Cumulative distributions of parameter values for behavioural and non-behavioural simulations for an insensitive parameter*

words: we want a model to predict the hydrograph peaks correctly (at least to within the magnitude of the errors associated with the observations), to predict the timing of the hydrograph peaks correctly, and to give a good representation of the form of the recession curve to set up the initial conditions prior to the next event. We may also require that, over a long simulation period, the relative magnitudes of the different elements of the water balance should be predicted accurately. The requirements might be somewhat different for different projects, so there may not be any universal measure of performance that will serve all purposes.

Most measures of goodness of fit used in hydrograph simulation in the past have been based on the sum of squared errors, or error variance. Taking the squares of the residuals results in a positive contribution of both overpredictions and underpredictions to the final sum over all the time steps. The error variance, σ_ε^2, is defined as

$$\sigma_\varepsilon^2 = \frac{1}{T-1}\sum_{t=1}^{T}(\widehat{y}_t - y_t)^2 \tag{7.2}$$

where $\widehat{y}_t$ is the predicted value of variable y at time step $t = 1, 2, \ldots, T$. Usually the predicted variable is discharge, Q (as shown in Figure 7.4), but it may be possible to evaluate the model performance with respect to other predicted variables so we will use the general variable y in what follows. A widely used goodness of fit measure based on the error variance is the modelling efficiency of Nash and Sutcliffe (1970), defined as

$$E = \left[1 - \frac{\sigma_\varepsilon^2}{\sigma_o^2}\right] \tag{7.3}$$

where σ_o^2 is the variance of the observations. The efficiency is like a statistical coefficient of determination. It has the value of 1 for a perfect fit when is $\sigma_\varepsilon^2 = 0$; it has the value of 0 when $\sigma_\varepsilon^2 = \sigma_o^2$, which is equivalent to saying that the hydrological model is no better than a one-parameter 'no-knowledge' model that gives a prediction of the mean of the observations for all time steps! Negative values of efficiency are indicating that the model is performing worse than this 'no-knowledge' model.

The sum of squared errors and modelling efficiency are not ideal measures of goodness of fit for rainfall–runoff modelling for three main reasons. The first is that the largest residuals will tend to be found near the hydrograph peaks. Since the errors are squared this can result in the predictions of peak discharge being given greater weight than the prediction of low flows (although this may clearly be a desirable characteristic for some flood forecasting purposes). Secondly, even if the peak magnitudes were to be predicted perfectly, this measure may be sensitive to timing errors in the predictions. This is illustrated for the second hydrograph in Figure 7.4 which is well predicted in shape and peak magnitude but the slight difference in time results in significant residuals on both rising and falling limbs.

Figure 7.4 also illustrates the third effect, i.e. that the residuals at successive time steps may not be independent but may be *autocorrelated* in time. The use of the simple sum of squared errors as a goodness of fit measure has a strong theoretical basis in

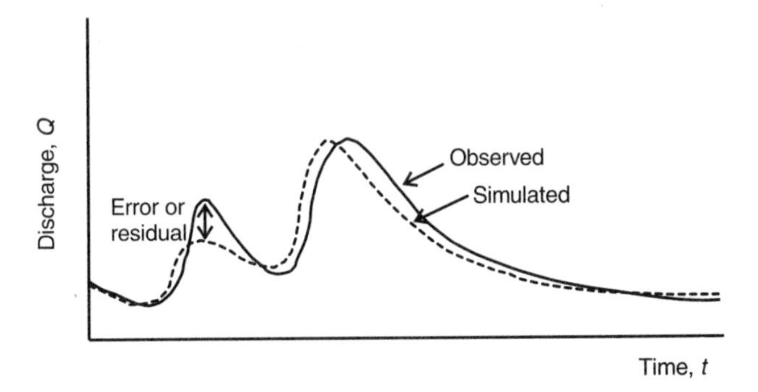

Figure 7.4 *Comparing observed and simulated hydrographs*

statistical inference, but for cases where the samples (here the predictions at each time step) can be considered as independent and of constant variance. In many hydrograph simulations there is also a suggestion that the variance of the residuals may change in a consistent way over time, with a tendency to be higher for higher flows. This has led to the use of measures borrowed from the theory of maximum likelihood in statistics, which attempt to take account of the correlation and changing variance of the errors (*heteroscedastic errors*, e.g. Sorooshian *et al.* 1983; Hornberger *et al.* 1985). Maximum likelihood aims to maximize the probabilty of predicting an observation, given the model. These probabilities are specified on the basis of a likelihood function, which is a goodness of fit measure that has the advantage that it can be interpreted directly in terms of such prediction probabilities. However, the likelihood function that is appropriate will depend on defining an appropriate structure for the modelling errors.

Underlying the development of the likelihood functions used in maximum likelihood approaches is the idea that there is a *correct* model, focusing attention on the nature of the errors associated with that model. Ideally, we would hope to find a model with zero bias, and purely random errors with minimum variance and no autocorrelation. For the relatively simple case of an additive error with a Gaussian distribution and single time step autocorrelation, the likelihood function is developed in Box 7.1. More complex error models will result in more complex likelihood functions (e.g. Cox and Hinkley 1974). In principle, the structure of the errors should be checked to ensure that an appropriate error model is being used. In practice, this must be an iterative process since, under the assumption that there is a correct model, it is the structure of the errors of that optimum model that must be checked, but finding the optimum depends on defining a likelihood function for an error structure.

Experience suggests that hydrological models do not, in general, conform well to the requirements of the classical techniques of statistical inference and that a more flexible and application oriented approach to model calibration is required. There are certainly many other performance measures that could be used. Some examples, for the prediction of single variables such as discharges in hydrograph simulation, are given in Box 7.1. It may also be necessary to combine goodness of fit measures for more than one variable, e.g. both discharge and one or more predictions of observed water table level. Again, a number of different ways of combining information are available, and some examples are given in Box 7.2. Some of the more interesting recent developments are based on a set theoretical approach to model calibration (see Section 7.6 below).

Remember that all these measures are aimed at providing a relative measure of model performance. That measure should reflect the aims of a particular application in an appropriate way. There is no universal performance measure and whatever choice is made, there will be an effect on the relative goodness of fit estimates for different models and parameter sets, particularly if an optimum parameter set is sought. The next section will examine the techniques for finding optimal parameter sets, after which a more flexible approach to model calibration will be discussed.

7.4 Automatic Optimization Techniques

A full description of all the available techniques for automatic optimization is well beyond the scope of this book, particularly since we have already noted that the concept of the optimum parameter set may not be a particularly useful one in hydrological modelling. In this section, we will give just a brief outline of the algorithms that are available. For more specifics, descriptions of different algorithms are available in Press *et al.* (1992) and Sen and Stoffa (1995), and a discussion of techniques in respect of hydrological models is given in Sorooshian and Gupta (1995).

7.4.1 Hill-Climbing Techniques

Hill-climbing techniques for parameter calibration have been an important area of research since the start of computer modelling in the 1960s. Hill climbing from any point on the response surface requires knowledge of the gradient of that surface so that the algorithm knows in which direction to climb. The available techniques may be classified into two basic types. The first are algorithms that require the gradient of the response surface to be defined analytically for every point in the parameter space. Mathematically, this requires that an analytical expression be available for the differential of the model output with respect to each parameter value. These gradient methods are not generally used with hydrological models since it is often impossible to define such differentials analytically for complex model structures. Much more commonly used are direct search algorithms, which search along trial directions from the current point with the aim of finding improved objective function values. Different algorithms vary in the search strategies used. Algorithms that have been widely used in rainfall–runoff modelling include the Rosenbrock method (Rosenbrock 1960) and the Simplex method (Nelder and Mead 1965). The latter is explained in Sorooshian and Gupta (1995).

Hill climbing is, of course, much easier on smooth response surfaces than on flat or jagged surfaces. Many hydrological models will not give smooth response surfaces but, as noted above, with three or more parameter values it may be difficult to evaluate or visualize the full shape of the surface. If a hill-climbing technique is used for parameter calibration, a *minimal* check on the performance of the algorithm in finding a global optimum is to start the algorithm from a number of very different (or randomly chosen) starting points in the parameter space and check the consistency of the final sets of parameter values found. If the final sets are close, then it may be implied that there is a single optimum. If not, then consider one of the algorithms in the sections that follow, which have all been developed to be robust with respect to complexities in the response surface.

7.4.2 Simulated Annealing

Another way of using random starting points to find a global optimum is simulated annealing. The name arises from an analogy between the model parameters included in the optimization and particles in a cooling liquid, which is the basis of the algorithm. If the particles are initially all in the liquid state they will be randomly distributed through the space occupied by the fluid. As the liquid is cooled to a lower temperature, annealing will take place in a way that minimizes the energy of the system. If the cooling is too fast, this energy minimization will occur locally; if very slow then eventually a global minimum energy state will result. The idea of simulated annealing is to mimic this cooling process, starting from randomly distributed sets of parameters in the parameter space to find a global optimum state with respect to the performance measure of the optimization problem.

There are a number of different variants on simulated annealing including very fast simulated reannealing and mean field annealing (see Tarantola 1987; Ingber 1993; Sen and Stoffa 1995). The essence of all of the methods is a rule for the acceptance of new parameter sets. Given a starting parameter set, a perturbation of one or more parameter values is generated and the new performance measure is calculated. If it is better than the previous one, the new model is accepted. If it is not better, it may still be accepted with a probability based on an exponential function of the difference in the performance measure value scaled by a factor that is equivalent to the temperature in the annealing analogy. As the temperature is gradually reduced over a number of iterations, this probability is reduced. This way of allowing parameter sets with worse performance to be accepted ensures that the algorithm does not get trapped by a local optimum, at least if the rate of cooling is slow enough. The choice of the cooling schedule is therefore important and will vary from problem to problem. The various simulated annealing methods differ in the ways that they attempt to increase the number of accepted models relative to those rejected and therefore increase the efficiency of the search. In hydrology, a recent application of simulated annealing may be found in Thyer *et al.* (1999).

There are similarities between simulated annealing and some of the Monte Carlo Markov Chain (MC^2) methods for parameter estimation that have seen a rapid recent development in statistics. Sen and Stoffa (1995) note that the Metropolis MC^2 algorithm is directly analogous to a simulated annealing method. This has been used in rainfall–runoff model parameter estimation by Kuczera and Parent (1998) and Overney (1998).

7.4.3 Genetic Algorithms

Genetic algorithm (GA) methods are another way of trying to ensure that a global optimum is always found, but are based on a very different analogy, that of biological evolution. A random population of 'individuals' (different parameter sets) is chosen as a starting point and then allowed to 'evolve' over successive generations or iterations in a way that improves the 'fitness' (performance measure) at each iteration until a global optimum fitness is reached. The algorithms differ in the operations used to evolve the population at each iteration, which include selection, cross-over and mutation. A popular description has been given by Forrest (1993), and more detailed descriptions are given by Davis (1991). Sen and Soffa (1995) show how some elements of simulated annealing can be included in a genetic algorithm approach. GA optimization has been used by Wang (1991) in calibrating the Xinanjiang model, Kuczera (1997) with a five-parameter conceptual rainfall–runoff model and Franchini and Galeati (1997) with an 11-parameter model.

One form of algorithm that has been developed for use in rainfall–runoff modelling, and which combines hill-climbing techniques with GA ideas, is the shuffled complex evolution (SCE) algorithm developed by Duan *et al.* (1992). In this algorithm, different simplex searches are carried out in parallel from each random starting point. After each iteration of the multiple searches, the current parameter values are shuffled to form new simplexes which then form new starting points for a further search iteration.

This shuffling allows global information about the response surface to be shared and means that the algorithm will generally be robust to the presence of multiple local optima. Kuczera (1997) concluded that the SCE algorithm was more successful in finding the global optimum in a five-parameter space than a classical crossover GA algorithm.

7.5 Recognizing Uncertainty in Models and Data: Reliability Analysis

The techniques of the last section are designed to find an optimum parameter set as efficiently as possible. A run of the model using that optimum parameter set will give the best fit to the observations used for the calibration, *as defined by the performance measure used.* It has long been recognized that different performance measures will generally result in different optimum parameter sets. Thus, as far as is possible, the performance measure should reflect the purpose of the modelling. The optimum parameter set alone, however, will reveal little about the possible uncertainty associated with the model predictions. There are many causes of uncertainty in a modelling study. Errors in initial and boundary conditions, errors in the calibration data and errors in the model itself, will all tend to induce uncertainty in the model predictions that should be assessed. As noted earlier, a review of sources of uncertainty in rainfall–runoff modelling and methods for uncertainty estimation is provided by Melching (1995). He includes methods based on Monte Carlo simulation; Latin Hypercube simulation; mean-value first-order second-moment estimation (MFOSM); an advanced first-order second-moment (AFOSM) method; Rosenblueth's point estimation method; and Harr's point estimation method. These are essentially all ways of sampling the response surface for the performance measure in the parameter space. Where enough runs of the model can be made, the Monte Carlo simulation technique will generally produce the most accurate results; the others are approximations to save computer time. However, a high dimensional parameter space will require many, many Monte Carlo samples, as explained in Section 7.2 above, so that the approximate methods still have value in practical applications.

The aim of uncertainty estimation is to assess the probability of a certain quantity, such as the peak discharge of an event, being within a certain interval but it is worth noting that different types of interval might be required. Haan and Meeker (1991), for example, distinguish three different types of interval. A confidence interval will contain the estimate of an unknown characteristic of the quantity of interest, e.g. the mean peak discharge of the event. Since we cannot estimate the peak discharge precisely from the sample of model runs available, then even the estimate of the mean will be uncertain. The confidence interval can then be used to define the mean estimate with specified probability. Most often, 5 and 95 percent limits are used to define a confidence interval (i.e. a 90 percent probability that the value lies within the interval). Confidence limits can also be calculated for other summary quantities for the distribution of peak discharge, such as the variance or even a quantile value.

A second type of interval is the tolerance interval. This is defined so as to contain a certain proportion of the uncertain model estimates of an observation used in model calibration. For the peak discharge example, tolerance intervals could be defined for the model predictions of a particular observed peak used in model calibration. Finally, the third type of interval is the prediction interval. In the rainfall–runoff context this could be defined as the interval containing a certain proportion of the uncertain model estimates of peak discharge (or any other predicted variable) for a future event. In rainfall–runoff modelling we are mostly interested in prediction intervals after calibration or conditioning of a model.

Uncertainty limits are related to the changes in the predicted variable in the parameter space or, more precisely, if a predicted variable (rather than the performance measure) is represented as a surface in the parameter space, to the *gradient* or slope of the surface with respect to changes in the different parameter values. If the slope is steep, then methods such as MFOSM will predict that the uncertainty in the predictions will be large. If the slope is quite small, however, then the methods will predict a small uncertainty since the predicted variable will change little if the parameter is considered to be uncertain. Recalling equation (7.1), the slopes are an indication of the local sensitivity of the predictions to errors in the estimation of the parameter values.

One question that then arises is where to calculate the values of the slopes in order to get a good estimate of the uncertainty limits. This is where the approximate methods must make certain assumptions. The classical assumption is to assume that the response surface is locally multivariate normal around the prediction of the optimum parameter set. The variance of the estimate of a variable Q will then be given by:

$$\mathrm{Var}(Q) = \sum_{i=1}^{p}\sum_{j=1}^{p} \frac{\partial Q}{\partial x_i}\frac{\partial Q}{\partial x_j} E\left[(x_i - \widehat{x}_i)(x_j - \widehat{x}_j)\right] \tag{7.4}$$

where the slopes (the differential terms) are evaluated close to the optimum, $E[\cdot]$ represents an expected value, the x values are the parameter values, and the $\widehat{x}$ values are the optimum parameter set. The term $E\left[(x_i - \widehat{x}_i)(x_j - \widehat{x}_j)\right]$ reflects the covariation of the parameters. If the parameters can be considered to be statistically independent, then

$$\mathrm{Var}(Q) = \sum_{i=1}^{p} \left[\frac{\partial Q}{\partial x_i}\sigma_i\right]^2 \tag{7.5}$$

where σ_i is an estimate of the variance of parameter x_i.

If the model response is linear then this may be an adequate approximation, but many rainfall–runoff models are highly nonlinear. Thus linearization around the optimum will not then provide accurate estimates of uncertainty in the predictions. Melching (1995) notes that this may be a particular problem in reliability analysis of engineering designs where the interest is often in the risk of a particular design failing under extreme conditions (such as a reservoir overflow channel or flood protection scheme in the case of rainfall–runoff modelling). Uncertainty then becomes important in evaluating risk, but for a nonlinear model it will be important to investigate the behaviour away from the best estimate or model response in the region of the more extreme responses. This is the purpose, for example, of the AFOSM method, which still uses linearization but does so around an estimate of a failure point or confidence limit rather than around the mean prediction. More details and references may be found in the Melching (1995) review.

7.6 Model Calibration Using Set Theoretic Methods

There is another approach to model calibration that relies much less on the idea that there is an optimum parameter set. It was noted in Section 1.8 that detailed examination of response surfaces reveals many different combinations of parameter values that give good fits to the data, even for relatively simple models. The concept of the optimum parameter set may then be ill-founded in hydrological modelling, carried over from concepts of statistical inference. A basic foundation of the theory of statistical inference is that there is a correct model; the problem is to estimate the parameters of that model given some uncertainty in the data available. In hydrology it is much more difficult to make such an assumption. There is no correct model, and the data available to evaluate different models may have large uncertainties associated with them, especially for extreme events, which are often those of greatest interest.

An alternative approach to model calibration is to try to determine a set of acceptable models. Set theoretic methods of calibration are generally based on Monte Carlo simulation. A large number of runs of the model are made with different randomly chosen parameter sets. Those that meet some performance criterion or criteria are retained; those that do not are rejected. The result is a set of acceptable models, rather than a single optimum model. Using all the acceptable models for prediction results in a range of predictions for each variable of interest, allowing an estimation of prediction intervals. This type of method has not been used widely in rainfall–runoff modelling (with the exception of the GLUE variant of the next section) but there have been a number of studies in water quality modelling (e.g. Klepper *et al.* 1991; Rose *et al.* 1991; van Straten and Keesman 1991).

A recent development in set theoretic approaches has been the multi-criteria calibration strategy of Yapo *et al.* (1998) and Gupta *et al.* (1998). Their approach is based on the concept of the Pareto Optimal Set, a set of models with different parameter sets that all have values of the various performance criteria that are not inferior to any models outside the optimal set on any of the multiple criteria. In the terminology of the method, the models in the optimal set are dominant over those outside the set. Yapo *et al.* (1998) have produced an interesting method to define the Pareto Optimal Set, related to the SCE optimization of Section 7.3. Rather than a pure Monte Carlo experiment, they start with N randomly chosen points in the parameter space and then use a search technique to modify the parameter values and find N sets within the optimal set (Figure 7.5). They suggest that this will be a much more efficient means of defining the Pareto Optimal Set.

They demonstrate the use of the model and the resulting prediction limits with the Sacramento ESMA-type rainfall–runoff model, used in the US National Weather Service River Forecasting System, in an application to the Leaf River catchment, Mississippi. The model has 13 parameters to be calibrated. Two objective functions were used in the calibration: a sum of squared errors and a heteroscedastic maximum likelihood criterion. To find the Pareto Optimal Set, 500 parameter sets were evolved, requiring 68 890 runs of the model. The results are shown in Figure 7.6, in terms of the grouping of the 500 final parameter sets on the plane of the two objective functions (from Yapo *et al.* 1998) and the associated ranges of discharges predicted by the original randomly chosen parameter sets and the final Pareto Optimal Set (from Gupta *et al.* 1998). A major advantage of the Pareto Optimal Set methodology is that it does not

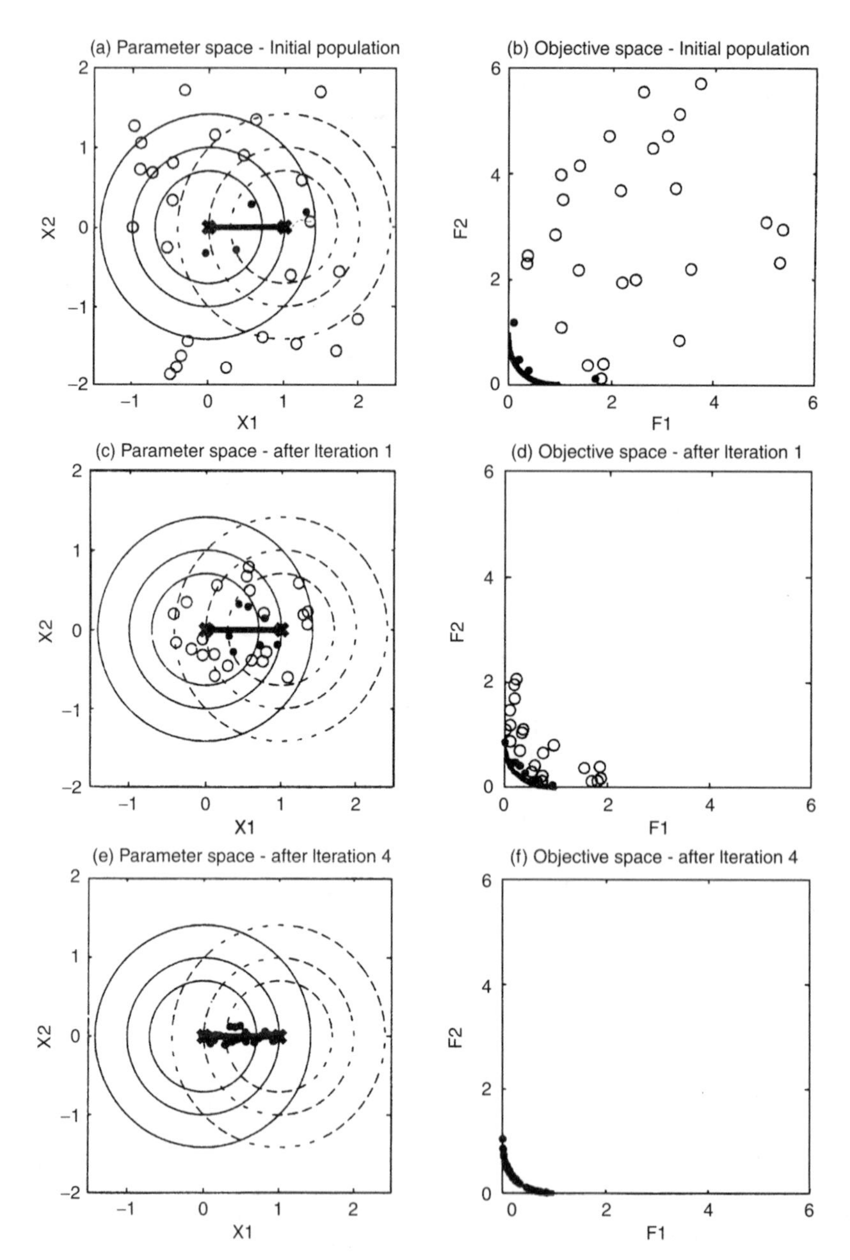

Figure 7.5 *Iterative definition of the Pareto Optimal Set using a population of parameter sets initially chosen randomly. (a) Initial parameter sets in a two-dimensional parameter space (parameters X1, X2). (b) Initial parameter sets in a two-dimensional objective function space (functions F1, F2) (c),(d) Grouping of parameter sets after one iteration. (e),(f) Grouping of parameter sets after iteration 4. After the final iteration, no model with parameter values outside the Pareto Optimal Set has higher values of the objective functions than the models in the Pareto Set (after Yapo et al. 1998). Reprinted from Journal of Hydrology 204: 83–97, copyright (1998), with permission from Elsevier Science*

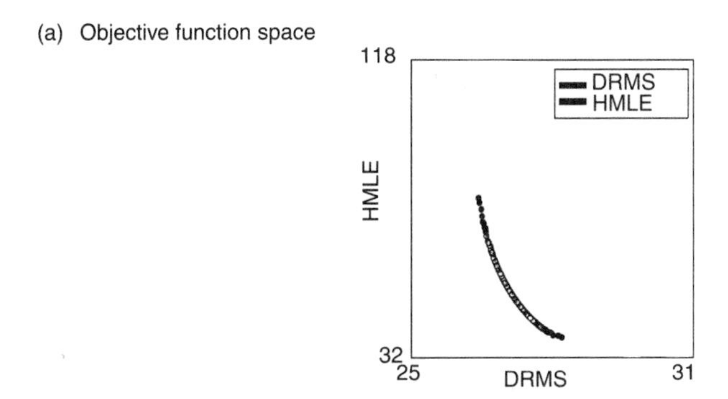

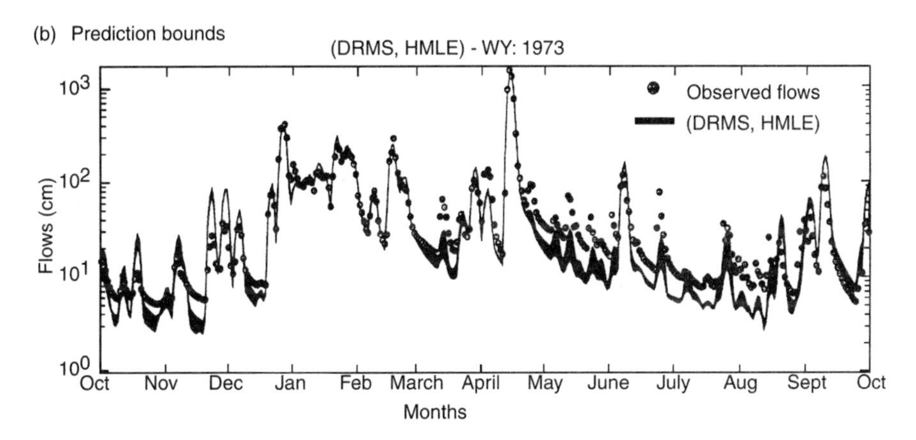

Figure 7.6 *Pareto Optimal Set calibration of the Sacramento ESMA rainfall–runoff model to the Leaf River catchment, Mississippi (after Yapo et al. 1998). (a) Grouping of Pareto Optimal Set of 500 model parameter sets in the plane of two of the model parameters. (b) Prediction limits for the 500 Pareto Optimal parameter sets. Reprinted from Journal of Hydrology 204: 83–97, copyright (1998), with permission from Elsevier Science*

require different performance measures to be combined into one overall measure. Gupta *et al.* (1999) suggest that this method is now competitive with interactive methods carried out by a modelling expert in achieving a calibration that satisfies the competing requirements on the model in fitting the data.

As shown in Figure 7.6(a), the set of models that is found to be Pareto Optimal will reflect the sometimes conflicting requirements of satisfying more than one performance measure. Figure 7.6(b), however, shows that this does not guarantee that the predictions from the sample of Pareto optimal models will bracket the observations since it cannot compensate completely for model structural error or discharge observations that are non-error-free. The original randomly chosen sets do bracket the observations, but with limits that are considerably wider (note the log discharge scale in Figure 7.6(b)). It must be remembered that the method is not intended to estimate prediction limits in any statistical sense, but one feature of this approach is that it does seem to result in an over-constrained set of predictions in comparison with the observations. The user would then have difficulty in relating the range of predictions to any degree of confidence that they would match any particular observation.

7.7 Recognizing Equifinality: The GLUE Method

If we accept that there is no single correct or optimal model, then another approach to estimating prediction limits is to estimate the degree of belief we can associate with different models and parameter sets: this is the basic idea that follows from recognizing the equifinality of models and parameter sets. Certainly we will be able to give different degrees of belief to different models or parameter sets, and many we may be able to reject because they clearly do not give the right sort of response for an application. The 'optimum', given some data for calibration, will have the highest degree of belief associated with it but, as will be seen below, there may be many other models that are almost as good. This can be seen in the dotty plots of Figure 7.7(a) which represent an application of TOPMODEL to the Maimai catchment in New Zealand.

The dotty plots are scatter diagrams of parameter value against some single objective function value. Each dot represents one run of the model from a Monte Carlo experiment using many simulations with different randomly chosen parameter values. They essentially represent a projection of a sample of points on the goodness of fit response surface onto individual parameter dimensions. In Figure 7.7(a) the good models are those that plot near the top. It will be seen that for each parameter there are good simulations across a wide range of parameters. We commonly find with this type of Monte Carlo experiment that good simulations go all the way up to the edge of the range of parameters sampled. There are generally also poor simulations across the whole range of each parameter sampled. Duan *et al.* (1992) show similar behaviour for a completely different model. Such behaviour is a good indication that whether a

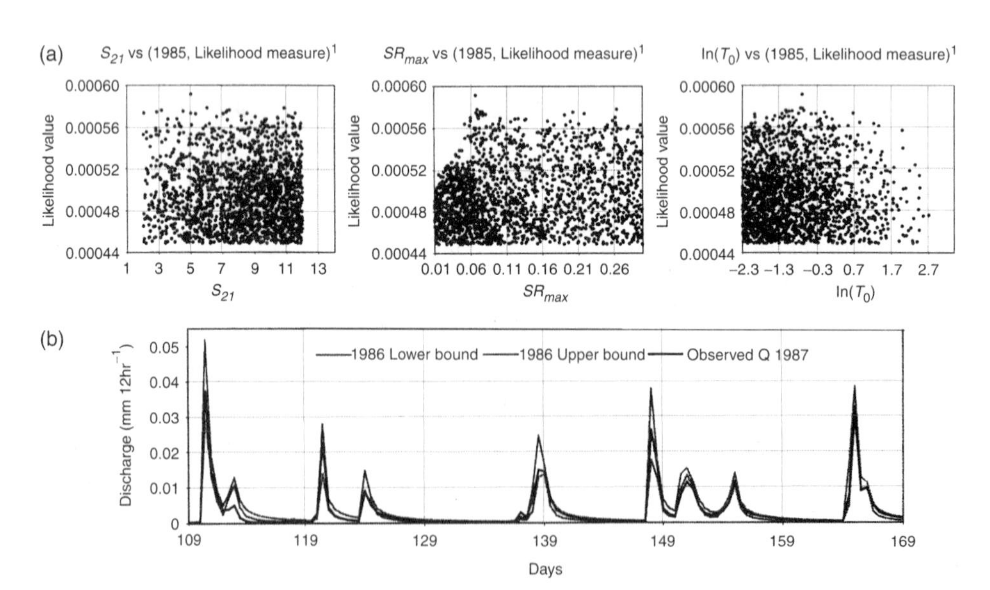

Figure 7.7 *An application of TOPMODEL to the Maimai M8 catchment (3.8 ha), New Zealand, using the GLUE methodology. (a) Dotty plots of the Nash–Sutcliffe model efficiency measure. Each dot represents one run of the model with parameter values chosen randomly by uniform sampling across the ranges of each parameter. (b) Discharge prediction limits for a period in 1987, after conditioning using observations from 1985 and 1986*

model gives good or poor results is not a function of individual parameters but of the whole *set* of parameter values and the interactions between parameters. As a projection of the response surface, the dotty plots cannot show the full structure of the complex parameter interactions which shape that surface. In one sense, however, that does not matter too much since we are really primarily interested in where the good parameter sets are *as a set*.

All of these good parameter sets will, however, give different predictions, but if we associate a measure of belief with each set of predictions (highest for the optimum, zero for those models that have been rejected) then we can estimate the resulting uncertainty in the predictions in a conceptually very simple way by weighting the predictions of all the acceptable models by their associated degree of belief. Such an approach allows the nonlinearity of the response of acceptable models using different parameter sets to be taken into account in prediction and uncertainty estimation.

This appears to lead quite naturally to a form of Bayesian analysis (see Lee 1989; Box and Tiao 1992). Bayesian statistics allow this type of subjective information to be used to estimate the probabilities of different outcomes (see Box 7.2). Here, prior distributions of models and predictions are assessed in terms of some likelihood measure (degree of belief or acceptability) relative to the available observations and a posterior distribution calculated that can then be used in prediction. This is the essence of the generalized likelihood uncertainty estimation (GLUE) methodology proposed by Beven and Binley (1992) which has now been used in a variety of hydrological modelling contexts with a variety of likelihood measures. Updating of the model likelihood distributions as new calibration data become available is handled easily within this Bayesian framework.

In the GLUE methodology, a prior distribution of parameter values is used to generate random parameter sets for use in each model using Monte Carlo simulation. An input sequence is used to drive each model and the results are compared with the available calibration data. A quantitative measure of performance is used to assess the acceptability of each model based on the modelling residuals. Any of the likelihood measures in Box 7.1 or Box 7.2 could, in fact serve this purpose. The only requirements are that the measure should increase monotonically with increasing goodness of fit and that 'non-behavioural' models should have a likelihood of zero. Different likelihood measures or combinations of likelihood measures will, however, lead to different estimates of the predictive uncertainty.

In using the model for predictions, all simulations with a likelihood measure greater than zero are then allowed to contribute to the distribution of predictions. The predictions of each simulation are weighted by the likelihood measure associated with that simulation. The cumulative likelihood weighted distribution of predictions can then be used to estimate quantiles for the predictions at any time step.

Implementation of the GLUE methodology requires a number of decisions to be made:

- a decision about the model or models to be included in the analysis
- a decision about the feasible range for each parameter value
- a decision about a sampling strategy for the parameter sets
- a decision about an appropriate likelihood measure

These decisions are all, to some extent, subjective but an important point is that they must be made explicit in any application. Thus the analysis can be reproduced if necessary, and the decisions can be discussed and evaluated by others.

Given a large enough sample of Monte Carlo simulations, the range of likelihood weighted predictions may then be evaluated to obtain prediction quantiles at any time step. This is most easily done if the likelihood values are renormalized such that $\sum L[M(\Theta_i)] = 1$, where $M(\Theta_i)$ now indicates the ith Monte Carlo sample, so that at any time step t:

$$P(\hat{Q}_t < q) = \sum_{i=1}^{N} L[M(\Theta_i) \mid (\hat{Q}_{i,t} < q)] \tag{7.6}$$

where $\hat{Q}_{i,t}$ is the variable of interest predicted by the ith Monte Carlo sample and N is the number of samples. The prediction quantiles, $P(\hat{Q}_t < q)$, obtained in this way (as shown for example in Figure 7.7(b)) are conditioned on the inputs to the model, the model responses for the particular sample of parameter sets used, the subjective choice of likelihood measure and the observations used in the calculation of the likelihood measure. They are, therefore, empirical, but note that in such a procedure the simulations contributing to a particular quantile interval may vary from time step to time step, reflecting the nonlinearities and varying time delays in model responses. It also allows for the fact that the distributional characteristics of the likelihood weighted model predictions may vary from time step to time step (see Freer *et al.* (1996) and the case study in Section 7.8 below).

Any parameter interactions in the model, and any effects of errors in the input data and observational data, will be implicitly reflected in the likelihood measure associated with each simulation and do not therefore have to be considered separately. This makes an assumption that these effects will be similar during a prediction, but avoids the problem that these effects are very difficult indeed to consider separately.

7.7.1 Deciding on Feasible Parameter Ranges

Deciding on feasible parameter ranges is not necessarily easy, even given some experience of previous applications of a model. The aim is to have a parameter space wide enough that good fits of the model are not excluded, but not so wide that the parameter values have no sense or meaning or that unnecessary non-behavioural model runs are made. It is often found, however, that even if the ranges are drawn quite wide, good fits are found right up to the boundary of some parameters (as in Figure 7.7). This may be because the model predictions are not very sensitive to those parameters, or it may be that the range has not been drawn wide enough since it implies that there will still be good fits beyond the edge of the range. The best suggestion is to start with quite wide ranges and see if they can be narrowed down after an initial sampling of the parameter space.

7.7.2 Deciding on a Sampling Strategy

The choice of a sampling strategy may be very important, since if a large number of parameters are included in the analysis a very large number of model runs would be required to define the form of the response surface adequately in a high dimension parameter space. The idea of using randomly chosen parameter sets is to at least get a large sample from this space but clearly a lot of computer time could be wasted if a large number of model runs are made in areas of the parameter space that give poor fits to the data. A number of strategies have been proposed in statistical parameter estimation with the aim of avoiding this problem. In most of the applications of GLUE to date, a uniform independent sampling of parameters in the parameter space has been used. This ensures the prior independence of the parameter sets before their evaluation using the chosen likelihood measure and is very easy to implement but can be a relatively inefficient strategy if large areas of the parameter space result in non-behavioural simulations.

The computational expense of making many thousands of simulations so that an adequate definition of the response surface is obtained is the major reason why Monte Carlo methods have not been more widely used in hydrological modelling. The greater the number of parameters and the greater the complexity of the response surface, then the greater the number of simulations that will be required. This constraint is becoming less, at least for relatively simple models, as computer power continues to increase and prices continue to fall. The recent development of low-cost, Ethernet-linked, parallel PC systems using off-the-shelf boxes will mean that Monte Carlo simulation will become increasingly feasible in both research and application projects. Such parallel systems are ideally suited to this type of calculation. If a single run of the model will fit in the memory of a single processor, then there is very little loss in efficiency in the parallel system: essentially each processor will be running at full calculation capacity during a run, except for some short periods in which the results from a run are passed to the master processor or written to disk and a new run is initiated.

However, with large models there will still be some advantages in trying to make the Monte Carlo sampling more efficient. In the GLUE methodology, for example, there is little advantage in sampling regions of the parameter space with low likelihood measure values once those regions have been established. It would be better to concentrate the sampling in the regions of high likelihoods. This is a subject that has been studied in a variety of different fields, resulting in an extensive literature on what is often called importance sampling. A number of methods have been developed to try to exploit the knowledge gained of the response surface in refining an adaptive sampling strategy. These include techniques such as some Latin Hypercube methods and Monte Carlo Markov Chain (MC^2) methods, which attempt to sample the response surface according to likelihood density, so that regions of high likelihood are sampled more frequently. The hope is that considerable savings in computer time will be made in defining the likelihood surface. Such methods may work well when there is a well-defined surface but for surfaces with lots of local maxima or plateaux, the advantages may not be so great. The efficiency of uniform sampling techniques can also be improved by running the model only in areas of the parameter space where behavioural models are expected on the basis of previous sampling. The tree-structured search of Spear *et al.* (1994) and the Guided Monte Carlo method of Shorter and Rabitz (1997) can both be used in this way.

There may, of course, be some prior information about parameters. This information may take a number of forms. The first would be some sense of expected distribution and covariance of the parameter values. Some parameter sets, within the specified ranges, may be known a priori as not being feasible on the basis of past performance or mechanistic arguments. Then each parameter set could still be formed by uniformly sampling the parameter space but could be given a prior likelihood (perhaps of zero). If the prior likelihood is zero, it will not be necessary to run the model; such a model will be considered as infeasible.

An interesting question arises when there are measured values available of one, some or all parameter values in the model. In some (rare) cases it may even be possible to specify distributions and covariances for the parameter values on the basis of measurements. These could then be used to specify prior likelihood weights in the (uniformly) sampled parameter space. Although it is often the case that such measurements are the best information that we have about parameter values, there is no guarantee that the values measured at one scale will reflect the effective values required in the model to achieve satisfactory functional prediction of observed variables. It might then be possible to feed disinformation into the prior parameter distributions but, if the parameter space is sampled widely enough to include suitable effective parameter values, the repeated application of Bayes equation or some other way of combining likelihood measures (see Box 7.2) should result in the performance of the model increasingly dominating the shape of the response surface relative to the initial prior estimates of parameter distributions.

7.7.3 Deciding on a Likelihood Measure

There are many measures that can be used to evaluate the results of a model simulation. These will in part depend on what observational data are available to evaluate each model, but even if only one type of data is available (such as observed discharges in evaluating a rainfall–runoff model) there are different ways of calculating a model error and using those model errors to calculate a likelihood measure. What is certain is that if we wish to rank a sample of models by performance, different likelihood measures will give different rankings and the same measure calculated for different periods of observations will also give different rankings.

The choice of a likelihood measure should clearly be determined by the nature of the prediction problem. If the interest is in low flows, then a likelihood measure that gives more weight to the accurate prediction of low flows should be used. If the interest is in water yields for reservoir design, then a likelihood measure based on errors in the prediction of discharge volumes would perhaps be more appropriate. If we are interested in predicting flood peaks, then a likelihood measure that emphasizes accurate prediction of measured peak flows should be chosen. In flood forecasting, a likelihood measure that takes account of accuracy in predicting the timing of flood peaks might be chosen. If an evaluation of a distributed model is being made then a likelihood measure that combines both performance on discharge prediction and performance on prediction of an internal state variable such as water table level might be appropriate. A summary of various likelihood measures is given in Box 7.1.

7.7.4 Updating Likelihood Measures

If more than one period of data is available for evaluating the model, or if new data become available, then the likelihood measures from each period can be combined in a number of different ways, as shown in Box 7.2. This can be viewed as an updating procedure. At each stage, including after the first period, there is a prior likelihood associated with each parameter set that is combined with the value of the likelihood measure for the period being used for evaluation to calculate a posterior value. Bayes equation is one well-known way of doing such calculations in statistical theory, but it is not the only one (Box 7.2). The posterior from one period then becomes the prior for the next application. The likelihood measures for a given parameter set for the periods may be correlated; indeed, one should hope it is the case that if a model performs well in one calibration period, it will continue to perform well in other periods. If this is not the case then its combined likelihood measure will be reduced.

It is possible that, in combining two measures from different observed variables during the same calibration period, there will be a correlation in model performance against different variables, i.e. a model that produces good simulations of an output variable might equally produce good simulations of an internal state variable (although it has to be said that this does not necessarily follow in many environmental models). If a model produces good simulations on both variables its relative likelihood will be raised; if it does not, it will be lowered.

The choice of method of combining likelihood measures may have implications for the choice of the measure itself, in particular if it is required that multiple combinations, for examples of measures from different periods of data, have the same result as treating the data as a single continuous period (where this is possible). Repeated application of Bayes equation would not lead to this end if the likelihood measure was a linear function of the inverse error variance. The successive multiplications would result in the most recent period of data having greater weight in the determination of the posterior likelihoods (which may, of course, give the desired effect if the system is thought to be changing over time). However, the use of a likelihood measure that is a linear function of the inverse exponential of the error variance, would result in an equivalence of final posterior likelihood (see Box 7.2).

7.7.5 The GLUE Software

The GLUE software package is a Windows program that has been designed to demonstrate the principles of the GLUE approach. As a demonstration it has some limitations in terms of the number of parameters, the number of predicted variables, and the number of likelihood measures that can be considered (see Appendix A), but the principles demonstrated are easily extended to other more general software environments such as MATLAB, MATHCAD or even an EXCEL spreadsheet. As noted in the last chapter, the TOPMODEL package will produce a file of results that can then be used directly in the GLUE package. An example data file for the application of TOPMODEL to the Slapton Wood catchment is included with the software. However, the program is general enough that it can analyse results from any Monte Carlo simulation experiments, subject to the limitations noted above. There are options in the program for making dotty plots for individual parameters for different likelihood measures or predicted output variables; analysing the sensitivity of individual parameters using a form of the HSY analysis described in Section 7.2; calculating the cumulative distribution of likelihood weighted predictions using equation (7.6); and listing the best and worst parameter sets on any likelihood measure (or giving highest and lowest predictions for any predicted output variable).

7.8 Case Study: An Application of the GLUE Methodology in Modelling the Saeternbekken MINIFELT Catchment, Norway

The approach of the GLUE methodology is probably best understood by means of an example. In this example, we will consider only one model, TOPMODEL, in an extension of the application to the small Saeternbekken MINIFELT catchment in Norway, used in the case study of Section 6.5. The version of TOPMODEL used was based on the original exponential transmissivity profile assumptions. It is worth pointing out that the choice of just a single model is equivalent to assigning a positive prior likelihood to parameter sets sampled for that model (my model!), and zero to all other models. This is, of course, quite common practice, but there is no reason why more than one model structure should not be included within the GLUE framework (apart from the computational expense of making even more Monte Carlo simulations).

The use of both discharge and borehole measurements in conditioning the uncertainty in the predictions of TOPMODEL for the Saeternbekken has been considered by Lamb *et al.* (1998b). They first studied the use of global (catchment-scale) parameter values. Five parameters were varied in the Monte Carlo simulations. The ranges chosen for each parameter reflect past modelling experience using TOPMODEL and an initial analysis of recession curves in setting the range for the m parameter (Table 7.1). An example of the TOPMODEL responses to the individual parameters has already been shown for a similar application in the dotty plots of Figure 7.7 which use the Nash–Sutcliffe efficiency as a likelihood measure.

Lamb *et al.* (1998b) chose to use a different likelihood measure, though also one based on the variance of the residuals, as:

$$L = \exp(-W\sigma_\varepsilon^2) \tag{7.7}$$

where W is a weighting coefficient. This will give values close to zero for larger error variance but, as with the Nash–Sutcliffe efficiency, will have a limit of one if the error variance is very small. In the case of the Saeternbekken catchment, a number of different measures of performance could be calculated using discharge and different borehole observations. This form of performance measure allows these to be combined easily using Bayes equation, since multiplying the likelihood measures is then equivalent to taking a weighted average of the different residual variances within the exponential (see equation (7.2.2) in Box 7.2). In this particular application, this allowed different weights to be used in the 1987 calibration period, when one of the recording boreholes did not appear to produce hydrologically meaningful responses.

An impression of the sensitivity of the individual TOPMODEL parameters can be gained by plotting the cumulative likelihood weighted distributions for each parameter after evaluating each parameter set on data from the 1987 simulation period (see Figure 7.8). For each parameter, a rather wide initial range of values was used. Because of its operation within the TOPMODEL structure, the transmissivity at saturation was sampled on a log scale as $\ln(T_o)$. Lacking any prior information about the covariation of

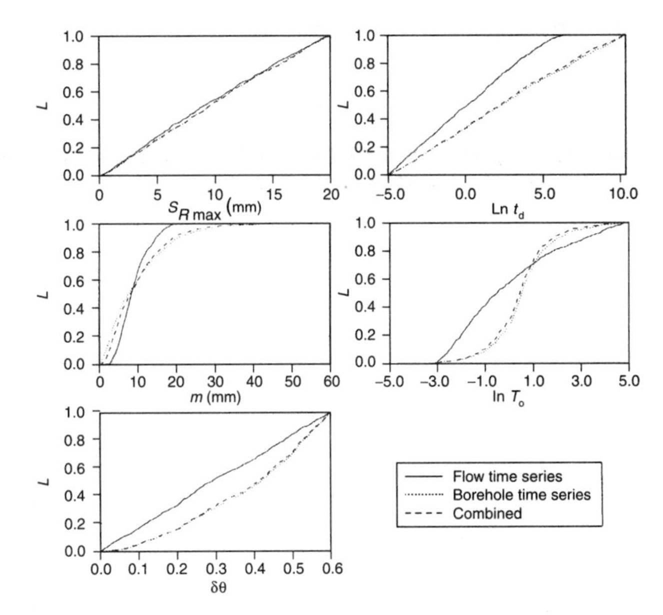

Figure 7.8 *Rescaled likelihood weighted distributions for TOPMODEL parameters, conditioned on discharge and borehole observations from the Saeternbekken MINIFELT catchment for the autumn 1987 period (after Lamb et al. 1998b). Reprinted from Advances in Water Resources 22: 305–317, copyright (1998), with permission from Elsevier Science*

the individual parameters, each was sampled independently from uniform distributions across the range shown. The prior distributions would be represented by a straight diagonal line on each plot. The strongest departures from the prior distributions are shown by the two parameters of the transmissivity profile, m and $\ln(T_o)$. The other three parameters are much less well conditioned by the observations. This is to be expected since the shallow wet soil in the catchment would not be expected to strongly limit evapotranspiration (controlled by the $S_{R\max}$ parameter), nor the time delay for vertical recharge (controlled by the t_d parameter). The relatively poor constraint on an effective storage capacity of the soil ($\delta\theta$) was not surprising in the conditioning on discharges alone but was more surprising when the borehole data were added as this parameter controls the calculation of a water table level from the storage deficits calculated by the model. It may be a reflection of the interaction in the model between this and other parameters in predicting the borehole levels.

It must be remembered that in the GLUE methodology it is the individual parameter sets that are being assessed in terms of the chosen likelihood measure. Figure 7.8 represents only a summary or marginal distribution for each parameter over all the parameter sets being considered. Prediction bounds for the 1987 simulation period are not shown here but showed that in predicting the catchment discharges, very little additional constraint was provided by adding the information from the recording borehole information.

The resulting prediction bounds are shown in Figure 7.9 for discharges and Figure 7.10 for the recording boreholes (see Figure 6.9). In both cases bounds are shown for this 1989 simulation period using parameter sets conditioned on the earlier

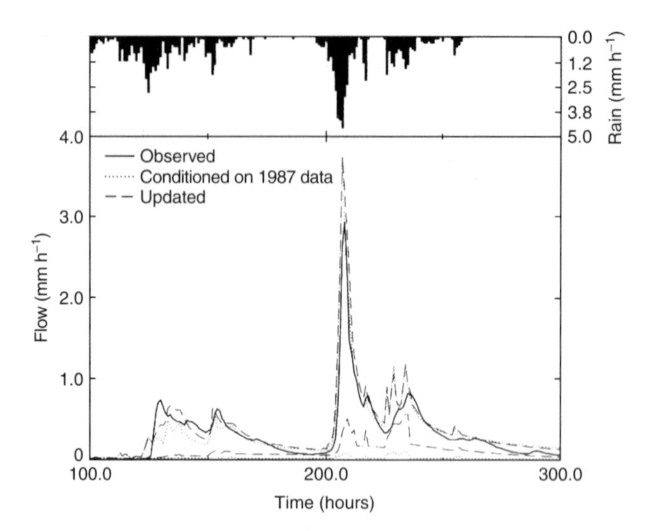

Figure 7.9 *Prediction bounds for stream discharge from the Saeternbekken catchment for the 1989 simulations, showing both prior bounds after conditioning on the 1987 simulation period and posterior bounds after additional updating with the 1989 period data (after Lamb et al. 1998b). Reprinted from Advances in Water Resources 22: 305–317, copyright (1998), with permission from Elsevier Science*

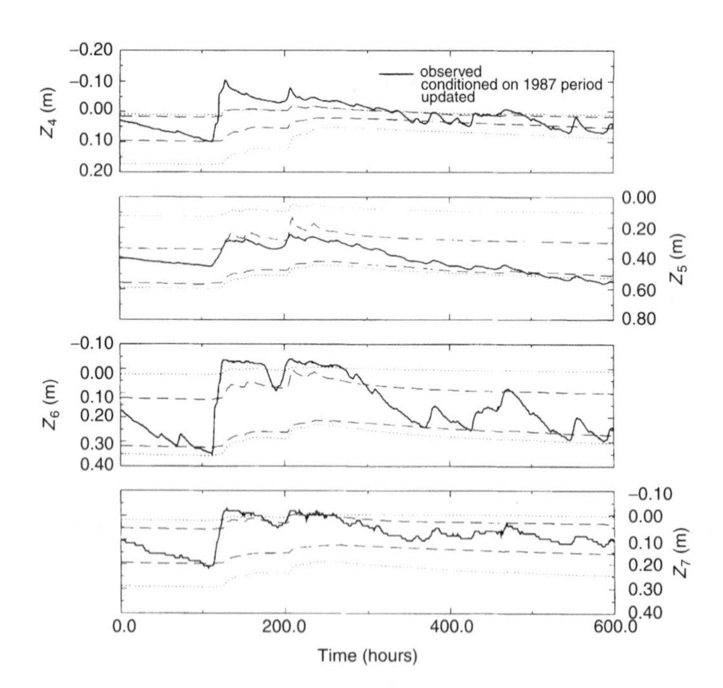

Figure 7.10 *Prediction bounds for four recording boreholes in the Saeternbekken catchment for the 1989 simulation period, showing both prior bounds after conditioning on the 1987 simulation period and posterior bounds after additional updating with the 1989 period data (after Lamb et al. 1998b). Reprinted from Advances in Water Resources 22: 305–317, copyright (1998), with permission from Elsevier Science*

1987 period, and also after combining with likelihoods calculated for the 1989 period itself. These therefore represent prior and posterior likelihoods for this period. In both cases it can be seen that adding the information from the 1989 period leads to a narrowing of the prediction bounds. For most of both sets of simulations the observations are bracketed by the prediction bounds but there are periods when the observations fall outside of the prediction bounds, indicating some deficiencies in the model structure used (or errors in the inputs or observed discharges). The predicted borehole responses do not show the same dynamic variation in the prediction bounds as the observations (although, as already seen in Figure 6.10, individual simulations can reproduce the dynamics much more closely).

Lamb *et al.* (1998b) also examined the prediction of the spatially distributed piezometer data which were available for five different discharges. All of the piezometers not indicating saturated conditions were used in the conditioning process. A weight was given to the residual variance for each set of measurements proportional to the number of points included. The results are shown in Figure 7.11. No account was taken of the possibility of local variations in transmissivity (see discussion of Section 6.5). Interestingly, the prediction bounds based on discharges and recording borehole measurements are much narrower than those based on the piezometer data themselves. Even the latter, however, are not wide enough to encompass a significant

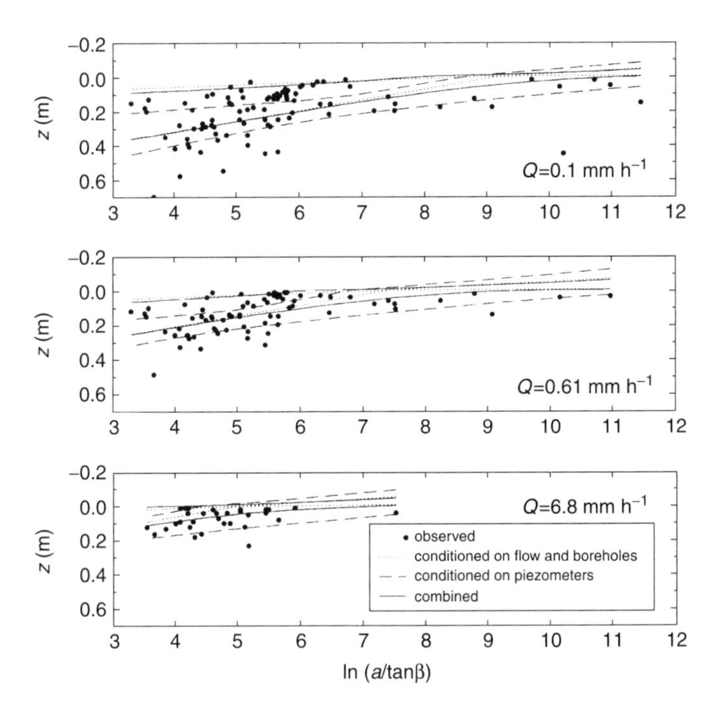

Figure 7.11 Prediction bounds for spatially distributed piezometers in the Saeternbekken catchment for three different discharges, showing prior bounds after conditioning on discharge and recording borehole observations, bounds based on conditioning on the piezometer data alone, and posterior bounds based on a combination of both individual measures (after Lamb et al. 1998b). Reprinted from Advances in Water Resources 22: 305–317, copyright (1998), with permission from Elsevier Science

number of the data points. This indicates that either the model dynamics cannot adequately reproduce the pattern of water tables in the catchment, or that local soil heterogeneity cannot adequately be represented by catchment-scale parameters. Certainly Lamb *et al.* (1997) have shown that improved predictions can be achieved by allowing a local effective transmissivity value for each piezometer site.

This study is of interest because of the way it has used both discharge and internal water table measurements to evaluate the model predictions and constrain the predictive uncertainty of TOPMODEL. It is, perhaps, representative in showing that we might not expect a rainfall–runoff model to reproduce all of the observations all of the time, even when the predictions are associated with some uncertainty bounds. It is also perhaps representative in revealing that there may be difficulties associated with using information from internal state measurements in model calibration or conditioning. First, the use of such data may require adding additional local parameter values; secondly, such local data may not have great value in conditioning the prediction bounds for the catchment discharges.

7.9 Dealing with Equifinality in Rainfall–Runoff Modelling

To summarize, the GLUE methodology provides one way (and not necessarily the only way) of recognizing the possible equifinality of models and parameter sets. The approach is conceptually simple, easily implemented, and leads quite naturally to an assessment of the uncertainty of model predictions. There are undoubtedly limitations of the method of which the most important seems to be computing constraints which will restrict the number of runs that can be made in sampling the parameter space. Thus, even with relatively simple models the number of parameters that can be varied will remain small. However, the type of calculations required are ideally suited to parallel computers. The GLUE methodology also involves a number of subjective decisions, so that any uncertainty bounds or predictions limits derived in this way will be intrinsically qualitative. However, as noted above, all the subjective decisions must be made explicit in the analysis so that they can be discussed or disputed and the analysis repeated, if necessary, with alternative assumptions. Thus, despite the qualitative nature of the procedure, there is some scientific rigour associated with it.

Other applications of the GLUE methodology to rainfall–runoff modelling have included Beven and Binley (1992), Beven (1993), Romanowicz *et al.* (1994), Freer *et al.* (1996), Seibert (1997), Franks *et al.* (1998), Dunn *et al.* (1999), Cameron *et al.* (1999) and Uhlenbrook *et al.* (1999). In most of these studies, likelihood measures have been calculated only using observed catchment discharges. In the study of Franks *et al.* (1998), however, some ground-based estimates of saturated area for a small part of the Naizin catchment in Brittany, France, together with the map of the TOPMODEL topographic index and satellite radar data, were used to estimate approximately the catchment-wide extent of saturation. This was then used as additional information to constrain the model predictions of discharge for the catchment. They showed that, with the addition of this information, there was some reduction in uncertainty in discharge predictions and that this was mostly a result of a dramatically reduced feasible range for the effective transmissivity parameter (in this case used as a catchment-wide value) after taking account of these contributing area estimates (Figure 7.12).

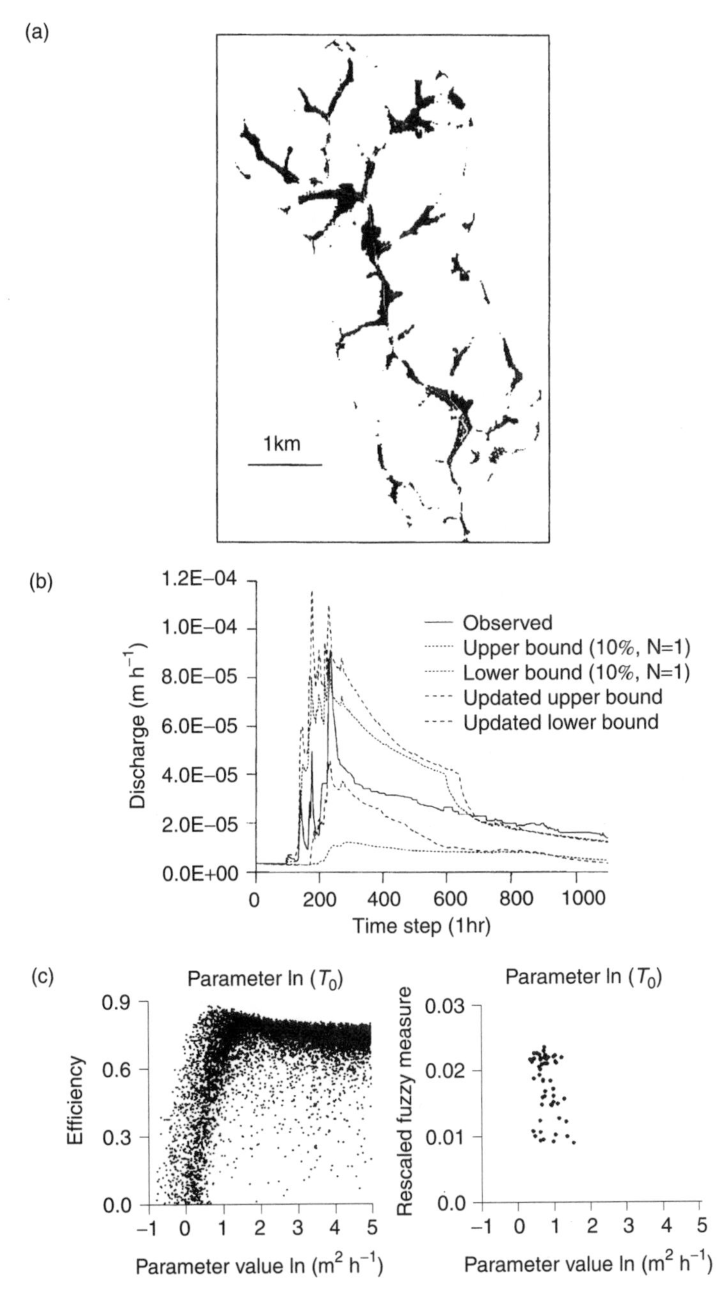

Figure 7.12 *Application of GLUE to the simulation of the Naizin catchment, Brittany, France (after Franks et al. 1998). (a) Best estimates of the maximum saturated contributing area in the catchment derived from observations on a small part of the catchment, topographic data and ERS-1 satellite radar data. (b) Prediction limits for discharge after conditioning on discharge alone, and on discharge and saturated area using the GLUE methodology. (c) Dotty plots of the TOPMODEL transmissivity parameter after conditioning on discharges alone and discharge and saturated area. Reproduced from Water Resources Research 34: 787–797, 1998, copyright by the American Geophysical Union*

In this study, the incorporation of the additional information allows the rejection of many of the model parameter sets that, based on fitting discharge observations alone, had previously been considered acceptable. The possibility of using additional empirical data in this way raises some interesting issues about the modelling process in general. If in the end we have to accept that the equifinality problem is endemic to modelling environmental processes, then an approach based on model falsification requires thoughtful and truly scientific strategies for defining hypotheses and data collection programmes for the most cost-effective refinement of the space of feasible models. In doing so, however, there will be an inevitable problem of compromise in what is accepted as a feasible model since, given the approximate nature of environmental models, and rainfall–runoff models in particular, it will generally be the case that we could reject all models if we look at their predictions in enough detail (see also Mroczkowski *et al.* 1997). It proves to be surprisingly difficult in rainfall–runoff modelling to bracket all the discharge observations (or other predicted variables) for at least 90 percent of the time steps, even using the GLUE approach. This is, in part, a result of data limitations as well as model structural limitations. A good example is provided by Freer *et al.* (1996) where an error in predicting the onset of snowmelt in one of the simulated years of data results in the observations being outside the prediction limits of the simulations for several weeks (see Figure 7.13). Thus, there would appear to be limitations on how far models can be validated as representations of the real runoff processes. This is not to imply, however, that the model predictions might not be sufficiently accurate to be useful for water resources management, flood forecasting or other purposes. We shall return to this discussion in Section 10.5.

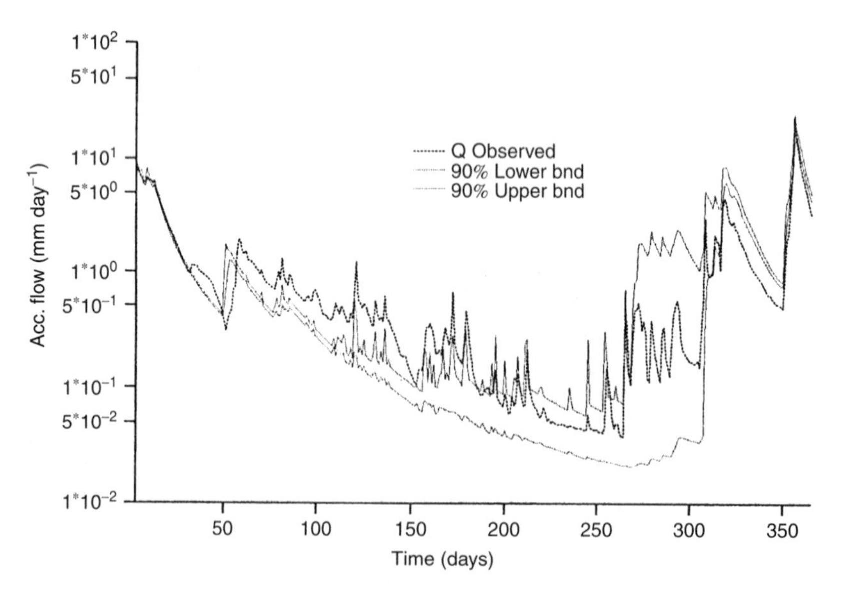

Figure 7.13 *Observed and predicted daily discharges simulated by a version of TOPMODEL for the small Ringelbach catchment (34 ha), Vosges, France. The model was run using an 18 minute time step. Prediction limits are estimated using the GLUE methodology (after Freer et al. 1996). Note the logarithmic discharge scale. Reproduced from Water Resources Research 32: 2161–2173, 1996, copyright by the American Geophysical Union*

7.10 Predictive Uncertainty and Risk

Let us assume that it has been possible to make a realistic assessment of the uncertainty associated with the predictions of a rainfall–runoff model, by whatever method (e.g. Figure 7.9 above). How best then to interpret and make use of those uncertain predictions? One interpretation is that the uncertainty represents the model error in representing the data but a better interpretation is in terms of the risk of a certain outcome in certain circumstances, given the model as a means of extrapolating knowledge and understanding to those circumstances. In essence, evaluating the risk of an outcome based on the model predictions is also an evaluation of the risk of the model predictions being wrong (as they may well prove to be). Risk, however, is something that is readily incorporated into modern decision-making processes, when it may also be necessary to take account of the costs in mitigating that risk, e.g. in enlarging a reservoir spillway or raising flood embankments. The important point is that the risk associated with the model predictions should be included in the decision analysis.

7.11 Key Points from Chapter 7

- Limitations of both model structures and the data available on parameter values, initial conditions and boundary conditions, will generally make it difficult to apply a hydrological model (of whatever type) without some form of calibration or conditioning of model parameter sets.
- In evaluating the goodness of fit of models to observations we often find that the response surface in the parameter space is very complex. There may be flat areas due to a lack of excitation of certain parameters, long ridges due to interactions between parameters, and many local optima as well as the point of global best fit. In general, carefully designed, simple models with a small number of parameters will avoid the problem of overparametrization and have smoother parameter response surfaces but this may be difficult to achieve in rainfall–runoff modelling.
- A review of automatic optimization techniques reveals that many have difficulties in finding the global optimum on a complex response surface. Recent techniques such as simulated annealing or genetic algorithms, such as the Shuffled Complex Evolution algorithm, have been designed to be more robust in finding the global optimum.
- Set theoretic techniques based on Monte Carlo simulation suggest, however, that the idea of an optimum parameter set might be illusory and would be better replaced by a concept of equifinality in simulating a catchment allowing that there may be many different model structures or parameter sets that could be considered as acceptable in simulation.
- The concept of the Pareto optimal set allows that multi-criteria parameter optimization might result in a set of models, each of which achieves a different balance between the different objective functions, but all of which are better than models outside the optimal set. This results in a range of different predictions from the different models in the set, but the range of predictions may not bracket the observations.

- Generalized likelihood uncertainty estimation (GLUE) is one technique for conditioning model parameter sets in the face of such equifinality. In the GLUE methodology many different model runs are made using randomly chosen parameter sets. Each run is evaluated against observed data by means of a likelihood measure. If a model is rejected it is given a likelihood value of zero. The likelihood measures are then used to weight the predictions of the retained models to calculate uncertainty estimates or prediction limits for the simulation. Likelihood values from different types of data may be combined in different ways or updated as more data are collected.
- This approach to modelling focuses attention on the value of different types of data in rejecting or falsifying models. Hypothesis tests may be formulated to refine the set of acceptable models in a truly scientific way. Some compromise in such tests will generally be necessary however, since if the criteria for acceptability are made too strict, all models will be rejected.
- The assessment of the uncertainty associated with a set of model predictions is also an assessment of the risk of a certain outcome that can be used in a risk-based decision analysis for the problem under study.

Box 7.1 Likelihood Measures for Use in Evaluating Models

Likelihood Measures and Likelihood Functions

The definition of a likelihood measure for use within the GLUE methodology requires a measure that is zero for non-behavioural models and increases monotonically as performance or goodness of fit to the available observational data increases. There are many different measures that could be used. The choice of measure should reflect the purposes of the study and the nature of the errors being considered but ultimately the choice of a likelihood measure is subjective.

This is not the case with the likelihood functions used in traditional statistics where the likelihood function expresses directly the expected probability of the data given the model. Statistical likelihood functions are therefore based on the idea that there is a true model and that the error is an *observational* error. This is readily appreciated if the simple example of linear regression is considered. A set of data is observed for two (or more) variables. A scatter plot suggests that there is a linear relationship between two of the variables. A linear regression is used to calculate the values of the intercept a and intercept b parameters of the model

$$Y = a + bX \tag{7.8}$$

The line does not go through all the data points but the line is calculated so as to minimize the sum of the squared errors between observed and predicted values of Y. The data points are then in error, relative to the model. The form of the likelihood function used then depends on assumptions about the nature of the errors (whether they are normally distributed, whether they are independent or correlated, etc.). There is a large body of likelihood function theory developed for different sets of assumptions.

In hydrology, however, we cannot assume that the model is correct. It is probably more appropriate to assume that the data are mostly correct (even though we know that measurement errors are sometimes large) and that it is the model that is in error. We also cannot often assume that the modelling errors conform to simple assumptions about their structure. This does not mean that the theory of likelihood *functions* cannot be used where appropriate, since treating the errors as model errors rather than measurement errors is essentially only a matter of interpretation. However, a wider view does open up the possibility of using likelihood *measures* based on different types of model evaluation, including qualitative and fuzzy measures.

Likelihood Measures Based on the Sum of Squared Errors

In rainfall–runoff modelling we are often evaluating the errors in simulating a time series of discharges or other observed data. A classical statistical measure for evaluating goodness of fit is then the sum of squared errors or error variance σ_ε^2:

$$\sigma_\varepsilon^2 = \frac{1}{NT}\sum_{t=1}^{NT}(Q_t - \hat{Q}_t\{\underline{\Theta}, \underline{Y}\})^2 \tag{7.9}$$

where Q_t is the observed value at time index t; $\hat{Q}_t\{\underline{\Theta}, \underline{Y}\}$ is the simulated variable given parameters $\underline{\Theta}$ and input data $\underline{Y}$, and NT is the number of time steps. Clearly σ_ε^2 becomes smaller as the fit of the simulation improves so this measure cannot be used directly as a likelihood measure. Various transformations can, however. Nash and Sutcliffe (1970) suggested a measure of the form

$$E = 1 - \frac{\sigma_\varepsilon^2}{\sigma_o^2} \tag{7.10}$$

where σ_o^2 is the variance of the observations. They called this measure the modelling *efficiency*. It has been widely used in fitting hydrological models to discharge data. It has the properties that for a perfect fit, it has the value 1. For a fit that is no better than assuming that the mean of the data is known (i.e. $\sigma_\varepsilon^2 = \sigma_o^2$), it has the value zero. For models that are worse than this, it can take on negative values (but since these are non-behavioural models they can be set to zero if the efficiency is treated as a likelihood measure).

Another measure based on the sum of squared errors is the inverse error measure suggested by Box and Taio (1992) in the form

$$L = (\sigma_\varepsilon^2)^{-n} \tag{7.11}$$

where n is a shaping parameter. In this case a large error variance will give a value of the measure close to zero, while as the error variance decreases the likelihood measure will increase. This measure has been used in rainfall–runoff modelling by Beven and Binley (1992) and Freer *et al.* (1997).

A third transformation is the exponential function:

$$L = \exp(-n\sigma_\varepsilon^2) \tag{7.12}$$

where again n is a shaping parameter. This will also give values close to zero for larger error variance but will have a limit of one if the error variance is very small.

All measures based on the error variance have some practical and theoretical limitations arising from the nature of the error series. In statistical theory the error variance is most suitable as a performance measure if the errors between the observed data and predictions are of mean zero, are distributed as a normal distribution with constant variance and are not correlated. Only if these assumptions are met would minimizing the sum of squared errors give unbiased estimates of model parameters. The assumptions are not normally met in rainfall–runoff modelling where the errors in predicting discharge may not have zero mean, they are almost certainly not normally distributed, the variance of the errors in predicting high flows will be different to the variance in predicting low flows, and they exhibit correlation from one time step to another (i.e. a positive error tends to be followed by another positive error at the next time step). This is especially true where a model does not get the timing of a hydrograph prediction correct (see Figure 7.4). The predicted form of the hydrograph might be perfect, but if it is too early or late then there will be a large sum of squared errors and the errors will be correlated.

In the rainfall–runoff modelling context, these problems were addressed by Sorooshian and his co-workers in a series of papers, summarized in Sorooshian and Gupta (1995). They suggested using likelihood measures chosen on the basis of assumptions about the structure of the errors, including a Heteroscedastic Maximum Likelihood Estimator (HMLE) which uses a form of weighted least-squares approach. The word heteroscedastic means that the variance of the errors changes over time. Here it is assumed that the variance of the errors changes with the magnitude of the flow being predicted. The HMLE is defined as

$$\text{HMLE} = \frac{\dfrac{1}{NT}\displaystyle\sum_{t=1}^{NT} w_t \left(Q_t - \hat{Q}_t\{\underline{\Theta}, \underline{Y}\}\right)^2}{\left[\displaystyle\prod_{t=1}^{NT} w_t\right]^{1/NT}} \tag{7.13}$$

where the w_t are weights defined by

$$w_t = Q_t^{2(\lambda-1)} \tag{7.14}$$

The HMLE criterion reduces to a simple error variance measure when all the weights are set equal to 1. The shaping parameter λ is designed to stabilize the variance of the errors between observed and simulated flows. Note that as with the error variance, the HMLE criterion gets smaller as the fit of the model improves. In optimization the parameter λ is normally varied along with the model parameters so as to minimize the HMLE. The HMLE has the effect of improving the convergence towards an optimum parameter set in comparison with a simple error variance criterion but it will tend to weight the fit of the model more towards recession periods than high-flow periods, whereas a simple error variance criterion will give more weight to higher flows. However, the HMLE does not take any explicit account of correlation in the errors.

All the above measures can also be calculated with respect to transformations of the observed and predicted values. One such transform is the log transform, so that rather than using the error $(Q_t - \hat{Q}_t\{\underline{\Theta}, \underline{Y}\})$, the difference between the log values, $(\ln(Q_t) - \ln(\hat{Q}_t\{\underline{\Theta}, \underline{Y}\}))$, is used. This will make allowance for the fact that the errors tend to be larger for larger flows and will also give more weight to the prediction of lower recession period discharges.

A very similar transform that is widely used in statistics with a view to normalizing the skewness of a data set is the Box–Cox transform (Box and Cox 1964). This is defined as

$$Q_t^* = (Q_t^{\xi} - 1)/\xi \qquad \text{for } \xi \neq 0 \tag{7.15}$$

$$= \ln Q_t \qquad \text{for } \xi = 0 \tag{7.16}$$

where Q_t^* is the transformed values of the variable Q_t, and ξ is a constant chosen so as to minimize the skewness of the transformed values.

Development of a Likelihood Function for Autocorrelated Gaussian Error

Consider the following traditional approach to likelihood estimation. We assume, perhaps after a suitable transformation, an error model of additive type, with errors with a normal or Gaussian distribution that exhibit correlation in time. As before, let Q_t be the observed time series, and $\hat{Q}_t(\Theta, Y)$ the model output at time t, given the times series of inputs Y and the set of parameter values Θ, so that

$$Q_t - \hat{Q}_t\{\underline{\Theta}, \underline{Y}\} = \varepsilon_t(\Phi) \tag{7.17}$$

where $\varepsilon_t(\Phi)$ represents the error model with parameters Φ. We now assume a model for the structure of the errors. For a nth order Gaussian autoregressive model AR(n),

$$\varepsilon_t = \mu + \sum_{i=1}^{n} \alpha_i(\varepsilon_{t-1} - \mu) + \sigma^2 \delta_t \tag{7.18}$$

where μ is a mean error, the α_i are correlation coefficients, σ^2 is a residual error variance, and δ_t is a random variable with the normal distribution N [0,1]. For the simplest first-order correlation case, $\Phi = (\mu, \sigma, \alpha)$ and the likelihood function is then given by

$$L(Q_t|\underline{\Theta}, \underline{Y}, \Phi) = (2\pi\sigma^2)^{-T/2}(1 - \alpha^2)^{1/2}$$

$$\exp\left[-\frac{1}{2\sigma^2}\left\{(1 - \alpha^2)(\varepsilon_1 - \mu)^2 + \sum_{t=2}^{T}(\varepsilon_t - \mu - \alpha(\varepsilon_{t-1} - \mu))^2\right\}\right] \tag{7.19}$$

where T is the number of time steps in the simulation. If the number of time steps is large then the power $T/2$ in this equation can give rise to some numerical difficulties in evaluation of the full likelihood function. The result of $T/2$ being very large is to greatly accentuate the peak values of likelihood in the parameter space. This is, of course, an advantage if an optimum is being sought, since essentially only the simulations having the minimum variance will survive an operation that may involve the power of $T/2$ having a value of hundreds or even thousands for long simulation periods. Thus, the concept of an optimum parameter set necessarily survives in this framework, and in finding the maximum likelihood solution the calculations can be carried out in log space so that $T/2$ becomes a multiplier and the numerical problems of using such large powers are avoided. The optimum of the log likelihood will be at the same point as the optimum of the likelihood.

Note that if the quantities Q_t and $\hat{Q}_t(\underline{\Theta}, \underline{Y})$ are taken to be log transformed observations and model outputs respectively, then this same model can be used for multiplicative errors, which allows for the effects of an error variance that changes with the magnitude of the observation (one possible heteroscedastic case). Note also that for any parameter set, having calculated the time series of errors between observed and predicted variables, the form of the error model should be tested to see if it is consistent with the assumed structure. In particular, it should be checked that the series of δ_t is indeed N[0,1] and uncorrelated in time. Such checks are not always carried out, even for an optimized model.

This likelihood measure has introduced (at least) three additional parameters, μ, σ and α, to take account of structure in the model errors. The cost is thus three additional dimensions over which the likelihood function should be evaluated; the benefit is that, as with the HMLE criterion, the peaks in the response surface will generally be better defined if the errors do indeed conform to the assumptions. This can be an advantage in optimization but can also have the effect of greatly exaggerating small differences in the fit of models that might be considered behavioural on the basis of, for example, the Nash–Sutcliffe efficiency measure. Again, underlying traditional likelihood functions is the idea that there is a true model and that it is the likelihood of the data given the model, $L(Q_t \mid \underline{\Theta}, \underline{Y}, \Phi)$, not the likelihood of the model given the observations, which would be written $L(\underline{\Theta} \mid Q_t, \underline{Y}, \Phi)$, that is being evaluated. The latter is more appropriate in the rainfall–runoff modelling case. The same measure can be used (if the error structure is appropriate) but it is clear that in many applications such a measure will lead to overconditioning towards the apparent optimum model.

Fuzzy measures for model evaluation

There have been some attempts to take a different approach to model evaluation based on non-statistical measures, partly because of some of the problems mentioned above. One approach is that of using fuzzy measures (e.g. Franks *et al.* 1998; Aronica *et al.* 1998). It has been used particularly in situations where observational data are scarce and statistical measures might be difficult to evaluate. The basic fuzzy measure can be defined as a simple function of the error between observed and predicted variables, such as discharge. If the error is zero (or within a certain range of zero) then the fuzzy measure is assumed to be at a maximum, say unity. The measure then declines to zero as the error gets larger in some defined way. A linear decline towards zero at some maximum allowable error is often assumed. The maximum allowable error might be different for overprediction compared with underprediction. It might also be different for different time steps or variables. The individual fuzzy measures for different time steps or variables may then be combined in some way (linear addition, weighted addition, multiplication, fuzzy union or fuzzy intersection; see Box 7.2) to create an overall measure for model evaluation. Such measures can be treated as likelihood measures within the GLUE methodology. They have no foundation in statistical theory but they can be used to express likelihood as a measure of belief in the predictions of a model.

Qualitative measures for model evaluation

Model evaluation need not be based solely on quantitative measures of model fit to observations. Some qualitative measures might also be useful. An example would be in evaluating the predictions of a distributed model. The distributed model might, with different parameter sets, be able to obtain good fits to observed discharges using different runoff generation mechanisms, e.g. infiltration excess runoff alone, saturation excess runoff alone, subsurface stormflow alone, or a mixture of these. If, however, observations suggest that there is negligible overland flow generated on the hillslopes on a catchment, then models that give good simulations based on such mechanisms should be rejected, even though those models might actually give the best simulations. An example is the application of an infiltration excess runoff generation, kinematic wave routing model to the Hubbard Brook catchment in New Hampshire reported in Loague and Freeze (1985). This catchment is forested, has high infiltration capacity soils, and does not produce infiltration excess overland flow (except perhaps in the most extreme events), suggesting that all the models based on this approach should be rejected, regardless of the fact that they might produce good simulations (in fact, the modelling efficiencies using prior estimates of parameter values were −0.32 for storm runoff volumes and −0.18 for peak discharges calculated over 26 storms). Most qualitative measures of performance can be set up as either a binary likelihood measure (1 if behavioural according to the qualitative measure; 0 if not) or as a fuzzy measure. They may be inherently subjective, in that different modellers might choose to emphasize different aspects of the behaviour in making such a qualitative assessment (see Houghton-Carr 1999).

Box 7.2 Combining Likelihood Measures

The need to combine different likelihood measures arises in a number of different circumstances, including the following:

1. combining likelihood measures for different types of model evaluation (such as one measure calculated for the prediction of discharges and one calculated for the prediction of water table or soil moisture levels);
2. updating an existing likelihood estimate with a new measure calculated for the prediction of a new set of observations.

Most cases can be expressed as successive combinations of likelihoods, where a prior likelihood estimate is updated using a new likelihood measure to form a posterior likelihood. This is demonstrated by one form of combination using *Bayes equation* which may be expressed in the following form:

$$L_p(\underline{\Theta} \mid \underline{Y}) = \frac{L_o(\underline{\Theta})L(\underline{\Theta} \mid \underline{Y})}{C} \tag{7.20}$$

where $L_o(\underline{\Theta})$ is the prior likelihood of parameters $\underline{\Theta}$, $L(\underline{\Theta} \mid \underline{Y})$ is the likelihood calculated for the current evaluation given the set of observations $\underline{Y}$, $L_p(\underline{\Theta} \mid \underline{Y})$ is the posterior likelihood, and C is a scaling constant to ensure that the cumulative posterior likelihood is unity. In the GLUE procedure this type of combination is used for the likelihoods associated with individual parameter sets so that for the ith parameter set,

$$L_p(\underline{\Theta}_i \mid \underline{Y}) = \frac{L_o(\underline{\Theta}_i)L(\underline{\Theta}_i \mid \underline{Y})}{C} \tag{7.21}$$

and C is taken over all parameter sets. Strictly, in the theory of Bayesian statistics, this form should be used only where a sampled likelihood (for a particular parameter set) can be assumed to be independent of other samples. In GLUE the parameter sets are chosen randomly so as to be independent samples of the parameter space.

The use of Bayes equation to combine likelihoods has a number of characteristics that may, or may not, be attractive in model evaluation. Since it is a multiplicative operation, if any evaluation results in a zero likelihood, the posterior likelihood will be zero regardless of how well the model has performed previously. This may be considered as an important way of rejecting non-behavioural models: it may cause a re-evaluation of the data for that period or variable, it may lead to the rejection of all models.

The successive application of Bayes equation will tend to gradually reduce the impact of earlier data and likelihoods relative to later evaluations. This may be an advantage if it is thought that the system is changing over time. It has the disadvantage that if a period of calibration is broken down into smaller periods, and the likelihoods are evaluated and updated for each period in turn, for many likelihood measures the final likelihoods will be different from using a single evaluation for the whole period. An exception is a likelihood measure based on an exponential transformation of an error measure E since for $j = 1, 2 \ldots m$ periods with $L_j = \exp(E_j)$:

$$L_p(\underline{\Theta}_i \mid \underline{Y}) = \exp(E_1)\exp(E_2)\ldots\exp(E_m) = \exp(E_1 + E_2 + \cdots E_m) \tag{7.22}$$

However, Bayes equation is not the only way of combining likelihoods. A simple weighted addition might be considered more appropriate in some cases such that for m different measures,

$$L_p(\underline{\Theta}_i \mid \underline{Y}) = \frac{W_o L_o(\underline{\Theta}_i) + W_1 L(\underline{\Theta}_i \mid \underline{Y}_1) + \cdots W_m L(\underline{\Theta}_i \mid \underline{Y}_m)}{C} \quad (7.23)$$

where the W_o to W_m are weights and the $\underline{Y}_1$ to $\underline{Y}_m$ are different evaluation data sets. A weighted addition of this form will have the effect of averaging over any zero likelihood values for any individual periods. Again C is calculated to ensure that the cumulative posterior likelihoods sum to 1.

Further forms of combination come from fuzzy set theory. Fuzzy operations might be considered appropriate for the type of fuzzy likelihood measures introduced in Box 7.1. A fuzzy union of several measures is effectively the maximum value of any of the measures, so that

$$L_p(\underline{\Theta}_i \mid \underline{Y}) = \frac{L_o(\underline{\Theta}_i) \cap L_1(\underline{\Theta}_i \mid \underline{Y}) \cap \ldots L_1(\underline{\Theta}_i \mid \underline{Y})}{C}$$

$$= \frac{\max(L_o(\underline{\Theta}_i), L_1(\underline{\Theta}_i \mid \underline{Y}), \ldots L_1(\underline{\Theta}_i \mid \underline{Y}))}{C}$$

The fuzzy intersection of a set of measures is the minimum value of any of the measures:

$$L_p(\underline{\Theta}_i \mid \underline{Y}) = \frac{L_o(\underline{\Theta}_i) \cup L_1(\underline{\Theta}_i \mid \underline{Y}) \cup \ldots L_1(\underline{\Theta}_i \mid \underline{Y})}{C}$$

$$= \frac{\min(L_o(\underline{\Theta}_i), L_1(\underline{\Theta}_i \mid \underline{Y}), \ldots L_1(\underline{\Theta}_i \mid \underline{Y}))}{C}$$

Thus, taking a fuzzy union will emphasize the best performance of each model or parameter set over all the measures considered, and taking a fuzzy intersection will emphasize the worst performance. In particular, if any of the measures is zero, taking a fuzzy intersection will lead to the rejection of that model as non-behavioural. All of these possibilities for combining likelihood measures are included in the GLUE software (see Appendix A).

8
Predicting Floods

Flood hydrology at the current time draws both on a microscale approach based on continuum mechanics and a macroscale approach based on the statistical study of large aggregates. Neither approach is entirely appropriate to catchment hydrology, which involves systems intermediate in size between the local scale of hydrologic physics and the global scale of a major geographical region.

Jim Dooge, 1986

There are two types of flood prediction required in hydrology. One is the prediction of flood discharges and extent of inundation during an event, particularly for decisions as to whether flood warnings should be issued. This type of 'real time' prediction is often called flood forecasting. In small catchments flood forecasting is primarily a problem of rainfall–runoff modelling, involving radar rainfall or telemetering raingauges sending data back to a flood operations office in real time for use with a forecasting model. In large catchments both rainfall–runoff modelling and hydraulic modelling of the channels may be involved: the former to determine how much water will contribute to the flood wave; the latter to allow predictions of inundation of the flood plain and flooding of property during the event. Methods for real-time prediction of flood discharges and inundation are dealt with in the first five sections of this chapter.

The second type of prediction required is the frequency of occurrence of floods of different magnitudes. The larger the flood, the lower the probability of exceedence that an event of that magnitude or bigger will occur in any one year. A lower frequency of occurrence may be represented as a longer 'return period' or 'recurrence interval', so that we expect the flood that occurs on average once in 100 years to be bigger than that with a return period of 50 years. These return periods are related to the inverse of the probabilities of exceedence. A 100-year event, for example, has a probability of exceedence of 0.01 in any single year; the 50-year event a probability of exceedence of 0.02. Defining an extreme event by its return period is a fairly natural way of expressing the probability of exceedence but it is important to understand properly the meaning of the return period. It is the expected average length of time between occurrences of an event of a given magnitude. Estimates of return periods are therefore very difficult to check with the lengths of record normally available at gauged sites.

Even a 10-year return period event would ideally require a century or more of flood data to obtain a robust estimate of peak magnitude, during which time the characteristics of most catchments have tended to change in a way that might have an effect on the flood frequency distribution. The frequency characteristics of extreme rainfalls are also known to have changed in the last 100 years or more, due to climatic fluctuations and possibly also climate change. Given such potential for significant change, what is really required is an estimate of the probability of a flood of a given magnitude under current (or future changed) conditions, and the expression of this probability in terms of a return period is somewhat misleading.

Thus flood frequency estimation is a very difficult problem. There are a variety of methods available to tackle the problem, including statistical estimation based on the sample of measured floods at a site, regionalization methods for catchments with no data and methods based on rainfall–runoff modelling. A full treatment of this interesting topic is beyond the scope of this book; we will look only at the rainfall–runoff modelling approach here – an approach that is currently attracting some interest.

There is speculation that in many areas of the world, extreme events will become more extreme as a result of both climate and land use changes. The argument follows the line that a warming of the atmosphere will cause more active circulation, leading to more extreme rainfalls, while deforestation and urbanization and other land use changes will tend to increase runoff coefficients. Additionally, sea level rise associated with melting of the polar ice caps and a rise in sea temperatures may make coastal and estuarine areas more vulnerable to flooding. If this does prove to be the case, the process of flood warning will become more and more important.

Historically, people tended to avoid frequently inundated flood plains as places to live. With increasing populations and increasing urbanization, however, there are more and more homes and commercial buildings in areas that are part of the natural flood plain. To some extent, flood protection works such as dams and dykes or levées can be used to reduce the probability of inundation and damage, and have often had the effect of encouraging flood plain development, but such works will not normally protect against the most extreme floods. Certainly the number of recorded floods with significant damage and the estimated annual costs of flood damage continue to rise, making floods one of the most costly natural hazards in terms of property damage and loss of life (see Smith and Ward 1998). People and businesses are generally resistant or unable for financial reasons to move from flood plain areas (although there have been examples of communities relocating businesses and residents from flood plain areas with Federal financial assistance following the Mississippi flood of 1993), so accurate and timely estimates of flooding will be a very practical application of rainfall–runoff modelling in the future.

8.1 Data Requirements for Real-Time Prediction

One of the most important ways of mitigating the costs of flood damage is the provision of adequate warnings, allowing people to act to protect their property and themselves. The responsibility for flood warnings varies between different countries. In most cases the system is based on local flood warning offices that become operational as soon as a potential flood-forming rainfall is forecast. The offices will then use rainfall–runoff modelling in real time to predict the likely flood discharges and stage in different areas as a basis for decisions about whether to issue flood warnings.

This process is very much easier in large catchments where the build-up of a flood, and the transmission of the flood wave downstream, can take days or even weeks. Recent examples are the Mississippi flood of 1993 (Chagnon 1996); the Meuse and Rhine floods of 1995 (Koopmans *et al.* 1995); the Northern California floods of 1997; the Oder floods in the Czech Republic, Poland and Germany in 1997 (Kundzewicz *et al.* 1999); the Red River flood of 1997 in the northern US and Canada; the Ch'ang Chiang (Yangtse) floods in China in 1998 and 1999; and the devastating floods on the Limpopo and Save rivers in Mozambique in 2000.

In small catchments, with short times to peak, real-time flood forecasting is much more difficult. The most extreme discharges in such catchments tend to occur as a result of localized convective rainfalls or high-intensity cells within larger synoptic weather systems. Even if there are telemetering raingauges or a rainfall radar monitoring the area, the response times may be too short to issue warnings in real time. The only option is then to issue warnings on the basis of the forecast rainfalls, but such warnings tend to be very general. The potential for a flood-producing rainfall might be recognized but it may be very difficult to specify exactly where. A good example of this occurs in southern France and northern Italy where, particularly in September, weather systems moving inland from the Mediterranean and subjected to orographic rise can give rise to very intense rainfalls and floods. There is at least a minor flood somewhere in this broad region virtually every year. Well-known examples are the 1988 flood in Nice and the 1992 flood on the 560 km^2 Ouvèze catchment at Vaison-la-Romaine in France.

In the case of Vaison-la-Romaine, the French Bureau de Météorologie issued a warning of the potential for flash flooding for the region. It is difficult in such a case for individual communities to react, since the rain may not fall on the catchment that would affect them. In Vaison, 179 mm were recorded in 24 hours, with higher amounts elsewhere on the catchment and intensities of up to 200 $mm\,h^{-1}$ during 6 minute periods. The time between the start of the rainfall and the peak of the flood was just 3.5 hours. The estimated peak discharge was 600–1100 $m^3\,s^{-1}$, with a peak stage in the town of Vaison of 21 m. The ancient Roman bridge in the centre of Vaison was overtopped but survived the flood, although there is impressive video footage of floating caravans being crushed beneath it. The majority of the fatalities in this event were tourists at a campsite situated in the valley bottom upstream of Vaison. There was much discussion after this event about whether development on the flood plain upstream of the town had made the depth of flooding and the impact of the flood much worse (Arnaud-Fassetta *et al.* 1993). What is clear is that many lives could have been saved if an accurate flood warning had been possible.

To make adequate warnings requires knowledge of the rainfalls as they occur, or, even better, accurate forecasts of potential rainfall intensities ahead of time. This would allow an increase in the forecast *lead time*, which might be important for small catchments and flash floods, such as that at Vaison-la-Romaine. A number of methods are currently under development for projecting a sequence of raingauge or radar rainfall data into the future (e.g. French and Krajewski 1994; Andrieu *et al.* 1996; Dolciné *et al.* 1998) but, whatever the method used, the forecasts tend to degrade very rapidly. In the future, there will be the technical possibility of using a fine-scale numerical weather forecasting model, nested within the grid of a larger circulation model, but accurate forecasts would also require improvements in the rainfall formation conceptualizations used in the current generation of models. As yet, the rainfall forecasts provided by mesoscale models are not sufficiently accurate. At the current time most flood warning systems are dependent on data from telemetering raingauges or radar rainfalls, which are transmitted back to the forecasting centre in real time.

The advantage of weather radar in these situations is that the radar will often pick up the most intense cells of rainfall in the weather system (e.g. Smith *et al.* 1996; see Section 3.1). Such cells may be smaller than the spacing between telemetering raingauges, and might therefore be missed by the ground-based systems. The problem with radar, as an input to predicting the resulting flood discharges, is that the relationship between the radar signal and the rainfall intensity may not always give an accurate estimate of the absolute intensity (see Section 3.1), particularly when there are attenuation effects due to heavy rainfall close to the radar masking the signal from further afield. Thus, for both radar and raingauge systems, it may be possible to recognize that intense rainfall is occurring over a catchment area but not exactly how intense.

This makes it important that any rainfall–runoff model used be capable of real-time adaptation to take account of any errors in the forecasts resulting either from errors in the inputs, whether from radar or rainfall, or from error in the model structure. This requires, however, that the flood warning centre also receives information about river stages in real time, at one or more gauging stations in a catchment, so that model forecasts can be compared with observed stages or discharges in real time and the model adapted to produce more accurate forecasts (at least until the gauge is washed away or the telemetering system fails). Some ways of doing this are discussed in the next section.

There is also another reason why discharge information might be useful in flood warning, particularly in larger catchments. Where the time delays in the channel system are sufficiently long compared with the required lead time for a forecast (in general 3–6 hours would be the minimum feasible lead time to allow a warning to be transmitted to the public), then a measurement of the discharge or stage upstream can be used as part of the system for predicting the stage and discharge and timing of the flood peak further downstream.

In general, flood warnings are issued in relation to the forecast stage of the river at a critical gauging point without modelling the detailed pattern of inundation upstream of that point. In many situations this may be adequate, since if flooding is predicted to occur somewhere in the flood plain, then a general warning can be issued. In large rivers, however, such as the Mississippi, the progress of the flood wave downstream may be very much controlled by the pattern of inundation during the flood, including the effects of embankment failures which are inherently difficult to predict ahead of time. Thus it may be necessary to use a hydraulic routing model in forecasting the expected depths downstream, continually revising the calculations as conditions change. This adds the additional requirement of knowing the channel and flood plain topography for use in the hydraulic model, together with parameters such as effective resistance coefficients. Topography is normally provided as a series of cross-sectional profiles surveyed across the flood plain and channel at different sites, but the increasing use of two-dimensional depth-integrated models will lead to the use of topographic data in the form of detailed digital elevation maps of the flood plain. Channel form can of course change during a flood event due to erosion and deposition. Models of sediment transport in rivers have not advanced to the stage where they can be used operationally and most current hydraulic flood routing models use a 'fixed bed' assumption.

8.2 Rainfall–Runoff Modelling for Flood Forecasting

Any rainfall–runoff model that has been calibrated for a particular catchment can be used in the prediction of flood discharges. The US National Weather Service River Forecast System (Burnash 1995), for example, is a development of the Sacramento model, a form of lumped explicit soil moisture accounting model with many parameters to be calibrated (e.g. Sorooshian *et al.* 1992; Gupta *et al.* 1999). Using methods such as those discussed in Chapter 7, the predictions may also be associated with an estimate of uncertainty in the predictions. Qualifying the estimates in this way may be important. Experience suggests that uncertainty in both measurements and predictions of flood peaks increases with peak magnitudes. In addition, even if a model has been calibrated for a certain range of discharges, uncertainty is bound to increase as predictions are made outside this calibration range for extreme events. Thus, it may not be possible to predict definitively whether a flood stage will be exceeded in an event ahead of time; it may, however, be possible to assess the risk that the flood stage will be exceeded by consideration of the distribution of (uncertain) predictions.

As noted earlier, the propensity for error in the predictions during extreme events also suggests that it would be advantageous to use an adaptive modelling strategy, so that if a comparison of observed and predicted discharges reveals that the model predictions are in error, then a strategy for adjusting the model predictions can be implemented. This is clearly only possible where discharge or river stage measurements can be made available in real time. Adaptation is also more easily implemented for simpler models. In complex models, with many different variables or components that could be adjusted, it might be difficult to decide what to adjust.

A comparison of real-time forecasting methods, including adaptive schemes, was carried out by the World Meteorological Organisation (WMO 1975) and a recent review of approaches has been given by Moore (1999). It was not clear from the WMO study whether using more complex models will have much advantage over using a greatly simplified approach to real-time modelling, since the real-time adaptation can account for much of the potential error in using a simplified model structure. In the real-time forecasting situation, the hydrologist is not necessarily concerned with getting a correct prediction of the flow processes upstream of a flood-prone area, but is very concerned about getting a good prediction of the peak flood stage in that area as many hours ahead as possible.

It is also possible to take an approach that makes no attempt to model runoff generation during flood forecasting at all. As noted in Section 4.9, neural net models have recently become popular as a means of estimating N-step ahead of flood discharges, using inputs that include rainfalls and previous values of discharge or water levels and a training set of historical events (see Figure 4.9). In discussing neural network approaches, however, it has already been noted that predictions made for events more extreme than those included in the training set may not be accurately estimated.

We will consider two simple model strategies for real-time forecasting here. The first is an adaptive deterministic method due to Lambert (1972) which is very difficult to beat in small to moderate catchments. This input–storage–output (ISO) model has been used in a number of UK flood forecasting schemes, particularly in the River Dee catchment in north Wales. The second is an adaptive form of the transfer function models considered in Section 4.3. Adaptive transfer functions can be used for both rainfall–runoff and upstream discharge (or stage) to downstream discharge (or stage), depending on what data are available. The example application presented in the case study of Section 8.5 is for an operational forecasting model for the town of Dumfries in Scotland, which uses both rainfall–flow and discharge–discharge transfer functions. Adaptation of such models can be implemented in a number of different ways. In the Dumfries model, a simple adaptive gain parameter is used, i.e. the transfer function is scaled up or down in real time without changing its form. This simple approach has proven very effective in this and other applications.

These types of models can be made part of a larger flood forecasting system that includes flood routing components. For the River Dee catchment, for example, ISO models were developed for all the gauged subcatchments and linked to a flood routing model. A general package incorporating a variety of rainfall–runoff and flood routing models has recently been implemented for the whole of the Yorkshire Region of the UK Environment Agency (see Moore *et al.* 1994).

8.3 The Lambert ISO Model

The idea behind Alan Lambert's ISO model is wonderfully simple. It is based on the development of a master recession curve for a catchment or subcatchment where a period of discharge measurements are available, by piecing together partial recession curves from individual events. In general, the shape of such a recession curve may not be easily represented by a simple mathematical function but Lambert suggested using simple linear and logarithmic functions to represent different parts of the discharge range (Lambert 1969, 1972). Then, at each time step during an event, average rainfall (less an estimate of evapotranspiration if necessary) is added to the relative catchment storage (the absolute storage does not need to be estimated), and using the representation of the storage–discharge function an incremental change in discharge is easily predicted. The discharge is subtracted from the storage, and the model is ready for the next time step. This is about as simple a rainfall–runoff model as is possible. Calibration of the model is simply a matter of deriving the master storage–discharge curve for as wide a range of discharges as possible. There are effectively no other parameters.

In continuous simulation, such a model would not be very accurate since we know that the relationship between storage and discharge is not simple. This is why we use more complex models that attempt to reflect the complexities of the rainfall–runoff processes more realistically. In real-time forecasting, however, the ISO model can be used in a way that reduces the impact of the errors inherent in using such a simple model. Its first advantage is in initializing the model at the start of an event. In general, the best index of the antecedent wetness of a catchment area is the discharge at the start of an event. The ISO model is easily initialized if the discharge is known at the first time step, since this can be used to infer the initial relative storage. The second advantage is that as soon as errors are detected between the observed and predicted discharges, the model can be reinitialized using the current measured discharge. This procedure can be implemented at every time step as soon as the measured discharges are received, and used to update the predictions into the future.

Thus this is also the very simplest possible adaptive modelling scheme. No complex mathematics is required for the adaptation, and it is easily understood, easily calibrated and easily implemented. It can be a very effective real-time forecasting scheme but clearly has some limitations. In particular, extrapolation outside the range of measured recession curves is uncertain. In the model, the relationships used for the upper and lower sections of the master recession curves are simply assumed to continue for more extreme conditions. In addition, adaptive use of the model to change the current relative storage invalidates any overall water balance for the model, but this will not be important in real-time forecasting if it leads to improved predictions.

In small catchments this is effectively the model to beat for real-time forecasting. More complex models that are less easy to adapt may not necessarily produce better real-time flood forecasts.

8.4 Adaptive Transfer Function Models for Real-Time Forecasting

It is worth noting that the simplest linear ISO model element is effectively a first-order transfer function model, equivalent to equation (4.9). The difference is that the inputs used with the ISO model are the rainfalls, whereas the transfer function models of Section 4.3 used the rainfalls filtered in some way to produce an effective rainfall. For a filter that is directly and only dependent on the current discharge, such as that used by Young and Beven (1994), simple ISO model type updating can still be used directly, but this will not be possible where effective rainfall components introduce additional storage elements. There is, however, a simple way of making such models adaptive, as noted above, by using an adaptive gain or multiplier. The best initial estimate of the gain parameter will normally be 1.0 but it is then allowed to vary as the event progresses to correct for any differences detected between the forecasts and the observations supplied to the flood warning system. If an underprediction is detected, the gain can be increased for the next time step; if an overprediction is detected, the gain can be reduced. The changes in the gain are filtered so that the changes from time step to time step stay relatively smooth. The adaptive gain approach is a simple way of compensating for any errors in the data or transfer function model structure that might affect the accuracy of the forecasts. It will generally lead to greatly improved forecasts. One example of a successful adaptive algorithm is given in Box 8.1.

There is one important limitation of transfer function models in flood forecasting applications. Transfer functions are designed to make predictions of output variables for which historical data are available for calibration. They are an empirical modelling strategy in this sense. Thus, although they can provide predictions, easily updated in real time, of river stage at specific sites in the catchment, they cannot predict the extent of flooding in the catchment, except in so far as this is possible from knowledge of the flood stage at the measurement sites. Quantitative prediction of flood extent would require a flood inundation model that provides distributed predictions, but such models have a great number of variables and parameters and are not easily updated in real time.

One strategy for combining transfer function models with distributed flood inundation models is considered in Section 8.6.

8.5 Case Study: A Real-Time Forecasting System for the Town of Dumfries

The adaptive algorithm outlined in Box 8.1 will be demonstrated by the flood forecasting system for the catchment of the River Nith (750 km^2) upstream of the town of Dumfries in Scotland, reported by Lees *et al.* (1994) (Figure 8.1). This application involved providing both rainfall–runoff models and flow routing models for the main channel, with data provided in real time from a number of rainfall and river stage measurement sites. The lower reach of the main channel also required that the state of the tide be taken into account in the flood forecasting procedure. Even though this is a fast-responding catchment, at least a 5-hour lead time for the forecasts at Dumfries was required to allow warnings to be issued in time for the public to respond. At the time of implementation in 1991, one of the requirements of the system was that if a flood stage was predicted, the software should automatically telephone the local police station with an appropriate computer-generated message. The police had responsibility

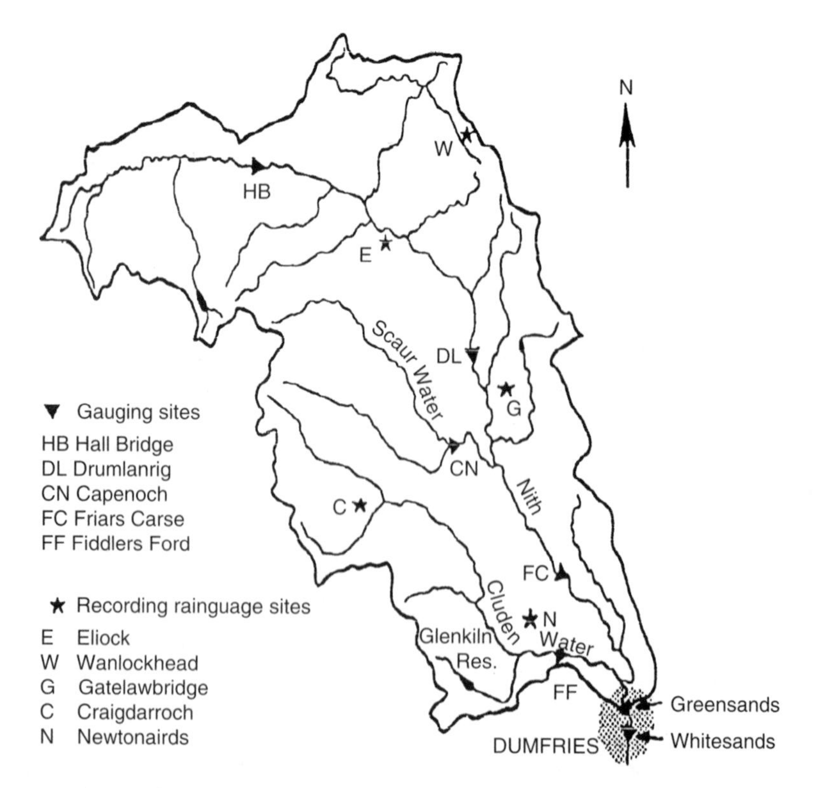

Figure 8.1 *The catchment of the River Nith above the town of Dumfries, Scotland (after Lees et al. 1994). Reproduced with permission of John Wiley & Sons Limited*

for distributing flood warnings to the public. It was important, therefore, that the system should be accurate in its predictions!

The models used in the Dumfries system were similar to the type of data-based mechanistic transfer function models described in Chapter 4, but with some additional features specific to the flood forecasting situation. Transfer functions were fitted both for rainfall–river flow modelling, and for flow routing in the main channel. The rainfall–flow model required a rainfall filtering which made use of the bilinear filter described in Section 4.3.2, so that effective rainfall at time t is proportional to $Q_t^n R_t$ where Q_t is the current discharge, R_t is the current rainfall and n is a parameter. The flow routing used stage-to-stage transfer functions for the upper, middle and lower parts of the mainstream channel, since it is stage that is the primary variable of interest in the flood forecasting case. All the transfer function models were implemented in such a way that if there were telemetry data available on-line for any of the prediction sites, these data could be used to update the model predictions in real time. This was achieved by introducing a time-variable gain parameter into the transfer function. The transfer functions calibrated for historical data were typically first-order models, with one or two b parameters (see Box 4.1), of the form

$$\widehat{y}_t = \frac{b_o + b_1 z^{-1}}{1 - a_1 z^{-1}} u_{t-\delta} \tag{8.1}$$

where $\widehat{y}_t$ is the predicted output variable (here river stage) at time t, $u_{t-\delta}$ is the input variable (upstream stage or effective rainfall) delayed by a time delay δ, z is the backward difference operator (see Box 4.1), and b_o, b_1 and a_1 are the calibrated parameters of the transfer function. The time-variable gain is introduced in the form

$$\widehat{y}_t = G_t \frac{b_o + b_1 z^{-1}}{1 - a_1 z^{-1}} u_{t-\delta} \tag{8.2}$$

where G_t is the gain at time t. Thus the best estimate of the gain will be a value of 1.0 but allowing it to be time variable is a way of compensating for any errors in the original calibrated model during an event. The time-variable gain is estimated in real time using a recursive least-squares algorithm with the parameter variation modelled as a random walk process with a long estimator memory (see Box 8.1). This damps down changes in the gain parameter from time step to time step but allows the model to adapt gradually to any drift of the observations away from the predictions of the original transfer function. The recursive algorithm also allows the variance of the forecasts to be estimated over time, so that if larger errors are detected, this can be reflected in estimates of uncertainty in the forecasts.

The resulting model, with its real-time adaptation, has proven to be very successful in predicting peak flood stages. Some example predictions for two large floods of the order of 450 $m^3 s^{-1}$, both of which caused flooding in the centre of Dumfries, are demonstrated in Figure 8.2, which shows both the simple transfer function forecasts and the adaptive forecasts. In both cases the adaptive forecast predicted the peak stage to within 5 cm, with only small errors in timing.

Downstream of the Greensands site the River Nith becomes tidal. Combined high river discharges and high tidal stages can greatly enhance the probability and impact of flooding in the town. Thus an additional component was added to the model to predict

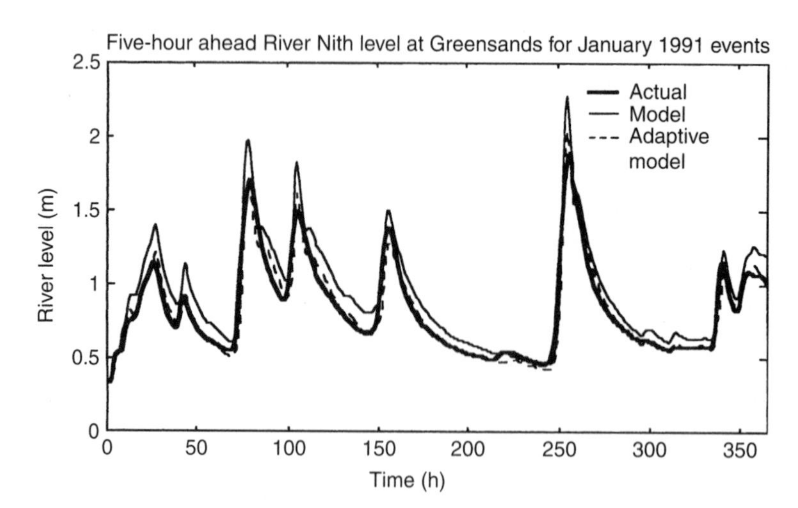

Figure 8.2 *Comparison of adaptive and non-adaptive 5-hour ahead flood forecasts on the River Nith (after Lees et al. 1994). Reproduced with permission of John Wiley & Sons Limited*

the effects of the tides on the water levels in the River Nith, also using a transfer function approach. More details may be found in the article by Lees *et al.* (1994).

The natural delays in the Nith system are less than the 5-hour lead time demanded. It was found that the lead time of the forecasts could be increased by using artificial delays in the transfer functions. Thus, the transfer functions derived from the historical data were calibrated with a delay longer than the natural time delays in the catchment. It was found that this gave less good fits in calibration, but better predictions in forecasting ahead of time, with lead times of the same order as the artificial time delay. The improvement was reinforced by the real-time adaptation of the gain parameter described above. This is a very simple method of increasing the effective lead time of forecasts that may not be useful everywhere (particularly where the rising limb of the hydrograph is steep) but the value of such an approach, relative to trying to forecast rainfalls ahead of time, can be checked for historical data sets.

8.6 Methods for Flood Inundation in Real Time

In rivers with extensive flood plains (generally catchments of areas 250 km^2 and larger) the timing and magnitude of a flood peak will be greatly affected by over-bank storage on the flood plain. Successful flood forecasting will then depend on either lumped (using transfer functions) or distributed (using hydraulic models) flood routing and flood inundation prediction (e.g. Bradley *et al.* 1995). It has been shown above how the transfer function approach can easily be made adaptive to improve real-time forecasts. This approach cannot, however, give any direct indication of the actual area that is expected to be flooded during a particular event. The hydraulic routing approach, implemented in either one (downstream) dimension or two (downstream and transverse) dimensions can give at least approximate distributed predictions of depths and areas of inundation as these change dynamically during flood events.

Both one- and two-dimensional approaches use either the full or simplified forms of the St Venant equations (see Box 5.5). As distributed models, with large numbers of parameters (primarily roughness coefficients for different elements representing the main channel and flood plain), these hydraulic models may be difficult to calibrate and implement within an adaptive updating framework. This will be particularly difficult where effective roughness coefficients for overbank flow change with stage (see Knight *et al.* 1994), and where only approximate information is available on the topography of the channel and flood plain system. However correct the hydraulic description incorporated into a flood routing model, accuracy is still going to be very dependent on the topographic description and parameter values used, leading to uncertainty in the predictions of flood plain storage and inundated areas. A recent paper by Romanowicz and Beven (1998) has shown how a simplified two-dimensional hydraulic model can be used within the GLUE framework of Chapter 7 to take account of such uncertainty within the type of conditioning methodology outlined in the last chapter. Model parameter sets are associated with a likelihood measure that is determined on the basis of predictions of downstream discharges and extents of inundated area measured on historical events. This results in a prior likelihood measure associated with each parameter set that can be used to weight the predictions of the models. In addition, however, a fairly simple adaptive step can be added by a further real-time conditioning on the likelihood measure according to how well that particular simulation is reproducing measurements or N-step ahead forecasts of the downstream discharges. The resulting procedure is simple because it does not require that the models be re-run (unless the upstream inputs are changed); it is only the likelihood measures associated with each model that are recalculated. Some results of such an approach, for three-hour ahead inundation forecasts in an application to a 12 km reach of the River Culm in Devon, England, are shown in Figure 8.3. Current research is investigating the possibility of using remotely sensed inundation in calibrating or constraining the predictions of such models (e.g. Bates *et al.* 1997).

8.7 Flood Frequency Prediction Using Rainfall–Runoff Models

In 1972, a seminal paper on the 'Dynamics of Flood Frequency' appeared in *Water Resources Research*, written by Peter Eagleson. In that paper, Eagleson outlined a method of 'derived distributions' for the calculation of flood frequency. Until that time, the primary methodology for estimating flood frequency was (and still is) by fitting a theoretical statistical distribution to available measurements of flood peak discharges, and using the distribution to estimate the frequencies of different magnitudes of events. This statistical approach has a number of limitations. One is that we do not know what the correct distribution should be. Log normal, Wakeby, generalised extreme value generalised logistic, log Pearson type III, and a number of other distributions have been used in the past (for recent discussions, see Vogel *et al.* 1993; IH 1999). Several different distributions might give acceptable fits to the data available, all producing different frequency estimates in extrapolation.

In addition, the calibration may not be very robust. Most discharge gauging sites do not have very long records available (very few are longer than 50 years). They

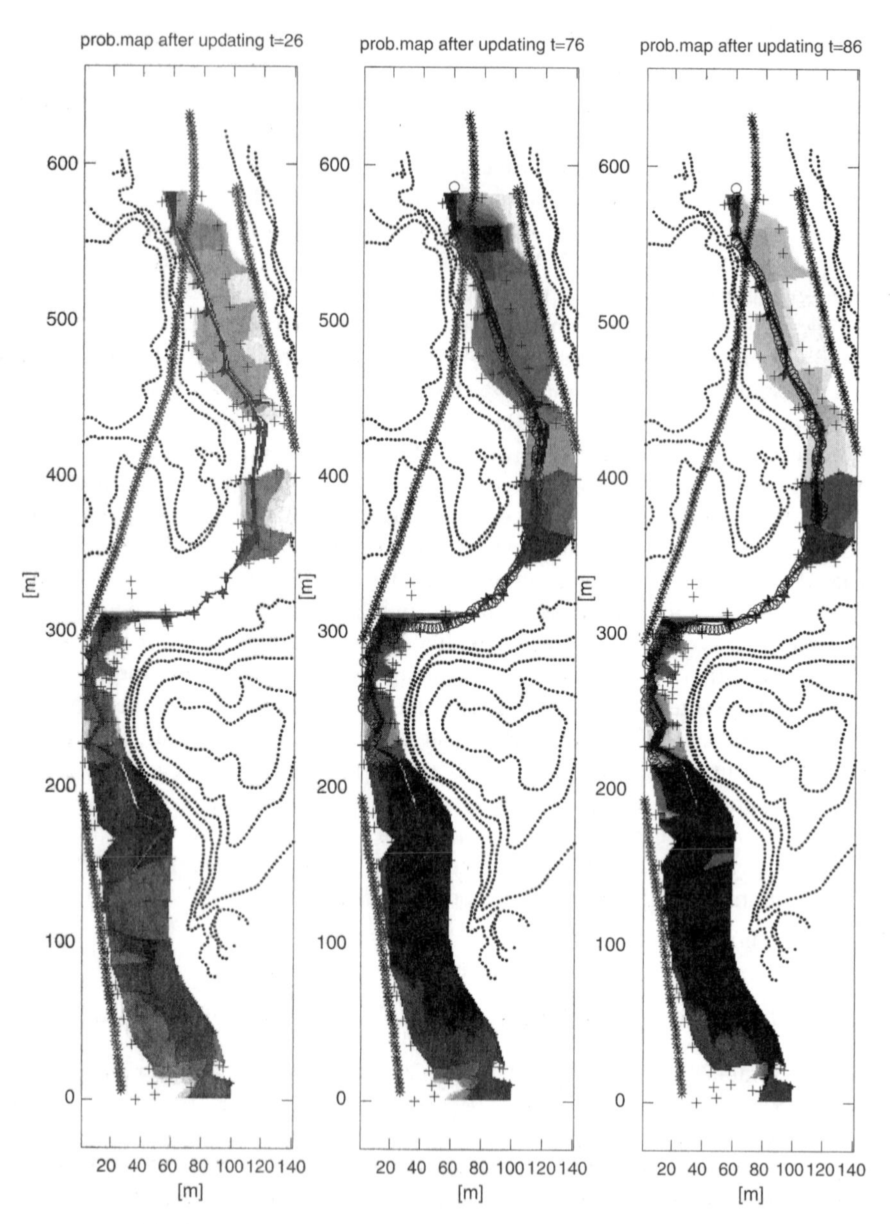

Figure 8.3 *Probabilities of flood inundation on the River Culm, Devon for three time steps during an event in January 1984 (after Romanowicz and Beven 1998). Dotted lines indicate contours, darker shading indicates increased probability of inundation*

therefore represent only a small sample from the possible distribution of floods at the site, so that the fitted distribution might be biased and the resulting frequency estimates uncertain, particularly in extrapolating to long return periods (smaller probabilities of exceedence). It is also known that both the frequencies of flood-producing rainfalls, and the land use characteristics have, in some cases, changed dramatically during the period of historical measurements (e.g. Walsh *et al.* 1982; Arnell 1989). Finally, the statistical distribution will take no explicit account of any changes in the nature of runoff generation processes for higher magnitude events (e.g. Wood *et al.* 1990).

Model structure and parameter estimation are also issues in trying to estimate flood frequencies by rainfall–runoff modelling. Although potentially a model might be able to represent any changes in the hydrological response of a catchment under more extreme conditions, there is no guarantee that a model calibrated by a prior parameter estimation or by calibration against a period of observed discharges will, in fact, produce accurate simulations for extreme flood peaks (e.g. Fontaine 1995). Lamb (1999), for example, has shown how the PDM model, calibrated over a 2-year continuous simulation period, can reproduce the magnitudes of the largest observed flood peaks when driven by a longer period of observed rainfall data, but not necessarily from the same events that produce the recorded peaks.

Peter Eagleson's idea was to look at the flood frequency problem as a transformation of the probability distribution of rainstorms into a probability distribution of flood peaks using a rainfall–runoff model. The advantage of this is that rainfall data are available for many more sites and generally much longer periods than are stream discharge data. The disadvantages are that rainstorms vary in intensity, duration and storm profile, so that if it is necessary to model a sequence of rainstorms, a more complicated stochastic model is required, while the results will be very dependent on the type of rainfall–runoff model and the set of parameter values used. As we have already seen, this will be an important source of uncertainty in the model predictions. Observed rainfalls had been used previously with conceptual rainfall–runoff models to simulate sequences of flood peaks (e.g. Fleming and Franz 1971) and shown to be competitive with statistical methods. Eagleson's initial proposal, however, was not so much concerned with practical prediction for individual sites as with providing a structure for thinking about the problem.

In the method of derived distributions, the probability density function of flood magnitudes is derived from the density functions for climate and catchment variables through a model of catchment response to a rainstorm. Eagleson managed to come up with a set of simplifying assumptions that allowed (in a paper of only 137 equations) the analytical integration of the responses over all possible rainstorms to derive the distribution of flood peak magnitudes. The method has been extended by later workers to include different stochastic characterizations of the rainstorms and different runoff-producing mechanisms at the expense of losing the mathematical tractability that allows analytical solutions. This is not perhaps so important given current computer resources when a simple rainfall–runoff model can easily be driven by stochastic inputs for periods of hundreds or thousands of years to derive a distribution of predicted flood magnitudes numerically. However, Eagleson's (1972) attempt is of interest in that it provided the framework that other later models have followed, i.e. that of starting with a stochastic rainstorm model that is used to drive, in either a probabilistic or a numerical way, a runoff generation model, which might then be linked to a routing model to predict the flood peak at a desired location. More recent developments in modelling flood frequency by continuous simulation are summarized in Table 8.1.

Table 8.1 *Selected studies of flood frequency estimation by rainfall–runoff modelling*

Study	Catchment	Rainstorm Model	Runoff Generation	Routing
Eagleson (1972)	44 US catchments	Eagleson Exponential I,D	Variable contribution area	Analytical KW
Cordova and Rodriguez-Iturbe (1983)	Querecual, Venezuela	Observed	Hortonian infiltration excess	GUH
Diaz-Granados *et al.* (1984)	Santa Paula Creek and Nashua, US	Observed rainfall	Hortonian infiltration excess	GUH
Hebson and Wood (1982)	Bald Eagle Creek, Davidson River, US	Eagleson Exponential I,D	Hortonian infiltration excess	GUH
Beven (1987)	Wye, Wales	Eagleson Exponential I,D,A	TOPMODEL	linear NWF
Calver (1993)	Tanllwyth, Wales; Larochette, Luxembourg	Observed	IHDM	IHDM KW
Troch *et al.* (1994)	Mahantango Creek, US	Observed	TOPMODEL	linear NWF
Calver and Lamb (1996)	10 UK catchments	Observed	PDM & TATE models	NWF and parallel TF
Franchini *et al.* (1996)	mid-west US catchments	Stochastic storm transposition	ARNO	ARNO parabolic
Onof *et al.* (1996)	Thames, UK	Bartlett Lewis	Hortonian infiltration excess	Linear reservoir
Blazkova and Beven (1997)	3 catchments, Czech Republic	Modified Eagleson model	TOPMODEL	linear NWF
Kurothe *et al.* (1997)	Davidson River, US	Gumbel model	Horton infiltration excess	GUH
Robinson and Sivapalan (1997)	Salmon Creek, Australia	Seasonal exponential/power law	Saturation excess/ subsurface	linear TF
Kilsby *et al.* (1998)	Tyne, UK; Cobres, Portugal; Broyce, Switzerland	Generalised Neyman-Scott	SHETRAN & ARNO models	Diffusion wave & parabolic
Hashemi *et al.* (1998)	Brue, UK	Modified Neyman-Scott	ARNO	parabolic
Lamb (1999)	40 UK catchments	Observed (hourly)	PDM	linear TF
Cameron *et al.* (1999)	Wye, UK	GPD extension to observed	TOPMODEL	linear NWF
Steel *et al.* (1999)	11 catchments in Scotland	Observed (daily)	IHACRES	linear TF

I, mean storm intensity; D, storm duration; A, inter-arrival time; GPD, Generalized Pareto distribution; IHACRES, identification of unit hydrographs and component flows from rainfall, evaporation and streamflow data; IHDM, Institute of Hydrology Distributed Model; KW, kinematic wave; GUH, geomorphological unit hydrograph; NWF, network width function; TF, transfer function.

8.7.1 Generating Stochastic Rainstorms

Eagleson's rainstorm model was based on independent distributions for storm duration, and mean storm intensity at a point. Both probability distributions were assumed to be of exponential form. In his study, Eagleson assumes that the storms are independent so that in calculating flood frequencies he only needs to know the average number of storms per year. Other later studies have taken account of the variation in antecedent conditions prior to different events and this requires the introduction of a third distribution of inter-arrival times between storms. To allow the analytical calculations, Eagleson's stochastic rainstorms are of constant intensity over a certain duration but, in using rainfall statistics from individual stations, he did take account of the difference between point rainfall estimates and the reduction in mean intensity to be expected as catchment area increases, using an empirical function derived by the US Weather Bureau.

Later studies that have used direct numerical simulation, either on a storm by storm basis or by continuous simulation, have not been so constrained in their rainstorm model component. Some have used long periods of observed rainfall having calibrated a model against a time series of observed discharges. Beven (1987b) added a rainstorm profile component to the Eagleson model based on the statistics of observed cumulative rainstorm profiles normalized to a unit duration and unit rainstorm volume. This was later adapted by Cameron *et al.* (1999) to use a database of observed normalized profiles from 10 000 real storms, classified by different duration classes. When a new storm profile is required, a normalized profile is chosen randomly from the database. An interesting variation on this technique of using normalized profiles, that will produce rainfall intensities with fractal characteristics, has been suggested by Puente (1997). Another way of using actual storm data in flood frequency analysis is to use the random transposition of real storms recorded in a region over a catchment of interest (Franchini *et al.* 1996a). Finally, a number of rainstorm models have been proposed based on pulse or cell models, where the time series of generated rainfalls is based on the superposition of pulses of randomly chosen intensity and duration by analogy with the growth and decay of cells in real rainfall systems. Thus rainstorms are not generated directly but arise as periods of superimposed cells separated by dry periods. The Bartlett–Lewis (Onof *et al.* 1996) and Neyman–Scott (Cowpertwait *et al.* 1996) models are of this type (see Table 8.1). Such models require a significant number of parameters, especially when the model is fitted to individual months of the year.

8.7.2 The Runoff Generation Model Component

Given a series of rainstorms, the next stage is to model how much of that rainfall becomes streamflow. Eagleson did this by estimating effective rainfalls using a simple Φ-index model (see Section 2.2), with a constant value of Φ, again to allow an analytical solution to the derived distribution of flood peaks. His model did also recognize that runoff might be generated only over part of the catchment, on a contributing area that would vary both between catchments due to different catchment characteristics of soils, vegetation and topography, and within a catchment due to variation in antecedent conditions. He modelled these variations by assuming a probability density function for the contributing area of triangular form, which 'provides the bias towards small fractions of the catchment areas observed by Betson (1964)' (Eagleson 1972, p. 885). Random selection of a contributing area for a storm is a way of effectively introducing the antecedent condition control on runoff generation into the predictions.

Eagleson then routes the effective rainfall produced in this way using kinematic wave routing for both overland and channel flows (see Section 5.5). The runoff generation component produces, for each rainstorm with $i_o > \Phi$, a constant effective rainfall of duration t_r for that storm. This very simple time distribution allows analytical solutions of the kinematic wave equation to be performed. Again, more recent studies using direct numerical simulation have not been so constrained and a variety of runoff generation models and routing methods have been used (see Table 8.1). Those that are based on event by event simulation will, as in the original Eagleson study, require some way of reflecting the effect of varying antecedent conditions on runoff production. Those that use continuous simulation over long periods of time, including the drying of the catchment during inter-storm periods, will account for antecedent conditions directly.

8.8 Case Study: Modelling the Flood Frequency Characteristics of the Wye Catchment, Wales

Few of the studies in Table 8.1 have compared the predictions of models to both predicted hydrographs and flood frequency characteristics at a site. One study which did is that of Beven (1987b) in an application to the 10 km^2 Wye catchment at Plynlimon in mid-Wales. He used a rainfall model very similar to that of Eagleson but a version of TOPMODEL as the rainfall–runoff model. This version allowed runoff generation by three different mechanisms: saturation excess using the topographic index theory developed in Box 6.2; infiltration excess on a spatially variable soil; and purely subsurface runoff generation. The results of this study were interesting in a number of ways. Only 14 years of observations were available for fitting the statistical distributions for the rainfall model and for estimating the observed flood frequencies. The model was run with an hourly time step, and analysis of the observations showed how the *form* of the apparent flood frequency distribution differed, in this small catchment, between hourly average discharges and instantaneous peak discharges. Models using hourly time steps should not be compared with instantaneous peak discharge observations, at least in small catchments. The study showed that, with the parameters used, even in this relatively wet catchment with frequent rainstorms, simulated contributing areas for surface runoff generation very rarely reached 100 percent. The variability in the apparent flood frequency characteristics generated by the model for different realizations (simulations) of 14 years was also demonstrated. The range of model predictions did, however, bracket the observed flood frequency curve derived from the actual 14 years of record.

This was an early attempt to take account of the realization effect of such Monte Carlo simulation techniques. The realization effect has a particular significance in flood frequency studies, especially in comparing model results with observed frequency estimates, since the observations themselves in effect represent only a single realization of the possible sequence of floods that could have occurred over the period of record. This could be important in evaluating different parameter sets within a model, since it is the combination of parameter values and a particular stochastic realization that will give a good fit (or not) to the observed peak magnitudes. This has been examined in a more recent study of the Wye flood data by Cameron *et al.* (1999). Again, TOPMODEL was used as the rainfall–runoff model but a new rainstorm generator was developed using generalized Pareto distributions to extend the upper tails of observed mean storm intensity distributions in different duration classes of storms. Storm profiles were chosen randomly from large samples of normalized observed intensity profiles in each duration class. This rainfall model was shown to give better reproduction of the extreme rainfall statistics than the simple Eagleson exponential distribution model outlined above, at the expense of adding additional parameters.

In this new study, the model predictions were compared with 21 years of hourly discharge observations as well as the annual maximum flood series. The continuous discharge predictions were made with TOPMODEL using observed mean catchment average rainfalls. The frequency predictions were made using 21 year realizations of the rainfall model. Model parameter sets were generated randomly and the resulting simulations evaluated within the GLUE framework of Chapter 7. In this way, the combined realization effect in predicting the flood frequency curve is taken into account. It was shown that parameter sets could be found that reproduced both the hydrograph data and the flood frequency curve (Figure 8.4). However, if the maximum annual floods of the 21 year simulations were compared with observations, the annual flood was not always generated by the same rainstorm while the ranking of the flood peaks was not always consistent with the observed ranking (as in the Lamb (1999) study). This reflects the limitations of current rainfall–runoff modelling. It should not, however, be altogether surprising since, even on a small experimental catchment like the Wye, flood-producing rainfall events can be subject to significant error in measured rainfalls and discharges (see the study of the 1973 and 1977 floods in this catchment by Newson 1980). What is not yet known is whether this source of error is more important than model structural errors.

8.9 Flood Frequency Estimation Including Snowmelt Events

The predictions of flood frequency for the Wye neglected any of the impact of snowmelt-generated floods but in many environments this is the dominant cause of the annual maximum flood, at least for larger catchments. The work of Cameron *et al.* has been extended to include a stochastic temperature model that can be used with the rainstorm model to predict the build-up and melt of the winter snowpack. Snowmelt is predicted using a form of the degree-day method (Box 3.4), and runoff generation using TOPMODEL. In an application to a small catchment in the Czech Republic, predictions from randomly chosen parameter sets and realizations were again evaluated within the GLUE framework, including an evaluation of the frequency distribution of maximum snowpack water equivalents (see Figure 8.5). The data available for model evaluation were, in this case, quite limited so fuzzy measures were used to calculate the weights associated with each model realization. Steel *et al.* (1999) have also estimated flood frequency for a number of Scottish catchments with significant seasonal snowmelt discharge components using the degree-day method of Schreider *et al.* (1997) to augment a daily observed rainfall series in driving the IHACRES rainfall–runoff model.

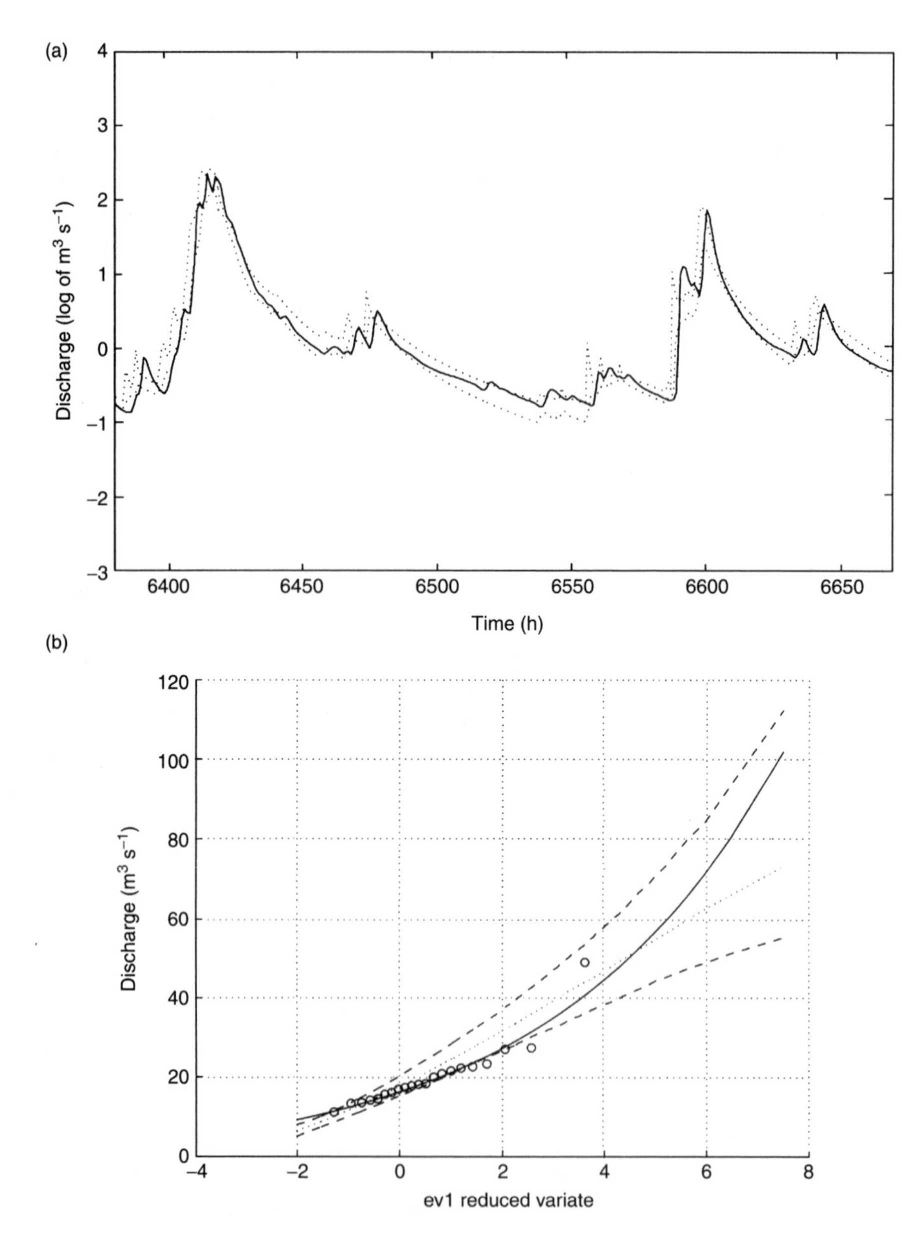

Figure 8.4 *Predictions of (a) flood hydrographs and (b) flood frequency curves for the River Wye catchment (10.6 km^2) at Plynlimon, Wales (after Cameron et al. 1999). Reprinted from Journal of Hydrology 219: 169–187, copyright (1999), with permission from Elsevier Science*

8.10 Hydrological Similarity and Flood Frequency Estimation

Sivapalan *et al.* (1990) produced a scaled flood frequency model based on the TOPMODEL concepts and showed that catchment runoff production could be compared on the basis of eight similarity variables. Their flood frequency curves

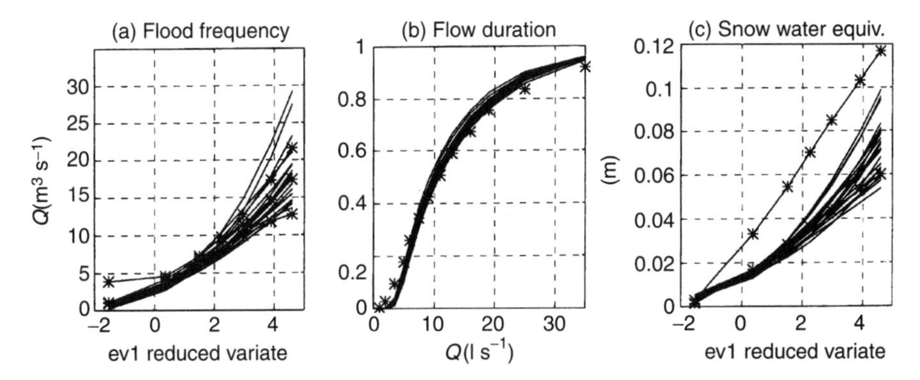

Figure 8.5 *Predictions based on 20 simulations of 100 years duration of (a) cumulative frequency of discharge, (b) flood peaks, and (c) cumulative frequency of maximum snow water equivalent for the ungauged Ryzmburk catchment (3.3 km²), Czech Republic. Lines with stars are regionalized estimates, except for snow water equivalent data which were observed at two sites in the region. Results calculated by Sarka Blazkova*

were derived from storm by storm simulations and showed a transition between saturation excess overland flow dominated flood peaks to infiltration excess overland flow dominated flood peaks for more extreme events. However, they presented only a sensitivity analysis and this similarity theory still has to be tested against real data sets, although a simplified version was shown to have value in differentiating the hydrology of seven small Australian catchments by Robinson and Sivapalan (1995).

8.11 Key Points from Chapter 8

- There are two uses of rainfall–runoff modelling in flood prediction; one in forecasting discharges in real time during flood periods; the other in predicting the frequencies of different flood peak magnitudes.
- Real-time forecasts will be very dependent on the accuracy of input data, particularly of spatial patterns of rainfall intensities. The availability of radar rainfall data has greatly improved the potential for forecasting flood peaks. Improvements in the lead time of forecasts are dependent on improvements in near-future forecasts of rainfalls, which at present are not very reliable.
- Flood forecasting is best carried out using adaptive models that can take account of prediction errors if discharge or stage data are available in real time by telemetry. This is most easily done with relatively simple models, such as the Lambert ISO model for small catchments or transfer function type models.
- Forecasts of flood inundation need at least a simplified distributed hydraulic model of flows in the channel and flood plain, together with detailed topographic data. Such models, having many parameters and variables, may be difficult to make adaptive for real-time forecasting. A way of updating the forecasts of inundation is described, based on the uncertainty estimation techniques of Chapter 7.

- Rainfall–runoff modelling is also being increasingly used for prediction of flood frequencies in combination with a rainstorm model as an alternative to a purely statistical estimation of flood frequency. This approach has the potential to take account of the changing nature of the hydrological response of a catchment with changing antecedent conditions and rainstorm volume or intensities, but at the present time is associated with significant uncertainties.

Box 8.1 Adaptive Gain Parameter Estimation for Real-Time Forecasting

Consider the simple first-order transfer function model:

$$\widehat{y}_t = G_t \frac{b_o + b_1 z^{-1}}{1 - a_1 z^{-1}} u_{t-\delta} \tag{8.3}$$

where $\widehat{y}_t$ is the predicted output variable, $u_{t-\delta}$ is the input variable (effective rainfall or upstream river flow), z is the backward difference operator (see Box 4.1), a_1, b_o, and b_1 are parameters fixed by previous calibration, and G_t is a time-variable gain.

The time-variable gain can be estimated in real time by the methods of time-variable parameter estimation developed in Young (1984), and used in the Dumfries flood forecasting system by Lees *et al.* (1994). The algorithm is a form of recursive filtering of the gain parameter G in that the gain at time t, G_t, depends on G_{t-1} at the previous time step, together with a function of the current prediction error $(y_t - \widehat{y}_t)$ where y_t is the observed flow at time t.

The algorithm has the form of predictor–corrector equations as follows:

Predictor step

$$\widehat{G}_{t|t-1} = G_{t-1} \tag{8.4}$$

$$P_{t|t-1} = P_{t-1} + C_{\mathrm{NVR}} \tag{8.5}$$

Corrector step

$$G_t = \widehat{G}_{t|t-1} + \frac{P_{t|t-1}\widehat{y}_t[y_t - \widehat{G}_{t|t-1}\widehat{y}_t]}{1 + P_{t|t-1}\widehat{y}_t^2} \tag{8.6}$$

$$P_t = P_{t|t-1} - \frac{[P_{t|t-1}\widehat{y}_t]^2}{1 + P_{t|t-1}\widehat{y}_t^2} \tag{8.7}$$

The degree to which the time-variable gain G_t is allowed to change between time steps is controlled by the value of the algorithm parameter C_{NVR}, which is a noise variance ratio (NVR). The higher the value of C_{NVR}, the faster the memory effect of previous values of G at times, $t-1, t-2, \ldots$ is allowed to decay away. Thus a large value of NVR means that there is a short memory and less filtering of the changes in G. In the Dumfries study of Lees *et al.* (1994), the value of C_{NVR} was allowed to vary with the magnitude of the prediction error so that the larger the prediction error at a time step, the more rapidly G was allowed to change.

One advantage of this type of methodology is that an estimate of the uncertainty in the predictions can be calculated and also updated recursively. If we define the n step ahead prediction error as

$$\varepsilon_{t+n} = (y_{t+n} - \widehat{y}_{t+n}) \tag{8.8}$$

then the variance of ε_n is given by

$$\mathrm{var}(\varepsilon_n) = \widehat{\sigma}_t^2[1 + P_{t+n}\widehat{y}_{t+n}^2] \qquad (8.9)$$

where $\widehat{\sigma}_t^2$ is current forecast error variance, which is estimated from the further recursion

$$\widehat{\sigma}_t^2 = \widehat{\sigma}_{t-1}^2 + P_t[e_t^2 - \widehat{\sigma}_{t-1}^2] \qquad (8.10)$$

where e_t is the scaled prediction error

$$e_t = \frac{y_t - \widehat{y}_t}{\sqrt{1 + P_t\widehat{y}_t^2}} \qquad (8.11)$$

The variance of ε_n can be used to estimate prediction limits for the forecasts based on the uncertainty in the time-variable estimates of G_{t+n} as reflected in the magnitude of P_{t+n}. Further details of this algorithm, and its extension to recursive updating of multiple parameters, may be found in Young (1984).

Note that in the predictor–corrector equations, only the error $(y_t - \widehat{y}_t)$ at the current time t can be known, but that this can be fixed in extrapolating the estimates of G and P forwards using the predictor–corrector equations to time $t + n$. In general, the magnitude of the prediction limits will increase with increasing discharge and with extrapolation further into the future. In the Dumfries forecasting system it was found that reasonable forecasts could be made for 5 hours ahead (see Figure 8.3), even though the natural time delay in the river network was of the order of 3 hours.

The time-variable gain in this type of model is a way of compensating for both data errors and any nonlinearities that are not properly represented in the model structure. Adaptive methods can also be used with nonlinear models. The most well-known technique is the extended Kalman filter (see Young 1984; Beck and Halfon 1991) which has been used widely in both hydrological and water quality modelling problems. Time-variable parameter estimation can, in fact, be used to investigate the nature of the nonlinearities of an input–output system (see Box 4.3).

9
Predicting the Effects of Change

Nowadays, model building has become a fashionable indoor sport

Ludwig von Bertalanffy, 1967

He who controls the future of global-scale models controls the future of hydrology

Peter Eagleson, 1986

Predicting the hydrological effects of changes in both climate and land use is certainly a currently fashionable indoor sport. Concerns about the impacts of global warming, deforestation and other changes have been enhanced by public and political awareness of the greenhouse effect, the ozone hole over Antarctica, the El Niño effect, satellite pictures of burned and burning forests in Amazonia, and the increasing demands of a growing global population on the world's freshwater resources. Hydrological modelling contributes only a small part of the global atmosphere–ocean–land modelling systems that are being used to predict changes into the next century. It is not, however, an insignificant part. Global circulation models (GCMs) are known to be sensitive to how the hydrology of the land surface is represented. There is also no doubt that both global and local changes will have feedback effects on hydrological processes in the new century, but the problem is how best to predict whether those impacts will be significant when hydrological systems are subject to so much natural variability.

Our abilities to predict the future are necessarily limited, even if we had perfect hydrological models (which, as we have seen earlier, we do not). At the local or catchment scale, there is a limitation due to our lack of knowledge of exactly what the boundary and auxiliary conditions will be in the future, in terms of both the forcing inputs and any changes in effective parameter values. At the global scale, there are limitations due to the grid resolution of global-scale models resulting from computing constraints, and the difficulties of formulating realistic representations of processes at the subgrid scale.

It follows that, as with models of present-day hydrological responses, we should be qualifying our predictions by estimating uncertainties of the impacts of change or, put in another way, the risk of seeing a certain degree of impact on flood peaks, minimum flows or the usable water resource in the future. Translating uncertainty into a future risk can provide a valuable contribution to the decision-making process. Risk is the more acceptable face of uncertainty. Rather than being faced with a prediction surrounded by a fuzzy cloud of uncertainty, the decision-maker is then faced with the more manageable problem of assessing an acceptable risk. The information provided by the simulations is essentially the same (see Section 7.10).

Understanding the nature of hydrological change is still very much in its early stages, limited by the short lengths of hydrological records which can make it difficult to detect change given the natural variability of hydrological systems and the small scale that is possible in most catchment land use manipulation experiments. There have been many studies that report predictions of the hydrological impacts of change but, as yet, I know of none where catchment-scale predictions made *before* a change have later been verified. In fact, the experience in groundwater modelling, where there have been a number of such post-audit studies, is not too promising in this respect. Most predictions of groundwater responses have been shown to be wrong (see Konikow and Bredehoeft 1992). The result has been a continuing discussion about the possibility of validation of groundwater models, much of which is also relevant for rainfall–runoff modelling (e.g. Oreskes *et al.* 1994).

The major problem in predicting the effects of change is that at least some of the model parameters should be expected to change. This is evidently the case if there is a change of land use, but changes may also be induced by a change in inputs as a result of climate change. It has been argued, for example, that an increase in CO_2 concentrations in the atmosphere will cause a decrease in the density of stomata on leaf surfaces because a plant can achieve the necessary exchanges of CO_2 for photosynthesis more efficiently. The result would be an increase in the effective canopy resistance, leading to smaller evapotranspiration losses as a result of climate change. The argument is plausible, and is supported by stomatal density measurements for some plants in the historical collections at Kew Gardens in England, but it is not clear if this will be a general response or a significant factor in the response of plants to climate change.

In general, therefore, if there is uncertainty associated with the estimation of parameter values for rainfall–runoff models under current conditions, the uncertainty is likely to be greater in trying to estimate the changes in parameter values under changed conditions. It has long been suggested that one of the most important reasons to develop a process-based distributed modelling capability is precisely that it might be easier to estimate such changes (e.g. Abbott *et al.* 1986a; Ewen and Parkin 1996; Dunn and Ferrier 1999). It is also rare for the whole of a catchment to change suddenly; the changes in characteristics are more likely to be gradual and often quite local. Thus a spatially distributed model has the advantage that it can implement any changes in parameter values in their correct spatial context. Some models include components that attempt to represent the growth of vegetation communities and their interactions with hydrological processes, e.g. RHESSys of Band *et al.* (1993), TOPOG-IRM of Dawes *et al.* (1997) and MACAQUE of Watson *et al.* (1999).

The advantages of process-based models in this context are still widely supported and for good reasons since the only alternative would seem to be to try and reason about potential changes in bulk catchment-scale parameters, such as the mean residence time of a transfer function, which necessarily integrate the impacts of change on a variety of interacting processes. This would appear to be much more difficult than reasoning about the impacts on the parameters of individual processes.

However, this belief might be a little naive since it has not yet been consistently demonstrated that the parameters of process-based models can be estimated a priori and produce successful simulations under current conditions. The attempt by Parkin *et al.* (1996) at prior estimation of feasible parameter values, in their *blind validation* of the SHE model to simulate the small Rimbaud catchment in the South of France, resulted in considerable uncertainty in the discharge predictions and a failure to meet most of their criteria for success (see the case study in Section 5.4). This could be considered to be a prerequisite of any belief in the predictions that might be made for changed conditions. And, at least in the case of land use, it might be possible by studying catchments where changes (such as urbanization or deforestation) have been observed in different environments, to structure some reasoning about the changes in effective parameters at the grid or catchment scale. The latter approach would be considered by many hydrologists to be less scientific, even if it might be equally effective at the scale at which predictions are mostly required. These two approaches will be discussed further in Chapter 10 when considering possible future developments in rainfall–runoff modelling.

9.1 Predicting the Impacts of Land Use Change

All catchments have a history. Some of this history will have a long time-scale, such as soils developed on till or fluvioglacial deposits from the last ice age, or deep lateritic soils resulting from weathering over even longer periods. Some changes will be the much more recent results of human activities, such as urbanization, deforestation, reforestation, reservoir or detention pond construction, the effects of wild fires, or the installation of field drainage systems. Both natural and anthropogenic changes in land use have affected catchments in the past and are continuing today. Some changes are documented, others may not be (I know of one experimental catchment in the UK where it was discovered, after the installation of the instrumentation, that the headwater hollows were drained by old field drains built of stone slabs and estimated to be more than 100 years old). In most gauged catchments, certainly larger gauged catchments, such changes have been ongoing during the period of historical records. However, hydrological analyses (e.g. flood frequency analyses) do not often attempt to take any account of the possible effects of such changes. Why not? Primarily because it may be very difficult to detect such changes in the historical record given the natural year to year and storm to storm variability in hydrological responses, especially where the changes are only gradual.

Techniques for the analysis of the effects of land use change on modelled hydrological responses are still very much at an early stage. The prediction of the effects of future change (and validation of those predictions) has hardly even started. In the remainder of this chapter we will examine some selected studies to demonstrate the state of the art in this important area, in the context of what we have already learned about the nature of the modelling process and predictive uncertainty.

9.1.1 Deforestation and Reforestation

The role of forests in controlling water yields and hydrograph peaks has long been debated in hydrology. It has been a major driving force behind many catchment experiments, notably those at Coweeta in the US (Swank and Crossley 1988), Hubbard Brook (Likens *et al.* 1977), the studies of Frank Law in the Forest of Bowland in the UK which led to the setting up of the UK Institute of Hydrology, and the Plynlimon (Wales) and later Balquhidder (Scotland) experimental catchments (Calder 1990), and the long-term monitoring of the Melbourne Water Supply District catchments in Australia. A review of studies of forest hydrology has been published by McCulloch and Robinson (1993).

This is a good example where both process-based modelling and catchment-scale parametrizations might find it very difficult to predict the effects of change since the evidence from field studies about the impacts of deforestation is mixed. In some catchment studies deforestation produces higher water yields and flood peaks; in others it produces lower discharges. This is the effect of a number of interacting processes that are associated with deforestation and reforestation. The process of logging often involves the building of forest roads and the movement of heavy machinery and trees over skid trails, with a consequent local impact on soil structure and runoff generation. The process of reforestation often includes the digging of ditches to improve the drainage of the soil which might either increase runoff or, by decreasing the antecedent catchment wetness prior to an event, decrease runoff. In fact, both might occur in the same catchment depending on the magnitude of an event and when it occurs. The impact will also depend on the nature of the growth following reforestation or deforestation. The vigour and stand densities of the regrowth of mountain ash eucalypts in the Melbourne water supply catchments, for example, are exceptional and cause a significant reduction in water yields following loss of a mature forest stand (see Section 9.2).

Thus predictions of the impact of forest deforestation or reforestation are likely to be associated with some uncertainty. In fact, many modelling studies of such impacts have concentrated on the changes to the vegetation itself and have ignored any of the ancillary changes in forest roads, soil structure or ditching. Most have also made purely deterministic predictions without any consideration of the uncertainty in estimating the parameters for both current and changed conditions. One exception to this was the study of Binley *et al.* (1991). In a precursor to the GLUE methodology of Section 7.5, they evaluated the prediction uncertainty of the Institute of Hydrology Distributed Model (IHDM4; see Section 5.4.3) in an application to the Gwy catchment in the headwaters of the River Wye at Plynlimon, Wales. This catchment is currently under upland pasture, but the adjacent headwaters of the River Severn are forested. Computer constraints at the time of the study meant that only single storm simulations could be carried out, which meant that initialization of the soil moisture distribution on each hillslope element prior to each event was an issue.

Parameters of the interception and Penman–Monteith evapotranspiration components were available from independent field studies and were fixed. By manually calibrating the model for a number of storms under current conditions, mean values for each of four parameters (hydraulic conductivity, porosity, initial soil capillary potential and overland flow roughness) were determined together with a covariance matrix. These

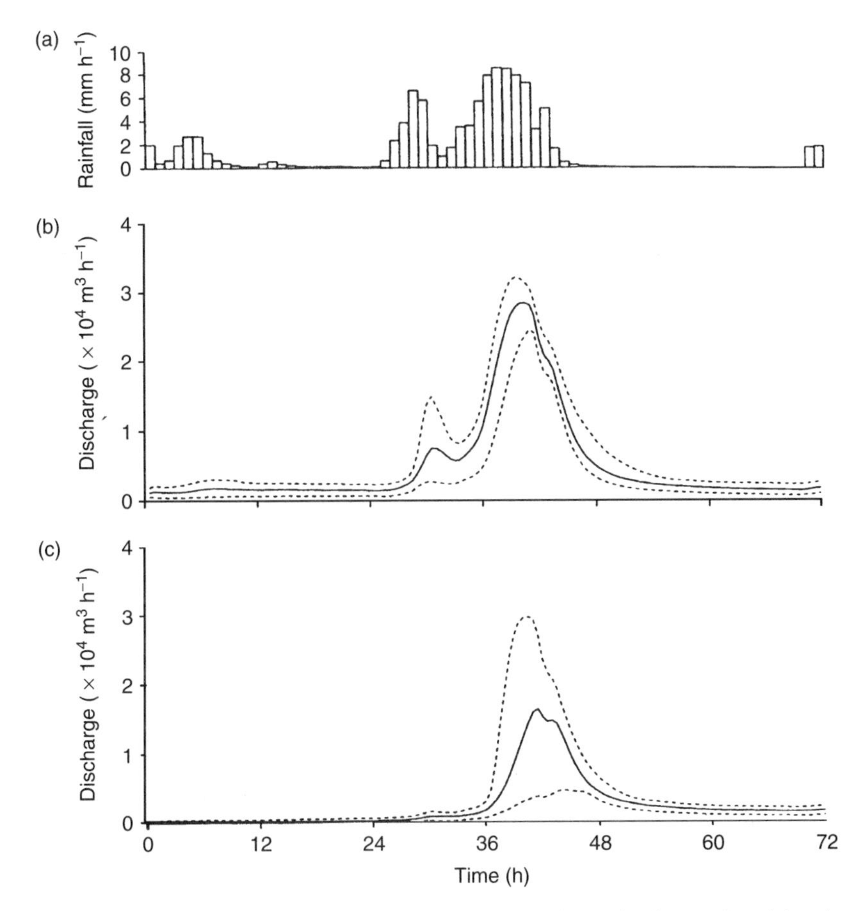

Figure 9.1 *Prediction bounds (based on the Rosenblueth method) simulated for the Gwy catchment at Plynlimon (3.9 km^2) using the Institute of Hydrology distributed model: (a) rainfall; (b) under current grassland land use; (c) for the same storm with estimated forest parameters (after Binley et al. 1991). Reproduced from Water Resources Research 27(6): 1253–1261, 1991, copyright by the American Geophysical Union*

values were then used to generate parameter sets from a multivariate normal distribution within a Monte Carlo procedure in simulating other 'validation' events, which included the largest storm on record. The prediction limits satisfactorily bracketed the observed discharges for these 'validation' events (Figure 9.1(b)).

The next part of the study was to estimate the change in discharge expected as a result of a change to coniferous forest cover. The interception and evapotranspiration parameters for forest had previously been shown to reproduce measured plot-scale data in the Severn catchment satisfactorily. The new set of vegetation parameters were therefore also assumed to be known without uncertainty. No changes were assumed in the means and variances of the hydraulic conductivity, porosity or overland flow roughness parameters. It was known from field measurements, however, that soil moisture levels prior to an event were generally drier under forest than under grass. A fixed change in the mean initial capillary potential was therefore imposed, while keeping the variance and covariances the same as for the current conditions. The results of the simulation for the same rainfall event are shown in Figure 9.1(b). The uncertainty in the predicted peak flow for the forested catchment was high but the differences in the distributions of the predicted peaks were shown to be statistically significant. Repeating the analysis without a change in the initial conditions, however, did not result in a significant difference between the grass and forest predictions.

Other studies have also shown that the effects of the initial conditions in a process-based distributed model can persist for considerable periods of time. Thus in any evaluation of the effects of change, great care should be taken in interpreting the results of short simulation periods. Fortunately computer constraints are now less and this type of study could be repeated using longer simulation periods to minimize the effects of the initial conditions on the results while still making a realistic assessment of the predictive uncertainty.

9.1.2 The Effects of Fire

Another important factor affecting runoff generation is fire, particularly where a major proportion of a catchment is affected. The effects on catchment discharges are generally much easier to detect and in some cases persist for decades. Many field studies have suggested that fire reduces infiltration rates and increases runoff coefficients, at least in the period immediately after a fire (e.g. Cerda 1998). Fire reduces protection of the surface resulting from the loss in vegetation cover and will tend to increase the water repellency or hydrophobicity of the soil surface (Soto and Diaz-Fierros 1998). An extensive fire in the Rimbaud catchment (which was modelled by Parkin *et al.* (1996) using SHE for a period prior to the fire; see the case study of Section 5.4) had the effect of increasing peak discharges dramatically in the following period (Lavabre *et al.* 1993). Prediction of the impacts of fire on runoff in the Melbourne water supply catchments is considered in the case study of Section 9.2.

9.1.3 Urbanization

Urbanization is known to have a significant effect on rainfall–runoff relationships but almost all the models we have considered up to this point take very little explicit account of urban areas in their rainfall–runoff predictions. This can often be justified on the basis that urban areas are only a small fraction of the area of a catchment, but small urban catchments will require specific account to be taken of the modifications to the hydrology made by humans. The prediction of runoff in urban areas for the design of storm sewer systems is an area of hydrology with an extensive literature of its own. Any urban rainfall–runoff model must take account of the patchwork of pervious and impervious areas and the network of natural and human-made drainage channels and pipes. There are a number of commercial software packages available for the prediction of rainfall–runoff relations in urban areas, including the US EPA Storm Water Management Model (SWWM; Huber 1995), the MOUSE package from the Danish Hydraulics Institute, and WALLRUS from Hydraulics Research and the Water Research Centre in the UK. These packages are designed specifically to take account of a specified network of pipes and channels in a catchment and the areas contributing to them. They generally include water quality modelling components as well as runoff prediction. Other general distributed modelling packages, such as the kinematic wave models HEC-1 and KINEROS, have components that can be used to represent the pervious/impervious mix of land uses in urban areas (Feldman 1995; Smith *et al.* 1995).

It is worth noting that virtually all urban hydrology models are based on the traditional idea that the major part of the hydrograph is due to surface runoff. This will often be a good approximation in urban areas where the most important storms for design purposes on a small catchment will be short intense convective rainstorms. However, it is worth remembering that many impervious surfaces in urban areas are not connected directly to storm drains or channels but rather to soakaways; and that many pervious surfaces will not generate surface runoff, even in high-intensity rainfall events. Such infiltration into the subsurface is generally treated as a loss from the model and is not considered further. The difficulty of characterizing all the different surfaces and drainage connections in urban areas, which may also change in flood conditions, is not a trivial task even in the design for small drainage systems. The process of modelling is, however, becoming easier as more urban drainage systems and land uses are characterized within GIS databases. Burges *et al.* (1998), however, still argue that continued monitoring of small catchments subject to change remains necessary to resolve some of the uncertainties associated with the prediction of the hydrological impacts of development.

9.1.4 Agricultural Drainage

The treatment of artificial field drains and ditches in hydrological models is a similar problem, although the records of where, when and how drains were installed in the past may not often be available to the modeller. Drains have an effect on the hydrology, otherwise farmers, foresters and others would not invest in them. There remains, however, some debate as to what effect drains might have on the magnitude of storm peaks. Two lines of reasoning are often presented. In the first, it is argued that since drains provide additional and relatively efficient pathways for the removal of runoff, they will tend to increase the magnitude of storm peaks. In the second, it is argued that since drains will tend to drain the soil between rainfall events, there will be more antecedent storage available prior to an event and so less runoff generation, so that, in spite of any increased efficiency of runoff, less runoff generation will lead to lower hydrograph peaks.

Analysis of runoff records from comparing drained and undrained experimental catchments has appeared to show both types of behaviour, even in the same catchment, under different conditions (see the review by Robinson 1986). Robinson showed that the drainage of heavy clay soils can result in a reduction in flood peaks by providing generally higher antecedent storage deficits, whereas the drainage of more permeable soils tends to speed up subsurface runoff, resulting in higher flood peaks. Thus modelling the effects of drainage at the catchment scale will not be simple, particularly where, even with a process-based distributed model, the spacing of tile or mole drains may be much smaller than the model grid scale. Thus some parametrization of these sub-grid scale effects will be necessary. Example parametrizations have been considered by, for example, Dunn and Mackay (1996), Kim and Delleur (1997) and Karvonen *et al.* (1999). The MIKE SHE model now has a component to represent the effects of field drains in predicting runoff as a linear function of local groundwater level above some drainage datum elevation (e.g. Al-Khudhairy *et al.* 1999).

9.2 Case Study: Predicting the Impacts of Fire and Logging on the Melbourne Water Supply Catchments

A particularly interesting example of predicting the effects of land use change comes from the water supply catchments of the city of Melbourne, Australia. The water yield of these catchments is sensitive to loss of old-growth mountain ash (*Eucalyptus regnans*) forest cover due to natural wild fires. Here, after a fire (or logging of a mature forest stand), catchment discharges are reduced. An annual yield of the order of 1000 mm under mature forest can be reduced to 500 mm at the peak of the regrowth due to the very high density and leaf area of the regrowth forest. The 'lost' water is returned back to the atmosphere as transpiration and evaporation of interception. The losses are at a maximum after about 25 years of regrowth but Kuczera (1987) has estimated that it may take over 100 years for the water yield to fully recover (see Figure 9.2).

The Kuczera water yield curve for these catchments was developed by an empirical regression at the catchment scale using long-term records of observed water yields following an extensive fire in 1939 and assuming a particular form for the nonlinear function. The hypothesis that it is due to regrowth vegetation densities has recently been tested using the process-based distribution function model MACAQUE by Watson *et al.* (1999). The model was calibrated for three different catchments (Watts, Grace Burn and Coranderrk; total area 145 km^2) under current conditions, including a comparison of some internal water table predictions. The model was shown to be more sensitive to assumptions about the spatial distribution of precipitation inputs than the topographic

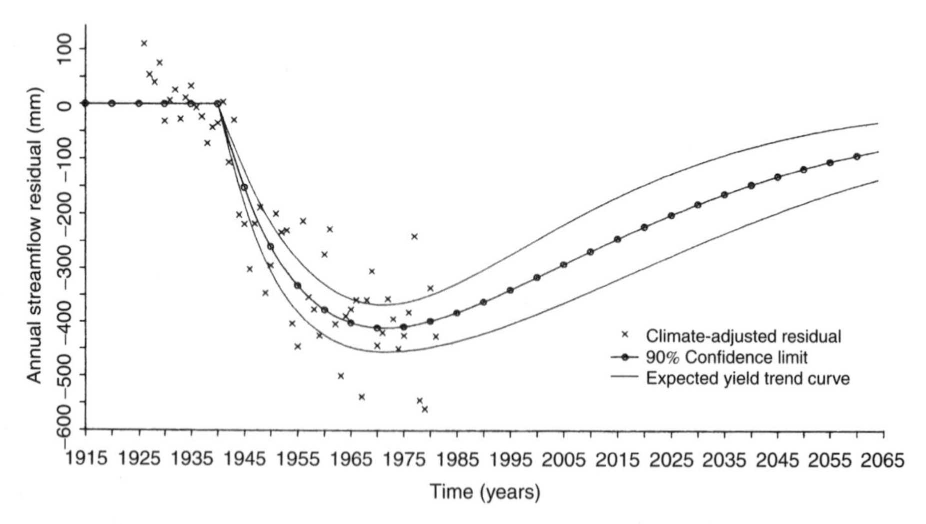

Figure 9.2 *Change in catchment water yield following forest fire, with approximate prediction limits, for regrowth of mountain ash in the Graceburn catchment, Australia (after Kuczera 1987). Reprinted from Journal of Hydrology 211: 69–85, copyright (1987), with permission from Elsevier Science*

distribution used. Vegetation density was simulated by a specified distribution of the leaf area index, according to an age map of the forest in the catchments. Long-term simulations were carried out, spanning the period 1910–1991 and varying the leaf area index and also a leaf conductance (or stomatal resistance) parameter to follow the regrowth of the forest after the 1939 fire. Parameter changes were based directly on relationships with stand age. The annual aggregated results of the different discharge predictions are shown in Figure 9.3.

The model in which both parameters were changed was the most successful in reproducing the long-term changes in discharge. However, even this simulation showed significant over- and under-prediction of observed discharges in particular years. The authors suggest that much of this error may be due to an inadequate representation of the spatial patterns of rainfall within the catchments. While MACAQUE is not a fully distributed process-based model, these results are perhaps representative of the type of accuracy that might be achieved in predicting the effects of land use change while representing the state of catchment as a single set of process related parameters, while the curve of Kuczera (1987) represents the catchment-scale approach to the same problem.

9.3 Predicting the Impacts of Climate Change

Predictions of the effects of climate change involve a change of time and space scales that has an important bearing on the predictive process. Current computational constraints on GCMs mean that the grid scale at which predictions are available is still very large (at best, of the order of 100 by 100 km, which is equivalent to a river basin of 10 000 km^2 such as the whole of the River Thames in the UK). Only a single prediction of average temperatures, precipitation or runoff is made for each grid square even though great sub-grid variability would be expected from measurements. This leads, for example, to the GCM 'drizzle' problem where the precipitation due to intense convective activity at scales much smaller than the grid square can only be represented in the model as an average precipitation of low intensity. It is possible now to refine local predictions of GCMs by imbedding a fine grid model within the normal coarser grid. The fine grid model then uses the predictions of the coarser grid as boundary conditions at the limits of its domain and is thus ultimately dependent on the coarser grid simulations.

All GCMs have a hydrological model to provide a lower boundary condition for the fluxes of energy and water vapour between the land surface and the atmosphere. These models are commonly known as soil–vegetation–atmosphere transfer models (SVATs) or land surface parametrization schemes (LSPs). In the past, these components have been necessarily very simple bucket-type models, similar to the soil moisture accounting models discussed in Section 2.4. We have seen that, with some calibration, such models can provide good simulations of measured soil moisture deficits at a point, but in the case of GCMs they are used not to represent a single point but the land surface of a whole grid square. The parameter values therefore have to be considered effective parameter values, but clearly in the case of a large GCM grid square there are no data against which to calibrate the effective values. Instead, values have generally been specified by intelligent guestimation.

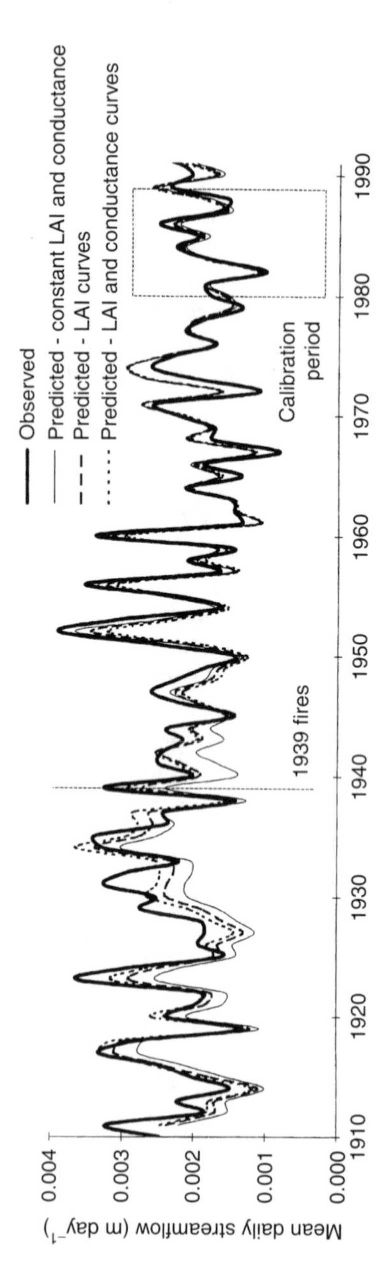

Figure 9.3 *Predictions of recovery of discharges in the Maroondah catchments, Australia (total area 145 km^2) following forest fire using the MACAQUE model (after Watson et al. 1999). Reproduced with permission of John Wiley & Sons Limited*

As computer power has increased, some of these constraints have started to be relaxed. In the latest generation of GCMs, bucket-type SVAT components are being replaced by more physically based parametrizations that treat the water and energy fluxes to and from the soil and vegetation, and the physiological controls of the vegetation on transpiration in much more detail. Examples are the MOSES scheme used in the UK Meteorological Office Unified Circulation Model (Cox *et al.* 1998) and the Simple Biosphere Model (SiB; Sellers *et al.* 1996) which, in their latest versions, also predict fluxes of carbon dioxide. SiB is a physically based, multi-layered vegetation, multi-layered soil, and multi-parameter model of a vegetation patch, and has been shown to produce good simulations of measured energy, water and carbon dioxide fluxes at the scale of a single vegetation stand. However, when coupled to a GCM, it is still only possible to use a single SiB model in each grid square. Thus the parameter values still have to be specified as effective parameters to represent the whole of the variability of processes and conditions within the large GCM grid square. There is currently no way of estimating such effective parameter values at the grid scale other than by intelligent guess work, as was the case for bucket models. In the case of more complex SVATs, however, there are many more parameter values to be estimated. It is not clear that the physical basis of the more complex SVAT models helps in estimating effective values at the grid scale and it has been shown that there may be considerable uncertainty in the estimation of SVAT parameters, even at the local patch scale (Franks *et al.* 1997).

Thus the hydrological component of GCMs has some important limitations. An ongoing intercomparison study of different SVAT formulations, known as PILPS (Project for the Intercomparison of Land-surface Parametrization Schemes), has shown that different schemes make significantly different predictions when parameters are estimated without calibration, particularly for predictions of runoff production (Lohmann *et al.* 1998c). This should perhaps be expected since, as we have seen, runoff production is not a simple one-dimensional process, but a three-dimensional process involving interactions between topography, soil horizons, surface vegetation, groundwater and the river network with dynamically changing contributing areas for both surface and subsurface contributions to runoff. The incorporation of more hydrologically acceptable land surface models into GCMs still seems some way off although there have been some recent interesting developments in what has been called macroscale hydrological modelling.

9.3.1 Macroscale Hydrological Models

One-dimensional SVAT models used at the GCM model grid scale with effective parameters values are one, particularly simple, example of a macroscale hydrological model formulation. A simple extension is to use multiple one-dimensional models to represent different tiles or patches of the landscape with different characteristics (e.g. Avissar and Pielke 1989; Koster and Suarez 1992; Jolley and Wheater 1997). This can be extended to a higher order scheme that attempts to take account of the joint distribution of different fluxes within the landscape (Avissar 1998). Another approach is to try to represent the distribution of responses within the landscape in some functional way, as in the Xinanjiang/ARNO/VIC model structure (see Box 2.2), versions of which have already been incorporated into GCMs (e.g. Dumenil and Todini 1992; X. Liang *et al.* 1994) or have used GCM-predicted variables as inputs for modelling large basins (Liston *et al.* 1994).

There have been some other developments in macroscale hydrological modelling that should be mentioned here, although these have (not yet) been incorporated directly into GCMs. In fact, the requirements of a macroscale hydrological model are really no different from any other catchment modelling exercise: they need a means of generating runoff that properly reflects the variability of hydrological processes and a means of routing the runoff to a point of interest (or in the case of GCMs, between different grid elements) that properly reflects the time-scales of the flow processes. The difference in macroscale modelling is the range of variability involved, the importance of the routing delays in forming the hydrograph of continental-scale rivers (e.g. Naden 1992; Naden *et al.* 1999) and the constraint that if a macroscale model is eventually to be coupled to a GCM it must still remain computationally fast to run. Some recent attempts to define a macroscale hydrological model that reflects the heterogeneity of runoff generation and that could be used within a GCM include the TOPLATS version of TOPMODEL (Famiglietti and Wood 1991; Stieglitz *et al.* 1997); the UP model of Ewen (1997); the SLURP model of Kite (1995); and the fuzzy disaggregation approach of Franks and Beven (1997) and Beven and Franks (1999). All have different ways of attempting to reflect the variability of responses to be found at the sub-GCM grid scale in a computationally efficient way.

TOPLATS (the TOPMODEL-based land–atmosphere transfer scheme) does so by using statistical distribution functions for soil, vegetation and topographic variability, with the TOPMODEL topographic index providing the primarily classificatory variable (see Section 6.4). The SLURP model similarly groups points within an area into a small number of calculation units within a model grid element, but does not use the topographic index as a classifying variable, only vegetation, soils and elevation. It has shown some success in modelling the hydrology of large basins using GCM output variables as inputs to SLURP (Kite and Haberlandt 1999).

The UP model is a particularly interesting approach to this problem in that it tries to maintain a purely physically based approach to large-scale predictions. The key to achieving this is to subsume the results of off-line simulations made using a detailed physically based 'parametrization' model, into a series of look-up tables for the fluxes between the major storages in the UP model (Figure 9.4). In an application to the 570 000 km^2 Arkansas–Red River basin, 1923 grid squares of size 17 by 17 km were classified into five different type domains (Ewen *et al.* 1999; Kilsby *et al.* 1999). UP model elements were developed separately for each of the five type domains by running simulations of the detailed model for representative catchment areas within the domains. Ideally, the look-up tables developed for each domain could be assumed to be independent of meteorological forcing, but in practice it was found that it was necessary to provide additional domains to allow for significant differences in precipitation inputs within the original classification. A comparison of observed and predicted discharges is shown in Figure 9.5.

In running simulations for the whole basin, each grid element has different rainfall and meteorological inputs. The water storage in each compartment is used with the look-up tables for that UP element type to determine the fluxes between compartments. Mass balance is maintained for each compartment in each grid element. The

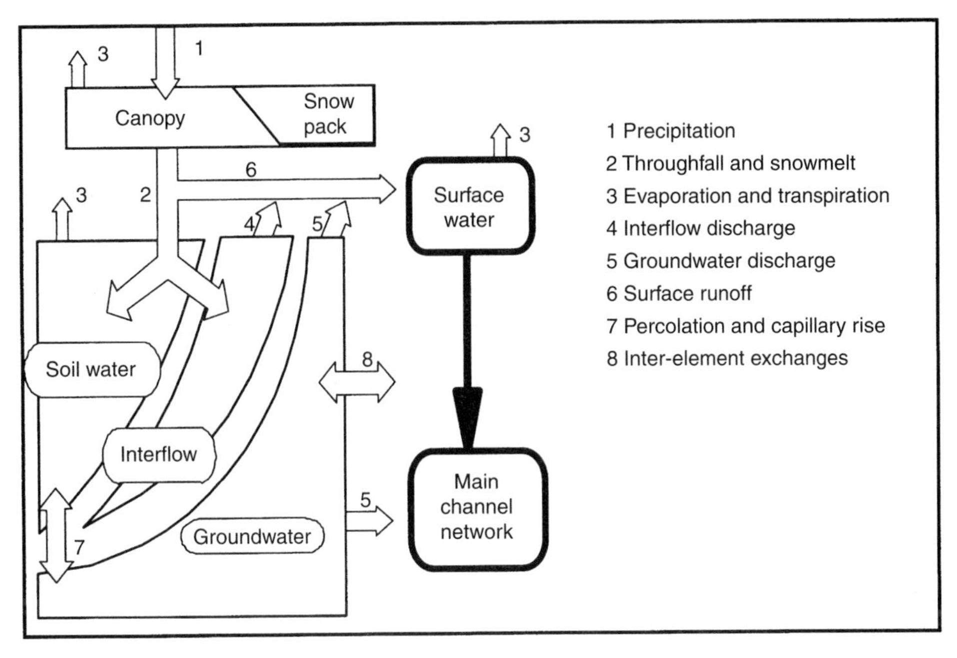

Figure 9.4 *The University of Newcastle UP macroscale rainfall–runoff model (after Ewen et al. 1999). Reprinted from Hydrology and Earth Systems Science with permission of European Geophysical Society*

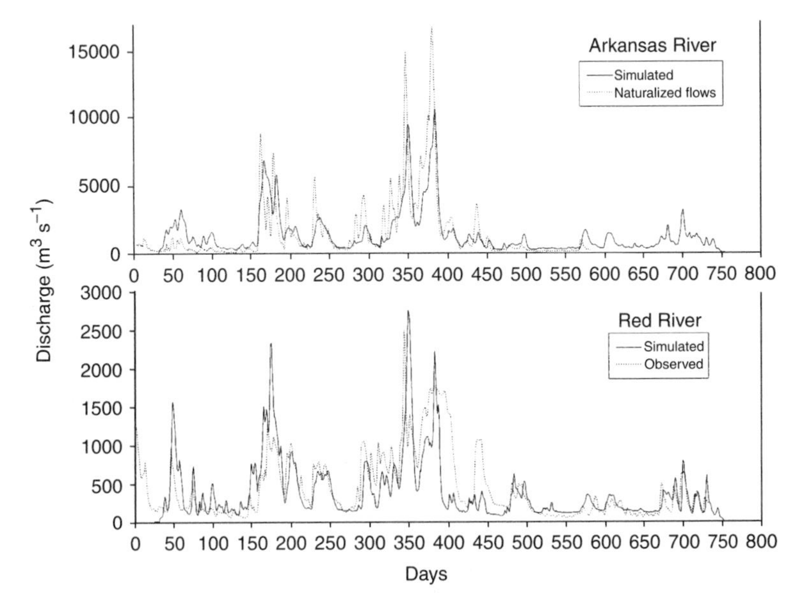

Figure 9.5 *Observed and predicted discharges at the outlets of the Arkansas and Red Rivers using the University of Newcastle UP macroscale rainfall–runoff model (after Kilsby et al. 1999). Reprinted from Hydrology and Earth Systems Science with permission of European Geophysical Society*

model is completed by the network width function runoff routing component of Naden *et al.* (1999). The simulations are therefore much more computationally efficient than a full physically based distributed model, but remain reasonably consistent with the predictions of the parametrization model used.

The results depend heavily on the realism of both the structure and parameter values of the underlying parametrization model since this is run in a purely deterministic way. This approach is therefore firmly founded in a belief that the a priori specification of all the model parameters required by the parametrization model is possible (note that the UP elements themselves require no parameters but that their look-up tables depend on the parameters of the underlying model). A comparison of observed and simulated discharges at different catchment scales from the application to the Arkansas–Red River basin (e.g. Figure 9.5) showed that significant errors remained, but it should be remembered that no calibration against observed discharges had been used in developing the simulation model.

A different approach to dealing with the problem of parametrization of ungauged areas in macroscale hydrological modelling has been taken by Abdulla *et al.* (1996) and Abdulla and Lettenmaier (1997) using the VIC-2L conceptual model of Box 2.2 in an application to the same Red–Arkansas system. They used a variety of gauged subcatchments within the Red–Arkansas basin as a training set to determine calibrated parameter values for the VIC-2L model. These were then regionalized by correlating the calibrated values against other land surface characteristics for the catchments. This then allowed parameters for all the other subcatchments to be estimated. The estimates were tested on six gauged catchments that were not included in the calibration set. No uncertainties in the estimated parameter values, for either gauged or ungauged subcatchments, were taken into account. The performance of the regionalization procedure was found to be adequate for humid and semi-humid subcatchments, but the approach did not perform as well for the semi-arid part of the Red–Arkansas system.

The fuzzy disaggregation approach of Franks and Beven (1997a) takes a third, quite different, approach to the same problem of trying to achieve computational efficiency while maintaining a realistic assessment of fluxes. This is also a two-stage approach, but is based more on the concept of equifinality of different models (see Sections 1.8 and 7.7) than on a belief in deterministic predictability. Franks and Beven point out that in trying to take account of the variability of hydrological response in the landscape (or GCM grid square), it is not the variability of parameters that we are interested in *per se* but in the functioning of different locations in terms of the partitioning of the water balance into evapotranspiration and discharge. Indeed, for GCM predictions, correctly partitioning of the fluxes will be more important than getting the distribution in time correct (which might be more important in hydrograph simulation). Using different parameter sets to represent different parts of the landscape is only one way of predicting the difference in function of different locations. However, specifying such parameter sets will inevitably be fraught with uncertainty, even where flux measurements are available for conditioning a particular model (see Franks *et al.* 1997).

They therefore adopt an approach of mapping the landscape into a model parameter space based on whatever information is available about feasible parameter sets to represent different locations in the landscape. Because of the inherent uncertainty, this is necessarily a fuzzy mapping and is achieved by using a large sample of feasible parameter sets and assigning each point a fuzzy weight of being associated with a particular simulation. The cumulative weights over all locations (which would normally be fine-scale grid elements and, in the case of Franks and Beven (1997a), remote sensing image pixels) allow a full mapping of the landscape into the model space.

The simplification of the calculations comes in the second stage from a realization that many different combinations of parameter sets in the model space will be functionally similar in terms of the partitioning of the water balance (or any other desired characteristic). A small number of simulations can therefore be used to represent the complete range of behaviours. This approach was used to predict the landscape-scale fluxes of the FIFE site in Kansas, taking account of the variability of the soils and topography and conditioning the mapping using remotely sensed surface temperature information. This idea of using landscape space to model space mapping is explored further in Chapter 10.

9.3.2 Uncertainty in Assessing the Hydrological Impacts of Climate Change

SVAT representations of the land surface are not the only sub-grid scale parametrizations required in a GCM and, in the current generation of GCM models, there are also questions about how best to represent cloud formation, rainfall generation, interactions with the ocean, and other processes at the grid scale. GCMs are in continual evolution as computer power increases, allowing finer grid scales and more detailed parametrizations to be used. What is clear is that, despite the recognized simplicity of many GCM components, a full uncertainty analysis of the predictions of climate change will not be feasible in the foreseeable future. The best that can be hoped for would be a limited number of runs assuming some different scenarios.

Thus, in predicting the hydrological impacts of change, we must accept the outputs of the latest GCMs as best estimates of conditions in the future, while remembering the limitations of the models that produced the estimates. They can certainly be accepted as possible scenarios for changed monthly averaged temperatures and rainfalls, but may be subject to considerable uncertainty. In order to use those estimates, however, except in the very largest drainage basins we must disaggregate the GCM predictions down to smaller time and space scales to regenerate some of the local sub-grid scale variability. This disaggregation process is currently the subject of a great deal of research that must consider, for example, how average rainfalls relate to local variability under different weather conditions, how variability in elevation might affect temperatures and precipitation as snow rather than rain, how changes in mean values might be related to changes in extreme conditions and so on. It is clearly a very complex problem that will, in itself, involve additional parameters in the disaggregation model components. Some disaggregation schemes have been reviewed by Wilby and Wrigley (1997), while recent examples are given by Conway and Jones (1998), Schnur and Lettenmaier (1998) and Wilby *et al.* (1998). Such disaggregation models will also be needed in the future as the basis for sub-grid parametrizations within GCMs that take more account of local variability in surface characteristics and responses in predicting grid-scale fluxes.

There is, however, a simpler approach based on modifying historical time series of input variables in a way that is consistent with the predictions of future conditions derived from GCMs. This is also a type of scenario modelling, using the GCM predictions to produce possible modifications in mean temperatures, mean precipitation, etc. This approach also has its limitations since it is not possible to produce completely consistent scenarios that include, for example, the feedbacks between changes in temperature, evapotranspiration and the humidity of the atmosphere, and any changes in variability associated with changes in the mean values. However, such an approach will not necessarily be more inaccurate than the direct use of disaggregated GCM output data.

Whatever the method used to create an input series under changed conditions, evaluation of the impacts will require a rainfall–runoff model to predict discharges and other hydrological variables such as soil moisture deficits, groundwater recharge or water table levels as necessary. The accuracy of such predictions will depend crucially on the chosen model to simulate the relevant processes under the changed conditions. The most obvious approach at this point is to calibrate a model on current-day data and then use the calibrated model to predict the response under the changed conditions. This approach has been used to make predictions of the hydrological impacts of climate change in, for example, the UK (Arnell 1996; Arnell and Reynard 1996), Belgium (Bultot *et al.* 1988; Gellens 1991), Switzerland (Bultot *et al.* 1992), Australia (Nathan *et al.* 1988), Greece (Panagoulia 1992), Canada (Ng and Marsalek 1992) and the USA (Schaake 1990; Cooley 1990; Lettenmaier and Gan 1990).

What do these predictions mean? They certainly do not mean that they are a true representation of what might happen in the future. They are, rather, conditional simulations; conditional on both the model used being a valid representation of the hydrology and on the scenarios of future climate used. Virtually all of these studies have been carried out without any assessment of predictive uncertainty and without any attempt to estimate how the parameters of the model might change with changes in climatic conditions.

There are, however, some exceptions. A little known study by Wolock and Hornberger (1991) was based on scenario modelling using a rainfall–runoff model, a version of TOPMODEL, in an application to the White Oak Run catchment in Virginia (as previously used in Hornberger *et al.* (1985) for current climate conditions). The scenarios consisted of modelled time series of precipitation and temperature inputs to the model for multiple periods of 60 years. Each series was given an imposed trend. After 60 years, they found that only some of the runs showed a resulting trend in the modelled runoff characteristics, implying that the chance of detecting a trend in an observational data set would be quite small. Trends in the annual maximum peaks were harder to identify than trends in annual runoff volume.

Prediction of the hydrological impacts of climate change will necessarily be uncertain. Jakeman *et al.* (1993b) have addressed this problem more directly by applying a version of the IHACRES model (see Section 4.3.1) to long periods of data from the French Broad River in North Carolina and the Tyfi River in Wales. The model was calibrated using observed rainfall, temperature and discharge data to find the best-fit parameter set on two- to three-year independent periods of data from each catchment. No strong trends in the resulting parameter values were found, although there was significant variability in all parameters from period to period on both catchments. This reinforces the difficulty of detecting trends in hydrological data, demonstrated by

Wolock and Hornberger, given the variability in the meteorological forcing data and the limitations of hydrological models.

All the fitted models were then used to predict the effects of raising the historical temperature data by a constant value ranging between 1 and 4 degrees. This led to a range of decrease in discharges for each temperature scenario, dependent on which parameter set was used, thereby allowing some assessment of the sensitivity of the predicted changes to uncertainty in the parametrizations. In this case, the dependence of the predicted impacts to the chosen model structure is also quite clear since the predicted discharges are directly dependent on a particular function relating the effects of evapotranspiration to temperature and catchment storage (a function that has changed in different versions of IHACRES). There is no physical justification for this function other than the fact that the model can be calibrated to give good simulations of observed discharges.

9.4 Case Study: Modelling the Impact of Climate Change on Flood Frequency in the Wye Catchment

In Section 8.8, it was shown that TOPMODEL could provide acceptable simulations of both continuous discharges and flood frequency estimates for the Wye catchment at Plynlimon, Wales. This work has been extended by Cameron (2000) in an assessment of uncertainty in the predicted hydrological impacts of climate change. The aim was to examine the possible impact of climate change on the frequency of high-magnitude events. The model calibration was undertaken within the GLUE framework (Section 7.7), resulting in a large sample of parameter sets considered as behavioural. The resulting uncertainty in the hydrograph predictions under current conditions is shown in Figure 8.4.

The historical record was then modified to be consistent with the predicted monthly changes in temperature and rainfalls issued by the Hadley Centre for the years 2020, 2050 and 2080 (HADCM2 scenarios; Hulme and Jenkins 1998). There are, of course, many different ways of interpreting an estimate of the mean monthly change when disaggregating down to an hourly simulation time step. A number of possibilities were tried but essentially made little difference to the results. Figure 9.6 shows the predicted change in the cumulative distribution of the estimated 100-year event. The distribution for current conditions reflects the likelihood weighted predictions from all the parameter sets retained as behavioural within GLUE. For the changed conditions, it is assumed that the likelihood associated with a parameter set does not change, only the input data. Figure 9.6 shows that the estimated change in the 100-year flood is much smaller than the current uncertainty in estimating the 100-year flood, but that the risk of a larger 100-year event is increased.

This is perhaps not an unexpected result but there have been very few such studies that have attempted to assess the uncertainty associated with the predictions of change. We have seen how the uncertainty can be estimated relatively easily using Monte Carlo simulation, at least in a scenario modelling framework. It should be done, since decisions about how to manage the impacts of future change can only be realistically based on an estimation of the risks of possible degrees of change. Figure 9.6 does

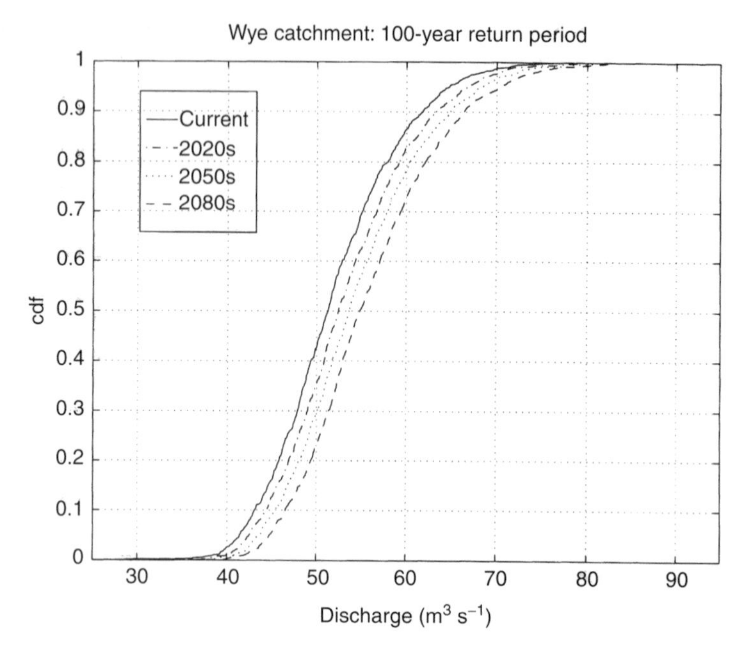

Figure 9.6 *Predicted change in the 100-year return period flood for the Wye catchment at Plynlimon (10.6 km^2) from current conditions to the year 2080, based on the Hadley Centre HADCM2 climate change scenarios, a stochastic rainfall model and the TOPMODEL rainfall–runoff model (after Cameron 2000)*

suggest that the risk of a higher magnitude 100-year return period flood will increase under these scenarios.

This also implies, however, that the risk estimates should be revised or updated using data that become available in the future. This emphasizes the continuing need for monitoring of systems subject to change, to allow conditioning of feasible models and uncertainty in the predictions of those models. The lack of a clear understanding of recent trends in hydrological data suggests that current predictions of hydrological impacts of change may not be very accurate and that the continued collection of data is essential.

9.5 Key Points from Chapter 9

- One of the most important uses of rainfall–runoff models is to predict the effects of future land use and climate change on catchment hydrology, particularly extreme events and water yields for water resource evaluation.
- The natural variability of hydrological systems in space and time, and the generally short periods of available observations, make it very difficult to study, understand and predict the effects of change, even using the most physically based distributed rainfall–runoff models currently available.
- The spatial complexities of land use change, e.g. in urban areas and in agricultural field drainage systems, make it particularly difficult to predict the impacts of change.

The availability of detailed spatial databases in geographical information systems may start to make this easier in the future.

- Climate change predictions based on GCMs are dependent on the representation of the land surface hydrology in the GCM. Computing constraints and the large scale of GCM grid squares still limit the possible hydrological models that can be coupled directly to GCMs.
- The impacts of climate change have to date been predicted using scenario simulations that are conditional both on the rainfall–runoff model being used and on the particular climatic scenario being used as forcing data. Such predictions should be associated with an estimate of uncertainty or risk, but there have been very few studies in which any attempt has been made to assess the predictive uncertainty of the impacts of change.

10

Revisiting the Problem of Model Choice

Remember that the computer is a tool for simulation, and what is simulated becomes the reality of the user. In a society like ours – the post-modern society – there are no 'great stories' to justify a specific perception of reality like there were in the 19th century. We should rather see the situation thus: communication is based on a number of language games which are played according to specific sets of rules. Each group of society can 'play a game', and thus it is the efficiency of each game that justifies it. The computer medium should be seen as a technical device that allows its owner to play particularly efficient games ... A good program is one that creates the reality intended by the sender in the most efficient way.

P. B. Andersen and L. Mathiessen, 1987

However the endeavour to produce such a theory [of flood hydrology] would be well worthwhile. It would improve our understanding of hydrologic phenomena, improve our decision making in relation to water resources, and improve our standing among geophysicists. To accomplish it, we require a broad background knowledge of our own subject and of cognate disciplines and a real capacity to think both imaginatively and to work hard.

Jim Dooge, 1986

10.1 Model Choice in Rainfall–Runoff Modelling as Hypothesis Testing

Despite the fact that we have paid little attention to a whole class of rainfall–runoff models (the conceptual 'bucket' or ESMA models), it should be apparent from the previous chapters that a very wide variety of models are available to any rainfall–runoff modelling application without any clear basis for making a choice between them. Recall the criteria that were established in Chapter 1 for model choice. These may be summarized as follows:

- Is a model readily available, or could it be made available if the investment of time (and money!) appeared to be worthwhile?

- Does the model predict the variables required by the aims of a particular project?
- Are the assumptions made by the model likely to be limiting in terms of what you know about the response of the catchment you are interested in?
- Can all the inputs required by the model, for specification of the flow domain, for the specification of the boundary and initial conditions and for the specification of the parameters values, be provided within the time and cost constraints of a project?

At this stage it should be apparent that these criteria essentially provide the basis for model rejection and, as pointed out way back in Chapter 1, it is all too easy to reject all the available models due to inadequate assumptions or infeasible demands for input data. This is not very helpful: in very many projects, the hydrologist is still required to make quantitative predictions of what might be expected in terms of flood peaks, reservoir inflows or other variables under different conditions. Thus one or more models must be retained.

This is, however, where the idea of conditioning of models becomes very important. We can, to some extent, overcome some of the limitations of the available models by conditioning their predictions on any available observations or prior knowledge about the catchment of interest. Traditionally, this has been done by the calibration or optimization of parameter values, though it has been suggested in Chapter 7 that a more general strategy of conditioning within an uncertainty framework will be a much more satisfactory approach for the future. In this framework, *any* model that predicts a variable of interest is a potentially useful predictor, until there is evidence (or a justifiable opinion) to reject it. The value of conditioning in this way is that the model or models retained must be consistent with the available data (at least to some level of acceptability); otherwise they would have been rejected. This will then give some basis for belief in the predictions when those models are used to extrapolate or predict responses for other conditions.

Try and picture the modelling process as this form of mapping of a particular (unique) catchment or element of a catchment into a model space. (Figure 10.1; for a more detailed discussion, see Beven 2000). Chapter 7 has shown how this mapping can be done using conditioning based on likelihoods or fuzzy weights. For many reasons, primarily associated with the limitations of model structure and parameter estimation, the mapping will be necessarily approximate. The mapping can be done for both gauged catchments and ungauged catchments based on whatever prior information or observations are available, but is likely to be more exact where observations of hydrological response are available and more approximate for the ungauged catchment case. There is, however, the potential to refine this mapping by the collection of more data, and this approach to model evaluation also has implications for the types of data collection exercises that might be used as the basis for a rainfall–runoff modelling study.

In fact, the process of model evaluation can be set up in terms of hypothesis testing. Given a set of models for a catchment, which have survived the initial selection and evaluation procedure, what hypotheses can be tested by the collection of data to allow some of those models to be rejected? It would seem that some types of data might be more valuable than others in testing models and in model rejection. For example, Lamb *et al.* (1998b) have considered the use of water table information in predicting the responses of the small Saeternbekken catchment in Norway (see Sections 6.5, 7.8). Their results highlighted the problem of using spatially distributed observations in

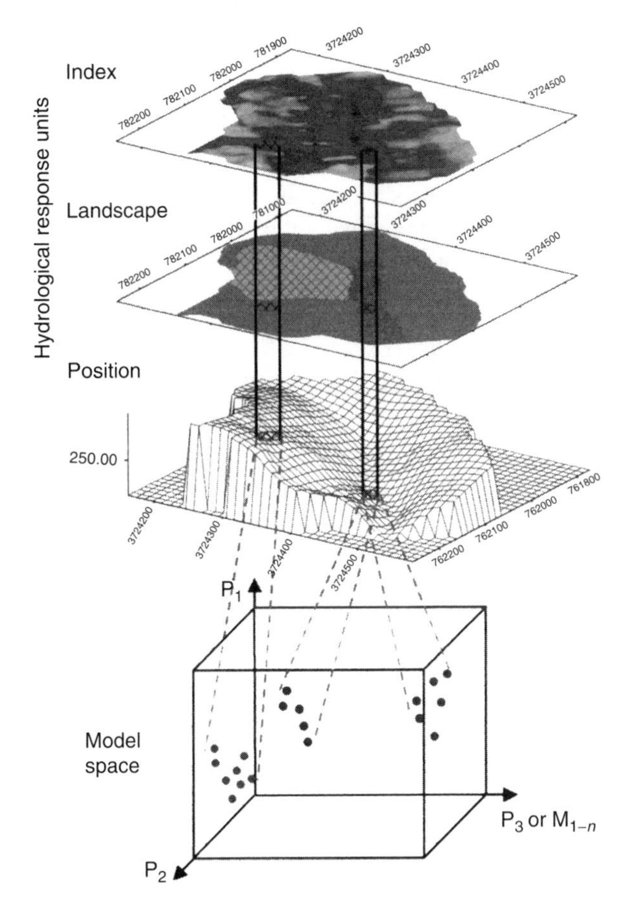

Figure 10.1 *Mapping a catchment space into a model parameter space*

model evaluation in that local water table responses will depend strongly on the local transmissivity and storage characteristics of the soil. Thus it is unlikely, given the heterogeneity of soil characteristics to be expected in a catchment, that even if the model structure were correct, an effective transmissivity and storage coefficient calibrated for a catchment would predict local variations in water table depth accurately. Local parameter values would be needed to make accurate local predictions. Essentially their conclusion was that knowledge of local water tables greatly improved the prediction of local water tables (although even then there were significant local anomalies) but did not help much in constraining the uncertainties in discharge predictions as estimated using the GLUE methodology.

Other types of data can also be used in multi-criteria model evaluation, if a model has the potential to predict the required variables. Different types of geochemical and tracer data, for example, have been used in several studies, including Hooper *et al.* (1986), Kleissen *et al.* (1990) and Mroczkowski *et al.* (1997). It should be noted, however, that in predicting geochemical variables it is also usually necessary to introduce additional parameter values or variables that may themselves not be easily identifiable (but see Robson *et al.* (1992) for an interesting test of TOPMODEL predictions using geochemical mixing).

The difficulty of making measurements remains an issue here. It is much easier (and very much cheaper) to make measurements at, or very close to, the surface. The types of measurements available for the investigation of subsurface flows are much more limited and we still do not have any adequate non-destructive way of assessing preferential flow pathways, whether they be due to natural piping or through mechanically induced cracks to mole drains in an agricultural field. Our perceptual model might allow for the possible importance of such processes, but if their effects must be inferred rather than measured then it is difficult to define an appropriate model description or condition a model on the basis of perceptions alone.

A view of model calibration as a process of hypothesis testing and model rejection would seem to be a positive one in moving ahead from a situation in which the limitations of a simple optimization approach in the face of parameter identifiability problems have become increasingly clear. It also brings together the variety of methods described in Chapter 7 in which multiple criteria are used in the model evaluation, from Pareto optimal sets to GLUE. Each has its own way of deciding on which models should be rejected and what weighting coefficients should be used in projecting the predictions of the retained models. In the Pareto optimal set method, all models that are not part of the set are rejected and each retained parameter set is given equal weight. In GLUE, different ways of combining likelihood measures can be chosen, and the weights are based on the current likelihood value associated with each set.

10.2 The Value of Prior Information

This type of conditioning framework does, however, allow for the use of prior information in the conditioning process. There are many types of prior information that might be used, for example prior degrees of belief in different types of model structure; prior estimates of ranges of parameter values that might be appropriate for different vegetation, soil or rock types; and the perceptual model of how a particular catchment might respond to rainfall. Such information can constrain the range of modelling possibilities to be considered in an application as a result of previous experience in a sensible way.

The use of prior information requires making choices, and rejecting possibilities in particular. As noted above, such choices need to be made with care. There is no shortage of examples in the literature of prior choices of inappropriate models, particularly the use of models based on infiltration excess concepts in applications where this may not be the dominant runoff mechanism (see discussions in Loague and Freeze 1985; Beven 1989; Grayson *et al.* 1992b). It might still be possible to calibrate or condition the model parameters to reproduce the observed discharges but should a model based on inappropriate concepts be used in extrapolation?

The fact that calibration can bring nearly all rainfall–runoff models into line with an observed discharge record can therefore be a problem as well as an advantage, since in many cases we will not know if an inappropriate model is being used. As noted in Chapter 7, there is a very natural tendency in making prior choices about model structures for every model developer to give a prior weight of 1 to his or her model and a prior weight of zero to all other models. This is, however, not a necessary choice:

multiple models may be included, as additional dimensions of the model space, within the modelling framework suggested here.

There is also a natural tendency to give a greater prior weight to models that are considered to be physically based relative to those that are more conceptual or data-based in formulation. Physically based models should, in principle, reflect our understanding of hydrological systems more closely and therefore be more robust in extrapolation to other conditions. This will only be true, however, if such models truly reflect the nature of the catchment processes and it is not clear whether we have really reached this stage, given the limitations of current physically based process descriptions in really representing the nonlinearities of hydrological processes at the scales of hillslopes and catchments, and the difficulties in estimating effective values of the model parameters.

10.3 The Ungauged Catchment Problem

What if there is limited information and no observed discharges for a catchment area for which predictions are required? This is the ungauged catchment problem. It is a common problem, since predictions are often required at sites with no gauging stations or at sites within a larger catchment that is gauged. The ungauged catchment problem is, as yet, not properly resolved due to the fact that it is very difficult to generalize about the nature of catchment responses in any quantitative way. It is a geographical aphorism to say that every catchment or site is unique. This uniqueness underlies all the difficulties inherent in hydrological theorizing and understanding. With the increasing availability of geographical information systems, some of the uniqueness of individual catchments can be reflected as variables in a spatial database. Ideally, this information could then be used to choose an appropriate model and specify the parameters of that model, perhaps with the additional investment of a few extra measurements in the field. This can already be done, but the uncertainties remain high (see for instance the study of Parkin *et al.* (1996), discussed in Section 5.4).

An alternative approach is to use a purely empirical model in which the response characteristics of gauged catchments are related to indices of the characteristics of those catchments. This is normally used within a regression framework with the catchment descriptors being treated as independent variables in predictive equations. An example of this type of study is the UK *Flood Studies Report* (NERC 1975), recently revised as the *Flood Estimation Handbook* (IH, 1999), in which such equations were provided to predict response variables such as time-to-peak, percentage runoff and mean annual flood. These equations can be thought of as empirical rainfall–runoff models, albeit that they are not in any way process-based.

In this type of empirical regression approach, the individuality of the response of individual catchments, as measured, appears as a residual from a model structure. Traditional statistical analysis, of course, assigns the residuals to purely stochastic effects but, from a physical viewpoint, there may still be information in the residuals. Why do some catchments tend on average to produce higher peak flows than the regional modal behaviour? Why do some catchments tend on average to produce lower peak flows, even after taking account of either catchment characteristics (in so far as they can be represented by indices) or network-scale transformations?

One reason is the uniqueness of individual catchments. Perhaps because of the assumptions of traditional statistical analysis, there appear to have been no detailed studies of the information content of such model residuals, but such a study might lead to some improved understanding or theory at the catchment scale. Significant residuals, however, also imply significant uncertainty that may need to be taken into account in predicting the responses of ungauged catchments. The regression framework does also allow uncertainty estimates for the dependent variables to be made, although this is not often done and one has to search hard for any mention of standard errors of estimation in the original *Flood Studies Report*.

10.4 Changing Parameter Values and Predictive Uncertainty

The ungauged catchment problem can be approached by the prediction of hydrological variables directly, without resort to an intermediate rainfall–runoff model. Indeed, if, as suggested in Chapter 7, it is difficult to identify a unique set of model parameters to represent a catchment, then the use of an intermediate model might be better avoided (see, for example, the mixed success in the attempts by Post and Jakeman (1996) and Sefton and Howarth (1998) to relate the parameters of the IHACRES model to catchment characteristics). However, the use of a model is more difficult to avoid in modelling the effects of change.

In all studies of the prediction of the effects of land use change, the major problem is how to speculate about the nature of changes in parameter values that might occur as a result of change, especially given an appreciation of the uncertainties associated with modelling discharges under current conditions. The classic approach to this problem is the scenario approach: given a model of a catchment under current conditions, create a scenario of inputs and parameter values for the changed conditions and analyse the difference in predicted responses. The results are then conditional on the assumptions made in the scenario (and model) but this can be repeated for different scenarios with different inputs or parameter values to give some idea of the sensitivity of the conclusions to different reasonable possible futures under the changed conditions.

It is but a simple extension of the scenario approach to a full Monte Carlo simulation using random choices of parameter values across a feasible range for the changed conditions in the same way as suggested in Chapter 7 for model conditioning under current conditions. If such an approach were adopted, the result would be two sets of uncertain predictions, one for the current conditions and one for the changed conditions. If the change being considered is a future change, the latter would be expected to be more uncertain since the modeller would have to define prior parameter distributions for the changed conditions without the benefits of conditioning. The question then arises as to whether it is possible to distinguish the predictions, given the uncertainty. Few studies have attempted to address this question. In some cases the answer is almost certainly no (e.g. the changing flood frequency characteristics of the River Wye analysed by Cameron 2000 in Section 9.4). In other cases, such as the dramatic effects of fire and regrowth on catchment responses (see the case study of Section 9.2) the answer may well be yes. In one sense, however, it does not actually matter since the comparison of the uncertain predictions before and after a change can still be considered in terms of the change in risk of certain conditions occurring, as discussed in Section 7.10, and those risks can be taken into account in any resulting decision analysis.

Another way of tackling the problem of change, to try to constrain the uncertainty in model predictions over both the short and long term, is through data assimilation. This is widely used in adaptive flood forecasting, using real-time observations of river discharge to correct model predictions over some forecast lead time as described in Chapter 8. It has been less widely used in other areas of rainfall–runoff modelling, but as more remote sensing and other spatially distributed data become available, then it will be possible to incorporate other types of data into assimilation algorithms. This is an area of active research that can be approached in a number of different ways (a recent example due to Houser *et al.* (1998) is described in the case study of Section 5.6), including the updating of model likelihood measures within the GLUE methodology (Box 7.2).

Long-term changes may be tackled in a similar way. Where predictions of future behaviour of a hydrological system are important but highly uncertain, there may be some justification for implementing a measurement program to monitor the impacts of change (including future climate change) within the context of the natural variability of hydrological systems. The assimilation of such monitoring data may, over time, allow the prediction of future behaviour to be made with greater certainty.

10.5 Predictive Uncertainty and Model Validation

What then are the implications of such an approach for model validation? The only validation test that is widely used in rainfall–runoff modelling is the split sample test in which one period of observations is used in calibration and another separate period is used to check that the model predictions are satisfactory. Klemeš (1986) has suggested that the split sample test commonly used in hydrology is a minimal test of model validity. He proposed stronger tests including differential split sample testing, proxy basin testing and proxy basin differential split sample testing. A proxy basin is a basin treated as if it is ungauged, but with some observations of discharges, and perhaps other variables, available for the evaluation of the model predictions. In all these stronger tests, some prior estimation of changes in parameter values following an initial calibration is required, to reflect either some changed condition, or the different characteristics of a proxy basin. Fundamental to this approach to model validation is the concept that a single 'true' model representation should exist.

Model comparisons based on split sample testing commonly reveal that it does not provide a strong test between models of different type and complexity (e.g. Naef 1981; Franchini and Pacciani 1991; Michaud and Sorooshian 1994; Refsgaard and Knudsen 1996). The study of Refsgaard and Knudsen (1996) is one of very few that have gone further in implementing the other tests in comparing three models in an application to catchments in Zimbabwe: a lumped conceptual model, a semi-distributed model based on hydrological response units, and the distributed MIKE SHE model. They qualify their conclusions by noting that they must be model system, climate and catchment dependent, but suggest that, with the exception of the lumped model application to proxy basins, there is little to choose between the results of the different models. All the models (but particularly the lumped model) are improved if a short period of data is made available on the proxy basins to allow some adjustment of the parameter values.

In this book, while recognizing that the single 'true' model is an ideal, it has been suggested that it is an unreachable ideal. We have neither the model structures nor the data necessary to identify that complex, unique, single realization that is the real catchment. Instead, given the limited information available, many models will be acceptable simulators or, put another way, no single model can be validated as the best representation of the catchment.

This should not, however, be unduly worrying. Using the model evaluation and rejection framework suggested here, individual models can be assessed, while whole classes of models might be rejected by collecting additional data to test a particular hypothesis constructed so as to differentiate between different models (e.g. whether saturated contributing areas have been observed in a catchment). Those that survive may be ranked in terms of their current degree of acceptability using one or more performance measures. This evolutionary approach to modelling has interesting elements of both relativism and Popperian falsification.

It also has one significant flaw (though one that will be shared with any other approach to rainfall–runoff modelling). As noted in Section 7.9, it might be all too easy to reject all the available models as, in one sense or another, inadequate representations of the catchment response when their predictions are examined in detail. This would certainly be the case if we required that model predictions were always within a realistic estimate of the measurement error for all the available observations. This does not imply, however, that the model predictions would not be useful. Thus, some degree of compromise in model rejection/acceptance may be necessary to retain some predictive capability. An uncertain or fuzzy catchment to model-space mapping allows this to be done within the limitations of the current state of knowledge in a scientifically rigorous way.

10.6 Final Comments: An Uncertain Future?

It is difficult to predict the future of rainfall–runoff modelling. There are no predictive techniques for methodological advances. Hydrological science is currently in a period of gradual development typical of 'normal' science in the sense of Thomas Kuhn. It was Kuhn who suggested that science often progresses by 'paradigm shifts' interspersed by periods of normal science. Computer advances have certainly made it easier to prepare GIS databases of spatially distributed inputs, to present space and time variable model outputs, and to calibrate and carry out sensitivity analyses of more complex models and parameter spaces, but there have been no major paradigm shifts in approach unless we count a shift from purely deterministic simulation to stochastic formulations (but of much the same model structures).

There were great hopes that a major advance might be forthcoming from the increasing availability of remote sensing images (see Engman and Gurney 1991). However, for hydrological science this has not been the case primarily because of the need for additional interpretative models before such images provide hydrologically useful information. It would appear that the improved sensors to be launched as part of the Earth Observation System programme will yield improved imaging but will also not lead to any radical shifts in approach. Areas in which remote sensing might be useful in rainfall–runoff modelling are in the estimation of surface soil moistures (see Houser *et al.* 1998) and in the estimation of spatial estimates of actual evapotranspiration rates (see e.g. Holwill and Stewart 1992; Bastiaanssen *et al.* 1994). At present, remote sensing estimates are highly uncertain (see Franks and Beven 1997) but could perhaps be greatly improved by new sensors and applications in conjunction with ground-level measurements such as large-scale scintillometer estimates of sensible heat fluxes (McAneny *et al.* 1995; Chehbouni *et al.* 1999). Direct estimates of actual evapotranspiration have a special value in that they could allow a check on closure of the water balance for an area and might be an important constraint on the performance of predictive models.

The major constraint on the utility of remote sensing is that in general it can only detect changes at or above the ground surface and the most interesting parts of hydrology take place underground. What is needed therefore to revolutionize hydrological thinking, theory development and model development is an easily applied technique to quantify water storage or visualize water flows beneath the surface. There are ongoing research experiments using tracers, ground-probing radar and subsurface electrical resistance and seismic tomography that are aimed at improving our understanding of the subsurface, but the routine operational use of any of these techniques seems a long way off. Thus, for the foreseeable future, there will continue to be considerable uncertainty about subsurface flow processes without the possibility of major advances on existing theory.

A major theme of this book has been the apparent equivalence in performance of different model structures or different sets of parameters within a model structure: what we have referred to as equifinality. Equifinality smacks of relativism, something that is not very acceptable to most scientists, including hydrological scientists. There should, after all, be an answer to aim for: a single, proper description of the catchment system that would be generally applicable and recognized as acceptable for predictive purposes. Perhaps such a description will develop in the future, but even if such a model were available, applying it would not necessarily avoid the equifinality problem unless techniques for the direct measurement of effective parameters were evolved at the same time. The unique nature of every catchment system will always require the measurement or estimation of parameter values to reflect that uniqueness. Without direct measurement, some form of conditioning on observed responses would be required and, with what would undoubtedly be a complex model, some equifinality of different parameter sets will certainly ensue.

It is also difficult to envisage that direct measurement of parameter values would be possible within a distributed framework for all locations within a catchment system, even for a small research catchment. This would imply that direct measurement might only be possible at a larger scale. What scale? Would it be easier, on the ground, to establish a nested set of discharge measurement sites as a way of assessing effective parameters for sub-units of a catchment? This would not allow direct measurement of parameters as such, but would allow the assessment of the variability of derived parameters within a catchment area. The spatially distributed description would then involve local subcatchment descriptions. This is, however, much easier to do for different headwater catchments than for slopes contributing to the interior links of a channel network (see Beven 1978) and the derivation of the effective parameters from subcatchment discharge measurements would still involve some uncertainty.

However, perhaps the scale could be refined somewhat. The tracing experiments of Huff *et al.* (1982) and Genereux *et al.* (1993) have shown that it is possible to assess the incremental discharges to a channel, at least approximately, by the dilution of a tracer supplied at a steady rate. If it were possible to assess the changing incremental discharges over time then this would allow an assessment of the differing responses associated with individual hillslope segments within a catchment area. The tracer technique does not seem practical or environmentally acceptable for this purpose but if, for example, a cheap ultrasonic tomography system for measuring velocities and hence discharges in any arbitrary channel shape were developed, then the deployment of multiple systems along a channel network might indeed become feasible.

This implies that a different type of model than the current generation of physically based models might need to evolve; one that was directed at using effective parameter values at the scale of the hillslope or sub-unit of the catchment. We have reached a stage where advances in modelling are likely to follow developments of measurement techniques, and what are needed are measurement techniques that allow an assessment of the variability of responses internal to a catchment at larger scales than at present.

It will undoubtedly continue to be necessary to associate model predictions with an assessment of uncertainty. Recalling the preface of this book, I suggested there that uncertainty estimation would normally be considered a topic for an advanced text rather than a primer in rainfall–runoff modelling. I hope that it is now apparent that uncertainty estimation should be an intrinsic part of the modelling process and associated with every prediction, and in Chapter 7 we have discussed a set of techniques for doing so in a simple, readily understood way. Uncertainty estimation does provide some protection against the possibility that a single deterministic prediction turns out to be wrong in a post-prediction audit (Konikow and Bredehoeft 1992). Having the capability of being wrong gracefully should be a convincing argument for practising hydrological modellers routinely to estimate the uncertainty associated with their predictions!

The future in rainfall–runoff modelling is therefore one of uncertainty: but this then implies a further question as to how best to constrain that uncertainty. The obvious answer is by conditioning on data, making special measurements where time, money and the importance of a particular application allow. It is entirely appropriate that this introduction to available rainfall–runoff modelling techniques should end with this focus on the value of field data.

Appendix A
Demonstration Software

A number of demonstration programs for the Windows environment are available to illustrate some of the concepts discussed in this book. All of the programs can be downloaded from the Web site at http://www.es.lancs.ac.uk/beven2000.html, together with example data sets and Windows help files giving detailed information about using the programs. All of the programs may be freely used in research and teaching.

The TFM program (Section A.1 below) illustrates the use of the lumped, catchment-scale, transfer function modelling described in Chapter 4. The TOPMODEL program (Section A.2), together with the DTM-ANALYSIS digital terrain analysis program (Section A.3), can be used to model the rainfall–runoff process at the catchment scale but allows the distributed predictions to be presented in map form, based on the pattern of the topographic index (see Chapter 6). The TOPMODEL program also has an option to run Monte Carlo sampling of different parameter sets, producing a file that can be used directly with the GLUE program (Section A.4). The GLUE program uses the concepts described in Chapter 7 to produce likelihood weighted estimates of predicted variables, and can evaluate the sensitivity of the predictions to different parameters values.

A.1 TFM

TFM is a program for the analysis of rainfall–discharge catchment data based on Transfer Function Model concepts, similar to those used in the IHACRES model of Jakeman *et al.* (1990, 1993; Jakeman and Hornberger 1994) and the bilinear power model of Young and Beven (1991, 1994) (see Chapter 4).

The opening screen of TFM reveals three buttons: one to QUIT the package, one to go to the Load File option and one to open a Log File. The Log File is used to record the data files used, any data transformations carried out and the results of Model Identification and Parameter Estimation options. The data are saved in ASCII text format for later editing or use in reports.

The basic options used in TFM are as follows.

A.1.1 Load File Option

The Load File option on the main TFM screen or the New Data option on the Plot Data screen allows the entry of file names for the input and output data sets to be used in the analysis. TFM requires input and output data at fixed time intervals to be available.

Conflicts between the input and output data files are identified and flagged by warning messages. Where the two data sets have different file lengths, the length is set to the length of the input data file. It is assumed that both files start at the same time.

Successfully loading the input and output data files brings up a Plot Data screen from which the other options available can be chosen.

Example data files can be downloaded with the program.

A.1.2 Transform Option

The Transform option in the Plot Data screen allows two nonlinear transformations to be made to the input data, corresponding to the bilinear power model and the storage or soil moisture index approaches (the latter is a simplification of the IHACRES approach). These transformations are required to create an effective rainfall input that may be more linearly related to the output than the original rainfall data. Each require a single parameter value (either the power in the bilinear power model or the time constant of the storage). These are easily optimized for any data set by making repeated runs of the model.

Other options allow an initial run of data to be excluded from the analysis, or the data file to be reduced by skipping every nth value. The transformed file can be saved or the original data can be restored.

A.1.3 The Identify Option

The Identify option allows a wide range of model structures to be evaluated quickly. The models are specified in terms of a range of numbers of a parameters, a range of numbers of b parameters, and a range of time delays.

The models are ranked in terms of the value of the Young Information Criterion or YIC (see Box 4.1). This should be as negative as possible, indicating both a good fit between observed and predicted outputs and well-defined parameter values. A value for the coefficient of determination R_t^2 is also given. This is 1 for a model giving a perfect fit, and 0 for a model that is no better than assuming a mean output value.

Underlying the model identification step is the idea that the data should be allowed to indicate which model structure is most appropriate rather than specifying a structure beforehand. There may be many models giving an acceptable fit to the data. The simplest model giving the most negative YIC value and a physically reasonable transfer function should then be chosen.

A double-click of the mouse on any of the models listed after the Identification process will take control directly to the Estimate option screen and will estimate the parameters of the model.

A.1.4 The Estimate Option

The Estimate option carries out the final estimation of the parameters of the chosen model. Results from the most recent estimation can be plotted. The most recent results are shown on the table of results shown on screen. If a Log File is open then the results are also written to the Log File.

Parameter estimation is carried out using a Simplified Refined Instrumental Variable (SRIV) technique which has been shown to be very robust to data errors (see Young 1984). The results may be evaluated using the YIC and R_t^2 criteria.

A number of other options allow graphical and statistical examination of the results.

- *Plot Model* gives a plot of the observed and predicted outputs, together with a plot of the observed error series. The modelled time series can be saved to a file.
- *Plot TF* gives a plot of the transfer function for the chosen model. This should be checked to ensure that the model transfer function is always positive and does not show any significant oscillatory behaviour or instabilities. The transfer function may be saved to a file.
- *Composite* gives a combined plot of the data, model fit and transfer function. This may be printed as a summary record of the analysis. The transfer function may be saved to a file.
- *Validate* allows the model to be tested against a new data set. The Validate option can only be chosen after the Estimate option has been performed and a model is available. It may be used in two ways depending on whether both input and output data are available. If both are available, the model is assessed in terms of fitting the new data for the validation period in terms of an R_t^2 value. If only input data are available, the model may be used to predict the missing flow data.

  To use the Validate option, fill in the file name fields (full path and extension required) and choose 'calculate'. Once the new outputs have been simulated, the data may be saved to a file.

A.2 TOPMODEL

This program is intended as a demonstration version of TOPMODEL for Windows and has been developed from versions used for teaching purposes over a number of years in the Environmental Science degree course at Lancaster University. Since 1974 there have been many variants of TOPMODEL developed at Leeds, Lancaster and elsewhere but never a 'definitive' version. This has been quite intentional. TOPMODEL is not intended to be a traditional model package but is more a collection of concepts that can be used where appropriate. It is up to the user to verify that the assumptions made are appropriate (see the discussions of limitations in Beven *et al.* 1995 or Beven 1997). This version of this program will be best suited to catchments with shallow soils and moderate topography which do not suffer from excessively long dry periods. Ideally, predicted contributing areas should be checked against what actually happens in the catchment (at least qualitatively).

The model supplied here has deliberately not been provided with an automatic optimization routine. This is for two reasons. First, the user is encouraged to view the output from the model and think about how the model is working. This is made possible, in part, by the fact that the results can be mapped back into space and viewed by the user in their correct spatial context. In this way, it may be concluded that this is not a good model to represent a particular catchment (but by thinking about *why*, it may be possible to improve the representation in some relatively simple way). This is why the distributed nature of the model predictions, combined with a simplicity of structure, is very important. Use it first as an aid to understanding before using it as a predictive tool.

Secondly, we do not believe that there is an optimum set of parameter values, even with a model that is as parametrically parsimonious as TOPMODEL, and do not want to encourage the practice of automatic optimization. At Lancaster we are now using the GLUE methodology (see Chapter 7) to carry out calibration/sensitivity analysis/uncertainty estimation based on many thousands of runs. This version of TOPMODEL provides an option for output of Monte Carlo simulation results for later use with the compatible GLUE package (see Section A.5 below).

Each catchment application requires a *project file*. This file has only four lines as follows:

1. Text description of application
2. Catchment Data filename
3. Hydrological Input Data filename
4. Topographic Index Map filename (may be left blank, but line must exist)

An example Project file and associated data files are also provided.

The following three options are available in the program.

A.2.1 The Hydrograph Prediction Option

The Hydrograph Prediction option allows the model to be run and hydrographs displayed. Parameter values can be changed on screen and the model run again. After each run, several indices of goodness of fit are given for evaluation.

If not all the time period of the simulation can be displayed at one time, the hydrograph can be scanned backwards and forwards using the arrow buttons.

If a Topographic Index Map file is available then a map button is displayed which allows the display of predicted simulation, either as a summary over all time steps or animated.

A.2.2 The Sensitivity Analysis Option

This screen allows the sensitivity of the objective functions to changes of one or more of the parameters being explored. An initial run of the model is made with the current values of the parameters. Then each chosen parameter is varied across its range, *keeping the values of the other parameters constant.* The results are displayed as small graphs. Any of the current parameter values or the minimum and maximum of the range to be included can be changed on screen.

A.2.3 The Monte Carlo Analysis Option

In this option a large number of runs of the model can be made (limited only by the storage capacity of the results file!) using uniform random samples of the parameters chosen for inclusion in the analysis. Values of the other parameters are kept constant at their current values. Check boxes can be used to choose the variables and objective functions to be saved for each run.

The Results file produced will be compatible with the GLUE analysis software package.

A.3 DTM-ANALYSIS

The DTM-ANALYSIS program is used to derive a distribution of $\ln(a/\tan B)$ values from a regular raster grid of elevations for any catchment or subcatchment using the multiple direction flow algorithm of Quinn *et al.* (1995). Output from the program is a histogram of the distribution of the $\ln(a/\tan B)$ values, and a map file of $\ln(a/\tan B)$ values that can be used for map output in the TOPMODEL programs.

There are three options in the program.

A.3.1 Topographic Index Distribution Calculation Option

This Topographic Index Distribution Calculation option is the very simplest version of the algorithm in that while pits and sinks in the elevation matrix are identified, no attempt is made to create continuous flow pathways to the catchment outlet. Sinks should be removed by modifying the elevation data using the Automatic Sink Removal option which uses successive averaging of surrounding elevations to resolve pits. This is simple but does handle the case of small river channels for which the elevation grid cannot resolve the continuous flow pathway.

This Topographic Index Distribution Calculation option also requires that only elevations of points within the catchment are supplied, all other values in the matrix being set to a value greater than 9999.0 (m). The Catchment Identification option can be used to cut out the catchment above a specified pixel using a hill-climbing algorithm.

It is recommended that the elevation data should be of 50 m resolution or better. It should be noted that the derived $\ln(a/\tan B)$ distribution will be dependent on the resolution of the elevation data used, and on the particular rules for distributing upslope areas and dealing with river channels that are smaller than the grid size. Different distributions may result in different effective parameter values for a given catchment.

A.3.2 Automatic Sink Removal Option

This option can be used to automatically correct for sinks and pits in an elevation array before running the Topographic Index Calculation option. The program will identify any local sinks (values surrounded on all nodes by equal or higher elevations) and will try to correct them by averaging the surrounding elevations. After each run through all the identified sinks, the program can check to see if there are any remaining sinks.

Note that this algorithm will not identify any large internal basins in the elevation map directly. However, such basins will always result in one local sink. Therefore by iterating through the program it should be possible to correct such features.

At the end of the option the new elevation data can be saved in a file. The output file is in exactly the same format as the input file. A separate file identifying all corrected sinks and changes in elevation values can also be saved.

A.3.3 Catchment Identification Option

The topographic index calculation requires that all elevation values outside the catchment be labelled as values of 9999.0 or greater. This option takes a raster elevation data file and identifies the catchment area draining to a specified point. The algorithm works by linking all points that are continuously upslope of the catchment outlet. At each iteration, the algorithm searches for points adjacent to those already identified as being in the catchment that are at an elevation greater than a specified threshold value. A value of zero may be specified for the threshold (i.e. adjacent points of equal elevation will be added to the catchment) but this can cause problems in some elevation arrays that contain flat benches extending around catchment divides. If this is the case the original elevation interpolation should be improved, or the resolution of the elevation matrix should be degraded.

A.4 GLUE

The GLUE package provides tools for sensitivity analysis and uncertainty estimation using the results of Monte Carlo simulations. The starting point for the GLUE concepts is the rejection of the idea of an optimum parameter set in favour of the concept of equifinality of model structures and parameter sets which implies that it is only possible to evaluate the relative performance of the range of possible models, either qualitatively or quantitatively, in terms of some likelihood measure. Note that likelihood is used here in a much wider sense than the likelihood functions of statistical estimation theory, but that traditional likelihood functions can be used within the GLUE framework. Note also that a fuzzy interpretation of the likelihood measures is also possible.

The predictions of the Monte Carlo realizations are then weighted by the likelihood measures to determine the prediction limits of the required variables. Thus, those parameter set realizations that perform well in the evaluation are given the greatest weight in prediction. No distribution assumptions are made in determining the prediction limits – they are based only on the available sample of predictions. There are facilities in the software for transforming the likelihood measures input to the package (e.g. by raising to a power to make the parameter response surface more peaked or by changing the threshold of acceptability). Parameter sets with likelihood measures below that threshold are then considered to be non-behavioural and have their likelihood values set to zero.

The GLUE methodology focuses attention on the subjective nature of model evaluation (e.g. choice of likelihood measure, choice of threshold value) but requires that those elements be defined explicitly and therefore made open to debate and justification.

Parameter interactions and nonlinearity in the model responses (which may be extremely complex and potentially even chaotic) are handled implicitly in the GLUE methodology. In essence, the nonlinear response of a particular model parameter set is summarized by the associated likelihood value, i.e. the performance of that particular model realization in reproducing the observations. Thus the analysis focuses on parameter *sets* rather than the behaviour of individual parameters and their interactions (although some facilities are provided in the Sensitivity Plots option to examine the sensitivity to individual parameters).

Errors in input data and the observation data are also handled implicitly. Thus the likelihood measure reflects the ability of a particular model to predict a particular series of observations (that may not be error-free) given a particular set of inputs (that may not be error-free). There is thus an implicit assumption that in prediction, error structures will be 'similar' in some broad sense to those in the evaluation period.

A major limitation of the GLUE methodology is the dependence on Monte Carlo simulation. For complex models requiring a great deal of computer time for a single run, it will not be possible to fully explore high-order parameter response surfaces. However, experience suggests that the upper limit of model performance is often well defined by a limited number of model realizations and that prediction intervals are reasonable in comparison with observations.

The options available in the program are as follows.

A.4.1 Dotty Plots

The Dotty Plot screen provides plots of any likelihood measure or predicted variable against value of each individual parameter. One dot is plotted for each Monte Carlo run in the input data file. The Dotty Plots are therefore projections of all the Monte Carlo samples onto single-parameter axes. As such they should be interpreted with care as they can conceal some of the structure in the N-dimensional parameter response surface. The chosen likelihood measure or variable can be changed or transformed by choosing from the menu line options.

A.4.2 Sensitivity Plots

The Sensitivity Plots screen provides plots of the cumulative distributions of parameter values group according to the ranking of each Monte Carlo run for a given likelihood measure or predicted variable. Strong differences between the cumulative distributions for a given likelihood measure or variable indicates sensitivity to that parameter. Distributions plotting close together indicate a lack of sensitivity. The chosen likelihood measure or variable can be changed or transformed by choosing from the menu line options.

A.4.3 Uncertainty Plots

This screen provides plots of the histogram and cumulative distribution of a predicted variable, weighted by a chosen likelihood measure. Sample quantiles (5 and 95 percent) are shown on the cumulative distribution plot. Both likelihood measure and variable can be changed by choosing from the menu line options.

A.4.4 List Simulations

This option provides an output of the top (or bottom) 20 simulations ranked according to either a likelihood measure or a predicted variable.

A.4.5 Transforming Likelihood Measures

This option provides a number of transform options, including applying a lower threshold limit, raising a likelihood measure to a power, taking the exponential of a likelihood multiplied by a coefficient, and taking the log of the measure. New likelihood measures derived in this way may be stored or edited.

A.4.6 Combining Likelihood Measures

The need to combine different likelihood measures arises in a number of different circumstances including combining likelihood measures for different types of model evaluation (such as one measure calculated for the prediction of discharges and one calculated for the prediction of water table or soil moisture levels); or updating an existing likelihood estimate with a new measure calculated for the prediction of a new set of observations. Four different types of combination operator are allowed in the GLUE software: Bayes multiplication, weighted addition, fuzzy union, and fuzzy intersection (see Box 7.2).

Appendix B
Glossary of Terms

Actual evapotranspiration The rate of evapotranspiration from a surface or vegetation canopy to the atmosphere under the prevailing meteorological conditions and water availability. [Section 3.3; Box 3.1]

Aerodynamic resistance Scaling parameter for sensible and latent heat fluxes in Penman–Monteith equation. [Section 3.3; Box 3.1]

Anisotropic Descriptive adjective for a porous medium in which hydraulic conductivity is preferentially greater in certain flow directions (see also *isotropic*). [Box 5.1]

Antecedent conditions The state of wetness of a catchment prior to an event or period of simulation. [Section 1.4]

Aquiclude A layer of soil or rock that is impermeable to water. [Section 5.1.1]

Atmospheric demand The rate of *potential evapotranspiration* for given atmospheric conditions of temperature, humidity and wind speed without any limit due to the availability of water. [Section 3.3]

Autocorrelated errors A time series of model residuals that are not independent at each time step, i.e. that exhibit statistical correlation at one or more time steps apart (see also *heteroscedastic errors*). [Section 7.3; Box 7.1]

Automatic optimization Calibration of model parameters using a computer algorithm to maximize or minimize the value of an *objective function*. [Sections 7.1, 7.4]

Baseflow That part of the discharge hydrograph that would continue after an event if there were no further rainfall. Sometimes taken to be equivalent to the total subsurface flow contribution to stream discharge, but environmental tracer measurements suggest that this is not good usage of the term since subsurface flow may be the dominant contribution to the hydrograph in many storms. [Section 2.2]

Baseflow separation A procedure associated with use of the *unit hydrograph* to separate the hydrograph into *storm runoff* and *baseflow* components. Many different methods are available, mostly without any firm basis. [Section 2.2]

Basis functions Interpolation functions used in representing the variation of a predicted variable within each element of a *finite-element* solution. [Box 5.3]

Bayes equation Equation for calculating a posterior probability given a prior probability and a likelihood function. Used in the GLUE methodology to calculate posterior model likelihood weights from subjective prior weights and a *likelihood measure* chosen for model evaluation. [Section 7.7.4; Box 7.2]

Behavioural simulation A simulation that gives an acceptable reproduction of any observations available for model evaluation. Simulations that are not acceptable are non-behavioural. [Sections 7.2.1, 7.7]

Big leaf model The representation of a vegetation canopy in predicting evapotranspiration as if it was a uniform surface. [Box 3.1]

Black box model A model that relates only an input to a predicted output by a mathematical function or functions without any attempt to describe the processes controlling the response of the system. [Sections 1.1, 4.1]

Blind validation Evaluation of a model using parameter values estimated before the modeller has seen any output data. [Section 5.4]

Boundary conditions Constraints and values of variables required to run a model for a particular flow domain and time period. May include input variables such as rainfalls and temperatures; or constraints such as specifying a fixed head (Dirichlet boundary condition), impermeable boundary (Neumann boundary condition) or specified flux rate (Cauchy boundary condition). [Sections 1.3, 5.1]

Calibration The process of adjusting parameter values of a model to obtain a better fit between observed and predicted variables. May be done manually or using an automatic calibration algorithm. [Section 1.8; Chapter 7]

Canopy resistance An effective resistance to the transport of water vapour from leaf stomata to the atmosphere. [Section 3.3]

Capillary potential The pressure, relative to atmospheric pressure, at which soil water is held in the pore space of a soil. In an unsaturated soil, capillary potential takes on negative values equivalent to the pressure drop across the curved air–water interfaces in the soil pores. [Section 5.1.1; Box 5.1]

Celerity or wave speed The speed with which a disturbance of pressure propagates through the flow domain. Important in the explanation of the large 'old water' component of storm runoff in many catchments. May vary greatly with different processes and with catchment wetness. [Sections 1.5, 5.5; Box 5.7]

Complementarity approach A method for the prediction of *actual evapotranspiration* based on the idea that the greater the actual evapotranspiration rate (and therefore the ambient humidity), the lower will be a measurement of evaporation from a free water surface or evaporation pan. [Section 3.3]

Conceptual model A hydrological model defined in the form of mathematical equations. A simplification of a *perceptual model*. [Section 1.3]

Contributing area A term used in a variety of ways in hydrology. Most often it refers to the part of the catchment contributing either surface or subsurface storm runoff to the hydrograph. [Section 1.4]

Data assimilation The process of using observational data to update model predictions (see also *real-time forecasting and updating*). [Section 5.6]

Degree-day method A method for predicting snowmelt as proportional to the difference between mean daily temperature and a threshold value. [Section 3.4]

Depression storage Water in excess of the *infiltration capacity* of the soil that is retained in surface hollows before significant downslope overland flow occurs. May later infiltrate into the soil after rainfall has ceased. [Section 1.4]

Deterministic model A model that with a set of initial and boundary conditions has only one possible outcome or prediction. [Section 1.7]

Diffusivity The produce of unsaturated hydraulic conductivity and the gradient of the curve relating capillary potential to soil moisture content. [Section 5.1.1; Box 5.1]

Distributed model A model that predicts values of state variables varying in space (and normally time). [Section 1.7]

Dotty plots A way of representing the results of Monte Carlo simulations in which an objective function from each simulation is plotted against the randomly chosen value of each parameter. Dotty plots therefore represent a projection of sample points on the response surface onto a single parameter axis. See also *objective function*, *response surface*. [Sections 1.8, 7.7]

Double mass curve A plot of the cumulative volumes associated with two measurement stations (either rainfalls or discharges). [Section 3.2]

Dynamic contributing area The area generating surface runoff that will tend to expand during a storm. [Section 1.4]

Eddy correlation method A technique for measuring *actual evapotranspiration* and sensible heat fluxes by integrating the rapid fluctuations in humidity and temperature associated with turbulent eddies in the lower boundary layer. [Section 3.3.3]

Effective rainfall A part of the storm rainfall inputs to a catchment that is equivalent in volume to the '*storm runoff*' part of the hydrograph (but note that the storm runoff may not be all rainstorm water). [Sections 1.3, 2.2]

Effective storage capacity The difference between the current soil moisture in the soil immediately above the water table and saturation. [Section 1.5]

Ephemeral streams Streams that are frequently dry between storm periods. [Section 1.4]

Equifinality The concept that there may be many models of a catchment that are acceptably consistent with the observations available. [Sections 1.8, 7.7, 7.9]

ESMA model See *explicit soil moisture accounting model.*

Evaluation See *validation.*

Explicit solution The independent calculation of predicted variables at a time step given values of the variables at the previous time step (see also *implicit solution*). [Section 5.1.1; Box 5.3]

Explicit soil moisture accounting model (or ESMA; sometimes called conceptual models). A hydrological model made up of a series of storage elements, with simple equations to control transfers between the elements. Mostly applied at the lumped scale, but some models use ESMA components to represent distributed hydrological response units. [Section 2.4]

Field capacity An imprecisely defined variable normally expressed as the water content of the soil when it is allowed to drain from saturation until rapid drainage has ceased (see *soil moisture deficit*). [Box 6.2]

Finite difference The approximate representation of a time or space differential in terms of variables separated by discrete increments in time or space. [Box 5.3]

Finite-element method The approximate representation of time or space differentials in terms of integrals of simple interpolation functions involving variables defined at nodes of an irregular discretization of the flow domain into elements. [Box 5.3]

Fuzzy logic A system of logical rules involving variables associated with a continuous fuzzy measure (normally in the range 0 to 1) rather than the binary measure (right/wrong, 0 or 1) of traditional logic. Rules are available for operations such as addition and multiplication of fuzzy measures and for variables grouped in fuzzy sets. Such rules can be used to reflect imperfect knowledge of how a variable will respond in different circumstances. [Sections 1.7, 5.2.2]

Gain A multiplier applied to a transfer function to scale inputs to outputs in a linear systems analysis; may be made adaptive in *real-time forecasting*. [Box 8.1]

Geomorphological unit hydrograph A *unit hydrograph* derived from the structural relationships of the catchment geomorphology, in particular the branching structure of the channel network. [Sections 2.3, 4.7.2]

Global optimum A set of parameter values that gives the best fit possible to a set of observations. [Sections 1.8, 7.2]

Head An expression of pressure as energy per unit weight, commonly used in hydrology and hydraulics since it has units of length. [Section 5.1.1]

Heteroscedastic errors A time series of model residuals that exhibit a changing variance over a simulation period (see also *autocorrelated errors*). [Section 7.3; Box 7.1]

Hortonian model Runoff production by an *infiltration excess* mechanism. Named after Robert E. Horton (see also Partial Area Model). [Section 1.4]

Hydrological response unit A parcel of the land surface defined in terms of its soil, vegetation and topographic characteristics. [Sections 1.7, 2.3, 3.8, 6.1, 6.3]

Hysteresis A term used to indicate that the relationship between soil water content and capillary potential or hydraulic conductivity is different when the soil is wetting compared with when it is drying. [Section 5.1.1; Box 5.1]

Implicit solution The simultaneous solution of predicted variables at a time step given values of the variables at the previous time step, usually by an iterative method (see also *explicit solution*). [Section 5.1.1; Box 5.3]

Incommensurate Used here to refer to variables or parameters with the same name that refer to different quantities because of a change in scale. [Section 1.8]

Infiltration capacity The limiting rate at which a soil surface can absorb rainfall. It will depend on factors such as antecedent moisture content, volume of infiltrated water, presence of macropores or surface crusting. [Section 1.4; Box 5.2]

Infiltration excess runoff Runoff generated by rainfall intensities exceeding the *infiltration capacity* of the soil surface. May be used either at the local point scale within a catchment (when the surface runoff may infiltrate further downslope), or at the catchment scale to represent that part of the storm hydrograph generated by an infiltration excess mechanism. [Section 1.4]

Initial conditions Values of storage or pressure variables required to initialize a model at the start of a simulation period. [Section 5.1]

Interception Rainfall that is stored on a vegetation canopy and later evaporated back to the atmosphere. [Section 3.3.2; Box 3.2]

Inverse method Model *calibration* by adjusting parameter values to reduce the differences between observations and predicted variables. [Section 5.1.1]

Isotropic Descriptive adjective for a porous medium in which the hydraulic conductivity is the same in all flow directions (see also *anisotropic*). [Box 5.1]

Land surface parametrization Hydrological model used to calculate water and energy fluxes from the land surface to the atmosphere in atmospheric circulation models. [Section 2.4]

Lead time The required time for a forecast ahead of the current time in *real-time forecasting*. [Section 8.1]

Learning set A set of observed data used in the calibration of a neural net model. [Section 4.9]

Likelihood measure A quantitative measure of the acceptability of a particular model or parameter set in reproducing the hydrological response being modelled. [Section 7.7; Boxes 7.1, 7.2]

Linearity A model is linear if the outputs are in direct proportion to the inputs. [Section 2.2; Boxes 2.1, 4.1]

Linear store A model component in which the output is directly proportional to the current storage value. The basic building block of the general linear *transfer function* model and the Nash cascade. [Section 2.3; Box 4.1]

Local optimum A local peak in the parameter response surface where a set of parameter values gives a better fit to the observations than all parameter sets around it, but not as good a fit as the *global optimum*. [Section 7.2]

Lumped model A model that treats the whole of a catchment as a single accounting unit and predicts only values of variables averaged over the catchment area. [Sections 1.5, 1.7]

Macropores Large pores in the soil that may form important pathways for infiltration and redistribution of water bypassing the soil matrix as a *preferential flow*. May result from soil cracking and ped formation, root channels and animal burrows. [Section 1.4]

Monte Carlo simulation Simulation involving multiple runs of a model using different randomly chosen sets of parameter values or *boundary conditions*. [Sections 7.5, 7.6, 7.7]

Network width functions A histogram of the number of reaches in the channel network at a given distance from the catchment outlet. Can be used as the basis for both linear and nonlinear flow routeing algorithms. [Section 2.3, 4.7.1]

Nomogram An empirical method for estimating runoff by using a series of graphs. [Section 2.1]

Nonlinear A model is nonlinear if the outputs are not in direct proportion to the inputs but may vary with intensity or volume of the inputs or with antecedent conditions. [Box 2.1]

Nonparametric method A method of estimating distributions without making any assumptions about the mathematical form of the distribution. [Section 7.2.1]

Nonstationarity A model in which the parameters are expected to change over time. [Box 2.1]

Objective function A measure of how well a simulation fits the available observations. [Sections 1.8, 7.3; Box 7.1]

Optimization The process of finding a parameter set that gives the best fit of a model to the observations available. May be done manually or using an *automatic optimization* algorithm. [Sections 1.8, 7.4]

Overland flow Downslope flow of water on the surface of the soil in excess of the *infiltration capacity* and *depression storage* capacity of the surface. [Section 1.4]

Parameter A constant that must be defined before running a model simulation. [Sections 1.6, 1.8]

Parameter space A space defined by the ranges of feasible model parameters, with one dimension for each parameter. [Section 1.8, 7.2]

Parsimony The concept, sometimes known as Occam's razor, that a model should be no more complex than necessary to predict the observations sufficiently accurately to be useful. [Box 4.1]

Partial area model Runoff production (by an *infiltration excess* mechanism) only over part of the hillslopes (the partial area) in a catchment. [Section 1.4]

Pedotransfer function A function for predicting soil hydraulic parameters from knowledge of soil texture and other more easily measured variables. [Sections 3.8, 5.1.1; Box 5.5]

Perceptual model A qualitative description of the processes thought to be controlling the hydrological response of an area. [Sections 1.3, 1.4]

Phreatophytes Plants whose roots extract water from below the water table. [Section 1.4]

Potential evapotranspiration Rate of evapotranspiration from a surface or vegetation canopy with no limitation due to water availability (see also *atmospheric demand*). [Section 3.3; Box 3.1]

Preferential flow Local concentrations of flow in the soil that may be due to the effects of *macropores*, local variations in hydraulic properties or fingering of a wetting front moving into the soil profile. May lead to rapid and deep infiltration of water bypassing much of the soil matrix. [Section 1.4]

Principle of superposition The addition of linear model responses to construct a total response. [Section 2.2; Box 2.1]

Procedural model A model represented as a computer program. May be an exact or approximate solution of the equations defining the *conceptual model* of a system. [Section 1.3]

Raster digital elevation model A gridded set of elevation values at regular spacing. [Section 3.7]

Rational method An empirical method, first used in the 19th century, for predicting peak discharges based on catchment area and a measure of average rainfall. [Section 2.1]

Real-time forecasting and updating Operational forecasting of flows during an event, usually to predict the possibility of flooding, often with adaptive updating of model parameters based on the errors between observed and predicted variables (see also *lead time*). [Sections 4.8, 8.4; Box 8.1]

Reliability analysis Evaluation of the uncertainty in model predictions arising from uncertainty in parameter values, often by assuming a certain shape for the response surface (see also *response surface*). [Sections 7.1, 7.5]

Response surface The surface defined by the values of an *objective function* as it changes with changes in parameter values. May be thought of conceptually as a surface with 'peaks' and 'troughs' in the multidimensional space defined by the parameter dimensions, where the 'peaks' represent good fits to the observations and the 'troughs' represent poor fits to the observations (see also *parameter space*). [Sections 1.8, 7.2]

Riparian area The part of the catchment that is immediately adjacent to the stream and is often the most important source of runoff, both surface and subsurface. [Section 1.4]

Runoff See *overland flow, storm runoff, surface runoff, subsurface stormflow.*

Runoff coefficient The proportion of the rainfall volume in a storm that appears in the storm hydrograph. The value will depend on how the *storm runoff* component of the hydrograph is defined. [Section 2.2]

Runoff routeing The translation of surface runoff and subsurface stormflow to a point of interest, usually the catchment outlet, taking account of surface, subsurface and channel flow velocities. [Sections 1.6, 4.4, 5.5, 5.6, 6.1]

Saturation excess runoff Runoff generated by rainfall onto saturated soil, even though the rainfall intensity may not exceed the normal infiltration capacity of the soil. May be used either at the local point scale within a catchment (when the surface runoff may infiltrate further downslope), or at the catchment scale to represent that part of the storm hydrograph generated by a saturation excess mechanism. [Section 1.4]

Similar media A method of scaling the *soil moisture characteristics* of heterogeneous soils by making assumptions about the structure of the media (e.g. that the geometry of the soil matrix is identical, different only in length scale between different samples). [Box 5.4]

Slope-area method Method of measuring a peak discharge after a flood event using one of the *uniform flow* equations by estimating the cross-sectional area, water surface slope and roughness coefficient at a site. [Section 3.2]

Snow course Transect where regular measurements of snow depth and density are made. [Section 3.1]

Soil moisture characteristics Curves or functions relating soil moisture content to unsaturated hydraulic conductivity and capillary potential. [Section 5.1.1; Box 5.2]

Soil moisture deficit (SMD) A state variable used in many hydrological models as an expression of soil water storage. SMD is zero when the soil is at *field capacity* and gets larger as the soil dries out. It is usually expressed in units of depth of water. [Sections 1.4, 3.1]

Specific moisture capacity The gradient of the curve relating unsaturated soil moisture to capillary potential. [Section 5.1.1; Box 5.1]

State variable A variable in the model that is part of the solution of the model equations and which varies with time during a simulation but which is not a flux or exchange of mass. May include storage and pressure variables, depending on the definition of the model. [Section 5.8]

Stemflow Rainfall that reaches the ground via the stems of plants. [Section 1.4; Box 3.2]

Stochastic A model is stochastic if, for a given set of initial and boundary conditions, it may have a range of possible outcomes, often with each outcome associated with an estimated probability. [Section 1.7]

Storm profile Time series of rainfall intensities during a storm. [Section 3.1]

Storm runoff There are many conflicting definitions of storm runoff. Here it is that part of the stream hydrograph due to a rainfall event over and above the discharge that would have occurred without a rainfall event and may involve both surface and subsurface flow processes and both event and 'old' water contributions. [Sections 1.4, 1.5, 1.6]

Streamline A line parallel to the direction of flow (see also *stream tube*). [Section 3.7]

Stream tube A part of a flow domain enclosed between two specified *streamlines*. [Section 3.7]

Sublimation Direct loss of water from a snowpack to the atmosphere by evaporation. [Section 3.1]

Subsurface stormflow The contribution to the stream hydrograph generated by purely subsurface flow processes. [Section 1.4]

Surface runoff The contribution to the stream hydrograph from overland flow. [Section 1.4]

Tessellation The discretization of space into a spatial grid or network of elements. [Section 3.7]

Throughfall Rainfall that reaches the ground either directly or indirectly by dripping from the leaves of plants. [Section 1.4; Box 3.2]

Throughflow Often used for rapid near-surface downslope subsurface flow in the soil profile. [Section 1.4]

Time compression assumption Treating the volume of infiltrated water during an event as if it had infiltrated at the *infiltration capacity* of the soil in order to calculate an equivalent *time to ponding*. [Box 5.2]

Time to ponding The time taken during a rainfall event to bring the soil surface just to saturation. [Box 5.2]

Transfer function A representation of the output from a system due to a unit input. [Section 2.2; Chapter 4]

Triangular irregular network A way of representing topography by a network of triangles between points of known elevation. [Section 3.7]

Uniform flow An open channel flow or overland flow in which the slope of the water surface is equal to the bed slope such that energy loss due to frictional shear stress is exactly matched by the potential energy gain as the water moves downhill. [Section 5.5.2; Box 5.6]

Unit hydrograph The *storm runoff* response from a unit of *effective rainfall*. [Sections 2.2, 2.3, 4.3]

Validation A process of evaluation of models to confirm that they are acceptable representations of a system. Philosophers of science have some problems with the concept of validation (see Section 1.8) and it may be better to use 'evaluation' or 'confirmation' rather than validation (which from its Latin root implies a degree of truth in the model). [Sections 1.8, 5.3, 10.5]

Vector digital elevation model A set of irregularly spaced elevation points defining contours of equal elevation. [Section 3.7]

Wave speed See *celerity*.

References

Abbott M. B. and Refsgaard J.-C. (1996) *Distributed Hydrological Modelling.* Kluwer Academic, Dordrecht.

Abbot M. B., Bathurst J. C., Cunge J. A., O'Connell P. E. and Rasmussen J. (1986a) An introduction to the European Hydrological System – Système Hydrologique Européen, SHE. 1. History and philosophy of a physically-based, distributed modelling system. *Journal of Hydrology* **87**: 45–59.

Abbot M. B., Bathurst J. C., Cunge J. A., O'Connell P. E. and Rasmussen J. (1986b) An introduction to the European Hydrological System–Système Hydrologique Européen, SHE. 2. Structure of a physically-based, distributed modelling system. *Journal of Hydrology* **87**: 61–77.

Abdullah F. A. and Lettenmaier D. P. (1997) Application of regional parameter estimation schemes to simulate the water balance of a large continental river. *Journal of Hydrology* **197**: 258–285.

Abdullah F. A., Lettenmaier D. P., Wood E. F. and Smith J. (1996) Application of a macroscale hydrologic model to the water balance of the Arkansas–Red River basin. *Journal of Geophysical Research* **101**: 7449–7460.

Abeliuk R. and Wheater H. S. (1990) Parameter identification of solute transport models for unsaturated soils. *Journal of Hydrology* **117**: 199–224.

Ahsan M. and O'Connor K. M. (1994) A simple non-linear rainfall–runoff model with a variable gain factor. *Journal of Hydrology* **155**: 151–183.

Al-Khudhairy D. H. A., Thompson J. R., Gavin H. and Hamm N. A. S. (1999) Hydrological modelling of a drained grazing marsh under agricultural land use and the simulation of restoration management scenarios. *Hydrological Science Journal* **44**: 943–971.

Al-Wagdany A. S. and Rao A. R. (1998) Correlation of the velocity parameter of three geomorphological instantaneous unit hydrograph models. *Hydrological Processes* **12**: 651–659.

Ambroise B., Perrin J. L. and Reutenauer D. (1995) Multicriterion validation of a semidistributed conceptual model of the water cycle in the Fecht catchment (Vosges Massif, France). *Water Resources Research* **31**: 1467–1481.

Ambroise B., Freer J. and Beven K. J. (1996a) Application of a generalised TOPMODEL to the small Ringelbach catchment, Vosges, France. *Water Resources Research* **32**: 2147–2159.

Amerman C. R. (1965) The use of unit-source watershed data for runoff prediction. *Water Resources Research* **1**: 499–507.

Amorocho J. and Brandstetter A. (1971) Determination of nonlinear functional response functions in rainfall–runoff processes. *Water Resources Research* **7**: 1087–1101.

Anderson E. A. (1968) Development and testing of snowpack energy balance equations. *Water Resources Research* **3**: 19–38.

Anderson M. G. and Burt T. P. (eds) (1990) *Process Studies in Hillslope Hydrology.* John Wiley, Chichester.

Andrieu H., French M. N., Thauvin V. and Krajewski W. F. (1996) Adaptation and application of a quantitative forecasting model in a mountainous region. *Journal of Hydrology* **184**: 234–259.

Arnaud-Fassetta G., Ballais J.-L., Beghin E., Jorda M., Meffre J.-C., Provensal M., Roditis J.-C. and Suanez S. (1993) La crue de l'Ouvèze à Vaison-la-romaine (22 septembre 1992). Ses effets morphodynamiques, sa place dans la fonctionnement d'un géosystème anthropisé. *Revue de Géomorphologie Dynamique* **42**: 34–48.
Arnell N. W. (1989) Changing frequency of extreme hydrological events in northern and western europe. In Roald L. N. K. and Hassel K. A. (eds) *FRIENDS in Hydrology*. IAHS Publication No. 187, Wallingford, pp. 237–249.
Arnell N. W. (1996) *Global Warming, River Flows and Water Resources*. John Wiley, Chichester.
Arnell N. W. (1999) A simple water balance model for the simulation of streamflow over a large geographic domain. *Journal of Hydrology* **217**: 314–335.
Arnell N. W. and Reynard N. S. (1996) The effects of climate change due to global warming on river flows in Great Britain. *Journal of Hydrology* **183**: 397–424.
Arnold J. G. and Williams J. R. (1995) *SWRRB – A Watershed Scale Model for Soil and Water Resources Management*. Water Resource Publications, Highlands Ranch, CO, pp. 847–908.
Arnold J. G., Srinavasan R., Muttiah R. S. and Williams J. R. (1998) Large area hydrologic modeling and assessment. Part 1. Model development. *Journal of the American Water Resources Association* **34**: 73–89.
Aronica G., Hankin B. G. and Beven K. J. (1998) Uncertainty and equifinality in calibrating distributed roughness coefficients in a flood propagation model with limited data. *Advances in Water Resources* **22**: 349–365.
Avissar R. (1998) Which type of soil–vegetation–atmosphere transfer scheme is needed for general circulation models: a proposal for a higher-order scheme. *Journal of Hydrology* **212/213**: 136–154.
Avissar R. and Pielke R. (1989) A parameterization of heterogeneous land surfaces for atmospheric numeric models and its impact on regional meteorology. *Monthly Weather Review* **117**: 2113–2136.
Bales J. and Betson R. (1982) The curve number as a hydrologic index. In Singh V. P. (ed.) *Rainfall–Runoff Relationship*. Water Resource Publications, Highlands Ranch, CO, pp. 371–386.
Band L. E., Peterson D. L., Running S. W., Coughlan J. C., Lammers R., Dungan J. and Nemani R. (1991) Forest ecosystem processes at the watershed scale: basis for distributed simulation. *Ecological Modelling* **56**: 171–196.
Band L. E., Patterson P., Nemani R. and Running S. W. (1993) Forest ecosystem processes at the watershed scale: incorporating hillslope hydrology. *Agric. Forest Meteorology* **63**: 93–126.
Bárdossy A., Bronstert A. and Merz B. (1995) 1-dimensional, 2-dimensional and 3-dimensional modelling of water-movement in the unsaturated soil matrix using a fuzzy approach. *Advances in Water Resources* **18**(4): 237–251.
Barling R. D., Moore I. D. and Grayson R. B. (1994) A quasi-dynamic wetness index for characterizing the spatial distribution of zones of surface saturation and soil water content. *Water Resources Research* **30**(4): 1029–1044.
Bastiaanssen W. G. M., Hoekman D. H. and Roebeling D. (1994) *A Methodology for the Assessment of Surface Resistance and Soil Water Storage Variability at Mesoscale Based on Remote Sensing Measurements*. IAHS Publication No. 2, Wallingford, UK.
Bastiaanssen W. G. M., Menenti M., Feddes R. A. and Holtslag A. A. M. (1998) A remote sensing surface energy balance algorithm for land (SEBAL). 1. Formulation. *Journal of Hydrology* **212/213**: 198–212.
Bates P. D., Anderson M. G., Baird L., Walling D. E. and Simm D. (1992) Modelling flood plain flows using a two-dimensional finite element model. *Earth Surface Processes and Landforms* **17**: 575–588.
Bates P. D., Anderson M. G. and Hervouet J.-M. (1995) An initial comparison of 2-dimensional finite element codes for river flood simulation. *Proceedings of the Institution of Civil Engineers, Water, Maritime, Energy* **112**: 238–248.

Bates P. D., Horritt M., Smith C. and Mason D. (1997) Integrating remote sensing observations of flood hydrology and hydraulic modelling. *Hydrological Processes* **11**: 1777–1795.

Bathurst J. C. (1986a) Physically-based distributed modelling of an upland catchment using the Systeme Hydrologique Europeen. *Journal of Hydrology* **87**: 79–102.

Bathurst J. C. (1986b) Sensitivity analysis of the Systeme Hydrologique Europeen for an upland catchment. *Journal of Hydrology* **87**: 103–123.

Bathurst J. C. and Cooley K. R. (1996) Use of the SHE hydrological modelling system to investigate basin response to snowmelt at Reynolds Creek, Idaho. *Journal of Hydrology* **175**: 181–211.

Bathurst J. C., Wicks J. M. and O'Connell P. E. (1995) The SHE/SHESED basin scale water flow and sediment transport modelling system. In Singh V. P. (ed.) *Computer Models of Watershed Hydrology*. Water Resource Publications, Highlands Ranch, CO, pp. 563–594.

Bazemore D. E., Eshleman K. N. and Hollenbeck K. (1994) The role of soil water in stormflow generation in a forested headwater catchment: synthesis of natural tracer and hydrometric evidence. *Journal of Hydrology* **162**: 47–75.

Beasley D. B., Huggins L. F. and Monke E. J. (1980) Answers: a model for watershed planning. *Transactions of the American Society of Agricultural Engineers* **23**: 938–944.

Beck M. B. (1987) Water quality modeling: a review of the analysis of uncertainty. *Water Resources Research* **23**: 1393–1442.

Beck M. B. and Halfon E. (1991) Uncertainty, identifiability and the propagation of prediction errors: a case study of Lake Ontario. *Journal of Forecasting* **10**: 135–161.

Becker A. and Braun P. (1999) Disaggregation, aggregation and spatial scaling in hydrological modelling. *Journal of Hydrology* **217**: 239–252.

Bell V. A. and Moore R. J. (1998) A grid-based distributed flood forecasting model for use with weather radar data. 1. Formulation. 2. Case studies. *Hydrology and Earth System Science* **2**: 265–298.

Bergkamp G. (1998) A hierarchical view of the interactions of runoff and infiltration with vegetation and microtopography in semiarid shrublands. *Catena* **33**: 201–220.

Bergstrom S. (1975) The development of a snow routine for the HBV-2 model. *Nordic Hydrology* **6**: 73–92.

Bernier P. (1985) Variable source areas and stormflow generation: an update of the concept and simulation effort. *Journal of Hydrology* **79**: 195–213.

Berod D. D., Singh V. P. and Musy A. (1999) A geomorphologic kinematic-wave (GKW) model for estimation of floods from small alpine watersheds. *Hydrological Processes* **13**: 1391–1416.

Betson R. P. (1964) What is watershed runoff? *Journal of Geophysical Research* **69**: 1541–1551.

Beven K. J. (1977) Hillslope hydrographs by the finite element method. *Earth Surface Processes* **2**: 13–28.

Beven K. J. (1978) The hydrological response of headwater and slideslopes areas. *Hydrological Sciences Bulletin* **23**: 419–437.

Beven K. J. (1979a) A sensitivity analysis of the Penman–Monteith actual evapotranspiration estimates. *Journal of Hydrology* **44**: 169–190.

Beven K. J. (1979b) On the generalised kinematic routing method. *Water Resources Research* **15**: 1238–1242.

Beven K. (1981) Kinematic subsurface stormflow. *Water Resources Research* **17**(5): 1419–1424.

Beven K. J. (1982) On subsurface stormflow: an analysis of response times. *Hydrological Sciences Journal* **27**(4): 505–521.

Beven K. J. (1984) Infiltration into a class of vertically non-uniform soils. *Hydrological Sciences Journal* **29**: 425–434.

Beven K. J. (1986a) Runoff production and flood frequency in catchments of order *n*: an alternative approach. In Gupta V. K., Rodriguez-Iturbe I. and Wood E. F. (eds) *Scale Problems in Hydrology*, Reidel, Dordrecht, pp. 107–131.

Beven K. J. (1986b) Hillslope runoff processes and flood frequency characteristics. In Abrahams A. D. (ed) *Hillslope Processes*. Allen and Unwin, Boston, pp. 187–202.
Beven K. J. (1987) Towards the use of catchment geomorphology in flood frequency predictions. *Earth Surface Processes and Landforms* **12**: 69–82.
Beven K. J. (1989) Changing ideas in hydrology: the case of physically-based models. *Journal of Hydrology* **105**: 157–172.
Beven K. J. (1991a) Hydrograph separation? In *Proceedings of the BHS 3rd National Hydrology Symposium*, Southampton, 1991.
Beven K. J. (1991b) Spatially distributed modelling: conceptual approach to runoff prediction. In Bowles D. and O'Connell P. E. (eds) *Recent Advances in the Modelling of Hydrologic Systems*. Kluwer Academic, Dordrecht, pp. 373–387.
Beven K. J. (1993) Prophecy, reality and uncertainty in distributed hydrological modelling. *Advances in Water Resources* **16**: 41–51.
Beven K. J. (1995) Linking parameters across scales: sub-grid parameterisations and scale dependent hydrological models. *Hydrological Processes* **9**: 507–525.
Beven K. J. (1996a) The limits of splitting: hydrology. *Science of the Total Environment* **183**: 89–97.
Beven K. (1996b) A discussion of distributed modelling. In Refsgaard J.-C. and Abbott M. B. (eds) *Distributed Hydrological Modelling*. Kluwer, Dordrecht, pp. 255–278.
Beven K. J. (1996c) Response to comments on 'A discussion of distributed modelling'. In Refsgaard J.-C. and Abbott M. B. (eds) *Distributed Hydrological Modelling*. Kluwer, Dordrecht, pp. 289–295.
Beven K. J. (1997a) TOPMODEL: a critique. *Hydrological Processes* **11**(9): 1069–1086.
Beven K. J. (1997b) *Distributed Hydrological Modelling: Applications of the TOPMODEL Concept*. John Wiley, Chichester.
Beven K. J. (2000) On uniqueness of place and process representations in hydrological modelling. *Hydrology and Earth System Science* **4**(2): 203–212.
Beven K. J. and Binley A. M. (1992) The future of distributed models – model calibration and uncertainty prediction. *Hydrological Processes* **6**(3): 279–298.
Beven K. J. and Clarke R. T. (1986) On the variation of infiltration into a homogeneous soil matrix containing a population of macropores. *Water Resources Research* **22**: 383–388.
Beven K. J. and Franks S. (1999) Functional similarity in landscape scale SVAT modelling. *Hydrology and Earth System Science* **3**(1): 85–94.
Beven K. J. and Freer J. (2000) Dynamic TOPMODEL. *Hydrological Processes* (in press).
Beven K. and Germann P. (1981) Water flow in soil macropores, II. A combined flow model. *Journal of Soil Science* **32**: 15–29.
Beven K. J. and Germann P. F. (1982) Macropores and water flow in soils. *Water Resources Research* **18**(5): 303–325.
Beven K. J. and Kirkby M. J. (1979) A physically based, variable contributing area model of basin hydrology. *Hydrological Science Bulletin* **24**(1): 43–69.
Beven K. J. and Wood E. F. (1983) Catchment geomorphology and the dynamics of runoff contributing areas. *Journal of Hydrology* **65**: 139–158.
Beven K. J. and Wood E. F. (1993) Flow routing and the hydrological response of channel networks. In Beven K. J. and Kirkby M. J. (eds) *Channel Network Hydrology*. John Wiley, Chichester, pp. 99–128.
Beven K. J., Warren R. and Zaoui J. (1980) SHE: towards a methodology for physically-based distributed forecasting in hydrology. IAHS Publication No. 129, Wallingford, UK, pp. 133–137.
Beven K. J., Kirkby M. J., Schoffield N. and Tagg A. (1984) Testing a physically-based flood forecasting model (TOPMODEL) for three UK catchments. *Journal of Hydrology* **69**: 119–143.
Beven K. J., Calver A. and Morris E. M. (1987) The Institute of Hydrology distributed model. Technical Report Report No. 98, Institute of Hydrology, Wallingford, UK.
Beven K. J., Wood E. F. and Sivapalan M. (1988) On hydrological heterogeneity: catchment morphology and catchment response. *Journal of Hydrology* **100**: 353–375.

Beven K. J., Lamb R., Quinn P., Romanowicz R. and Freer J. (1995) Topmodel. In Singh V. P. (ed.) *Computer Models of Watershed Hydrology*. Water Resource Publications, Highlands Ranch, CO, pp. 627–668.

Binley A. M. and Beven K. J. (1992) Three dimensional modelling of hillslope hydrology. *Hydrological Processes* **6**: 347–359.

Binley A. M., Beven K. J. and Elgy J. (1989) A physically-based model of heterogeneous hillslopes. II. Effective hydraulic conductivities. *Water Resources Research* **25**(6): 1227–1233.

Binley A. M., Beven K. J., Calver A. and Watts L. G. (1991) Changing responses in hydrology: assessing the uncertainty in physically based model predictions. *Water Resources Research* **27**(6): 1253–1261.

Blackie J. R. and Eeles C. W. O. (1985) Lumped catchment models. In Anderson M. G. and Burt T. P. (eds) *Hydrological Forecasting*. John Wiley, Chichester, pp. 311–346.

Blazkova S. and Beven K. (1997) Flood frequency prediction for data limited catchments in the Czech Republic using a stochastic rainfall model and topmodel. *Journal of Hydrology* **195**: 256–278.

Blöschl G., Kirnbauer R. and Gutnecht D. (1991) Distributed snowmelt simulations in an alpine catchment 1. Model evaluation on the basis of snow cover patterns. *Water Resources Research* **27**: 3171–3179.

Boorman D. B., Hollis J. M. and Lilly A. (1995) Hydrology of soil types: a hydrologically-based classification of the soils of the United Kingdom. Technical Report 126, Institute of Hydrology, Wallingford, UK.

Borah D. K., Prasad S. N. and Alonso C. V. (1980) Kinematic wave routing incorporating shock fitting. *Water Resources Research* **16**: 529–541.

Bouchet R. J. (1963) Evapotranspiration Réelle et Potentielle, Signification Climatique. IAHS Publication No. 62, Wallingford, UK, pp. 134–142.

Box G. E. P. and Cox D. R. (1964) An analysis of transformations. *Journal of the Royal Statistical Society* **B26**: 211–243; Discussion 244–252.

Box G. E. P. and Jenkins G. M. (1970) *Time Series Analysis, Forecasting and Control*. Holden-Day, San Francisco.

Box G. E. P. and Tiao G. C. (1992) *Bayesian Inference in Statistical Analysis*. Wiley-Interscience, New York.

Bradley A. A., Cooper P. J., Potter K. W. and Price T. (1995) Floodplain mapping using continuous hydrologic and hydraulic simulation models. *Journal of Hydrological Engineering ASCE* **1**: 63–68.

Brakensiek D. L. and Rawls W. J. (1994) Soil containing rock fragments: effects on infiltration. *Catena* **23**: 99–110.

Bras R. E. (1990) *Hydrology*. Addison-Wesley. Reading, MA.

Braun L. N. and Lang H. (1986) Simulations of snowmelt runoff in lowland and lower alpine regions of Switzerland. In *Modelling Snowmelt-Induced Processes*. IAHS Publication No. 155, Wallingford, UK, pp. 125–140.

Bronstert A. (1999) Capabilities and limitations of distributed hillslope hydrological modelling. *Hydrological Processes* **13**: 21–48.

Bronstert A. and Plate E. (1997) Modelling of runoff generation and soil moisture dynamics for hillslopes and microcatchments. *Journal of Hydrology* **198**: 177–195.

Brooks R. H. and Corey A. T. (1964) Hydraulic properties of porous media. Technical Report Hydrology Paper No. 3, Colorado State University, Fort Collins, CO.

Bruneau P., Gascuel-Odoux C., Robin P., Merot Ph. and Beven K. J. (1995) The sensitivity to space and time resolution of a hydrological model using digital elevation data. *Hydrological Processes* **9**: 69–81.

Brutsaert W. and Sugito M. (1992) Application of self-preservation in the diurnal evolution of the surface energy budget to determine daily evaporation. *Journal of Geophysical Research* **97**(D17): 18 377–18 382.

Bultot F., Coppens A., Dupries G. L., Gellens D. and Meulenberghs F. (1988) Repercussions of a CO_2 doubling on the water cycle and on the water balance – a case study for Belgium. *Journal of Hydrology* **99**: 319–347.
Bultot F., Gellens D., Spreafico M. and Schadler B. (1992) Repercussions of a CO_2-doubling on the water balance – a case study in Switzerland. *Journal of Hydrology* **137**: 199–208.
Burges S. J., Wigmosta M. S. and Meena J. M. (1998) Hydrological effects of land-use change in a zero-order catchment. *Journal of Hydrological Engineering ASCE* **3**: 86–97.
Burnash R. (1995) The NWS river forecast system – catchment modeling. In Singh V. P. (ed.) *Computer Models of Watershed Hydrology*. Water Resource Publications, Highlands Ranch, CO, pp. 311–366.
Calder I. R. (1977) A model of transpiration and interception loss from a spruce forest in Plynlimon, Central Wales. *Journal of Hydrology* **33**: 247–265.
Calder I. (1986) A stochastic model of rainfall interception. *Journal of Hydrology* **89**: 65–71.
Calder I. R. (1990) *Evaporation in the Uplands*. John Wiley, Chichester.
Calder I. R. (1996) Dependence of rainfall interception on drop size: 1. Development of the two-layer stochastic model. *Journal of Hydrology* **185**: 363–378.
Calder I. R., Harding R. J. and Rosier P. T. W. (1983) An objective assessment of soil moisture deficit models. *Journal of Hydrology* **60**: 329–355.
Calder I. R., Hall R. L., Rosier P. T. W., Bastable H. G. and Prasanna K. T. (1996) Dependence of rainfall interception of drop size: 2. Experimental determination of the wetting functions and two-layer stochastic model parameters for five tropical tree species. *Journal of Hydrology* **185**: 379–388.
Calver A. (1988) Calibration, sensitivity and validation of a physically-based rainfall–runoff model. *Journal of Hydrology* **103**: 103–115.
Calver A. (1993) Flood generation modelling. *Institution of Wat. Environ. Management J.* **7**: 614–620.
Calver A. and Cammeraat L. H. (1993) Testing a physically-based runoff model against field observations on a Luxembourg hillslope. *Catena* **20**: 273–288.
Calver A. and Lamb R. (1996) Flood frequency estimation using continuous rainfall–runoff modelling. *Phys. Chem. Earth* **20**: 479–483.
Calver A. and Wood W. L. (1995) The Institute of Hydrology distributed model. In Singh V. P. (ed.) *Computer Models of Watershed Hydrology*. Water Resource Publications, Highlands Ranch, CO, pp. 595–626.
Cameron D. (2000) Estimating flood frequency by continuous simulation. PhD thesis, Lancaster University, Lancaster.
Cameron D. S., Beven K. J., Tawn J., Blazkova S. and Naden P. (1999) Flood frequency estimation by continuous simulation for a gauged upland catchment (with uncertainty). *Journal of Hydrology* **219**: 169–187.
Campolo M., Andreussi P. and Soldati A. (1999) River flood forecasting with a neural network model. *Water Resources Research* **35**: 1191–1197.
Cappus P. (1960) Bassin versant expérimental d'Alrance: études des lois de l'écoulement. Application au calcul et à la prévision des débits. *La Houille Blanche A*: 493–520.
Carsel R. F., Mulkey L. A., Lorber M. N. and Bakin L. B. (1985) The pesticide root zone model (PRZM): a procedure for evaluating pesticide leaching threats to groundwater. *Ecological Modeling* **30**: 49–69.
Cerda A. (1998) Changes in overland flow and infiltration after a rangeland fire in a Mediterranean scrubland. *Hydrological Processes* **12**: 1031–1042.
Chagnon S. A. E. (1996) *The Great Flood of 1993*. Westview Press, Boulder, CO.
Chapman T. G. (1996b) Common unitgraphs for sets of runoff events. Part 2. Comparisons and inferences for rainfall loss models. *Hydrological Processes* **10**: 783–792.
Charbeneau R. J. (1984) Kinematic models for soil moisture and solute transport. *Water Resources Research* **20**: 699–706.

Chaubey I., Haan C. T., Grunwald S. and Salisbury J. M. (1999) Uncertainty in the model parameters due to spatial variability of rainfall. *Journal of Hydrology* **220**: 48–61.

Chehbouni A., Kerr Y. H., Watts C., Hartogenesis O., Goodrich D. *et al.* (1999) Estimation of area-average sensible heat flux using a large-aperture scintillometer during the semi-arid land–surface–atmosphere (SALSA) experiment. *Water Resources Research* **35**: 2505–2511.

Chiew F. H. S., Stewardson M. J. and McMahon T. A. (1993) Comparison of six rainfall–runoff modelling approaches. *Journal of Hydrology* **147**: 1–36.

Chow V. T. (1964) *Handbook of Applied Hydrology*. McGraw-Hill, New York.

Chow V. T. and Kulandaiswamy V. C. (1971) General hydrologic systems model. *Journal of the Hydraulics Division, ASCE* **97**(HY6): 791–804.

Chutha P. and Dooge J. C. I. (1990) The shape parameters of the geomorphologic unit hydrograph. *Journal of Hydrology* **117**: 81–97.

Ciarapica L. (1998) TOPKAPI – un Modello Afflussi–Deflussi Distribuito Applicabile Dalla Scala Di Versante Alla Scala Di Bacino. PhD thesis, Università di Bologna, Italy.

Clapp R. B. and Hornberger G. M. (1978) Empirical equations for some hydraulic properties. *Water Resources Research* **14**: 601–604.

Clapp R. B., Hornberger G. M. and Cosby B. J. (1983) Estimating spatial variation in soil moisture with a simple dynamic model. *Water Resources Research* **19**: 739–745.

Clapp R. B., Timmins S. P. and Huston M. A. (1992) Visualising the surface hydrodynamics of a forested watershed. In Russell T. F. *et al.* (ed.) *Proceedings of the IX International Conference on Computational Methods in Water Resources, V.2 Mathematical Models of Water Resources*. Elsevier, New York, pp. 765–772.

Clark C. O. (1945) Storage and unit hydrograph. *Transactions of the American Society of Civil Engineers* **110**: 1416–1446.

Clarke R. T. (1973) A review of some mathematical models used in hydrology, with observations on their calibration and use. *Journal of Hydrology* **19**: 1–20.

Clarke R. T. (1998) *Stochastic Processes for Water Scientists: Developments and Applications*. John Wiley, Chichester.

Clausnitzer V., Hopmans J. H. and Nielsen D. R. (1992) Simultaneous scaling of soil water retention and hydraulic conductivity curves. *Water Resources Research* **28**: 19–31.

Cluckie I. D. (1993) Hydrological Forecasting: Real Time. In Young P.C. (ed.) Encyclopaedia of Environmental Systems, Pergamon Press, Oxford, pp. 291–298.

Colbeck S. C. (1974) Water flow through snow overlying an impermeable boundary. *Water Resources Research* **10**(1): 119–123.

Connolly R. D., Silburn D. M. and Ciesiolka C. A. A. (1997) Distributed parameter hydrology model (Answers) applied to a range of catchment scales using rainfall simulator data. III. Application to a spatially complex catchment. *Journal of Hydrology* **193**: 183–203.

Conway D. and Jones P. D. (1998) The use of weather types and air flow indices for GCM downscaling. *Journal of Hydrology* **212/213**: 348–361.

Cooley K. R. (1990) Effects of CO_2-induced climatic changes on snowpack and streamflow. *Hydrological Science Journal* **35**: 511–522.

Cooper D. M. and Wood E. F. (1982) Parameter estimation of multiple input–output time series models: application to rainfall–runoff processes. *Water Resources Research* **18**: 1352–1364.

Cordova J. R. and Rodriguez-Iturbe I. (1983) Geomorphoclimatic estimation of extreme flood probabilities. *Journal of Hydrology* **65**: 159–173.

Corradini C., Melone F. and Smith R. E. (1997) A unified model for infiltration and redistribution during complex rainfall patterns. *Journal of Hydrology* **192**: 104–124.

Corradini C., Morbidelli R. and Melone F. (1998) On the interaction between infiltration and Hortonian runoff. *Journal of Hydrology* **204**: 52–67.

Cosby B. J., Hornberger G. M., Clapp R. B. and Ginn T. R. (1984) A statistical exploration of the relationship of soil moisture characteristics to the physical properties of soils. *Water Resources Research* **20**: 682–690.

Cowpertwait P. S. P., O'Connell P. E., Metcalf A. V. and Mawdsley J. (1996) Stochastic point process modelling of rainfall. I. Single-site fitting and validation. II. Regionalisation and disaggregation. *Journal of Hydrology* **171**: 17–65.

Cox D. R. and Hinkley D. V. (1974) *Theoretical Statistics*. Chapman and Hall, London.

Cox P. M., Huntingford C. and Harding R. J. (1998) A canopy conductance and photosynthesis model for Se in a GCM land surface scheme. *Journal of Hydrology* **212/213**: 79–94.

Crago R. D. and Brutsaert W. (1996) Daytime evaporation and the self-preservation of the evaporation fraction and the Bowen ratio. *Journal of Hydrology* **178**: 241–255.

Crawford N. H. and Linsley R. K. (1966) Digital simulation in hydrology: Stanford Watershed Model IV. Technical Report 39, Department of Civil Engineering, Stanford University, CA.

Cunge J. (1969) On the subject of a flood routing method (Muskingum method). *Journal of Hydraulics Research* **7**: 205–230.

Daluz Vieira J. (1983) Conditions governing the use of approximations for the St. Venant equations for shallow surface water flow. *Journal of Hydrology* **60**: 43–58.

Darcy H. (1856) *Les Fontaines de la Ville de Dijon*. Dalmont, Paris.

Datin R. (1998) Outils Opérationnels pour la prévision des crues rapides: traitements des incertitudes et intégration des prévisions météologique. Développements de TOPMODEL pour le prise en compte de la variabilité spatiale de la pluie. PhD thesis, INP Grenoble.

Davie T. (1996) Modelling the influence of afforestation on hillslope storm runoff. In Anderson M. and Brooks S. M. (eds) *Advances in Hillslope Processes, V*. John Wiley, Chichester, pp. 149–184.

Davis L. (1991) *Handbook of Genetic Algorithms*. Van Nostrand Reinhold, New York.

Dawdy D. R. and O'Donnell T. (1965) Mathematical models of catchment behaviour. *Journal of the Hydraulics Division, Proceedings of the American Society of Civil Engineers* **91**(HY4): 123–127.

Dawes W. R., Zhang L., Hatton T. J., Reece P. H., Beale G. T. H. and Packer I. (1997) Evaluation of a distributed parameter ecohydrological model (TOPOG-IRM) on a small cropping rotation catchment. *Journal of Hydrology* **191**: 64–86.

Dawson C. and Wilby R. (1998) An artificial neural network approach to rainfall–runoff modelling. *Hydrological Science Journal* **43**: 47–66.

De Bruin H. A. R., Van den Hurk B. J. J. M. and Kohsiek W. (1995) The scintillation method tested over a dry vinyard area. *Boundary Layer Meteorology* **76**: 25–40.

De Troch F. P., Troch P. A., Su Z. and Lin D. (1996) Application of remote sensing for hydrological modelling. In Abbott M. and Refsgaard J.-C. (eds) *Distribute Hydrological Modelling*. Kluwer, Dordrecht, pp. 165–191.

Diaz-Granados M. A., Valdes J. B. and Bras R. L. (1984) A physically-based flood frequency distribution. *Water Resources Research* **20**: 995–1002.

Dickinson R. E. and Henderson-Sellers A. (1988) Modelling tropical deforestation: a study of GCM land-surface parameterizations. *Quarterly Journal of the Royal Meteorological Society* **114**: 439–462.

Dietterick B. C., Lynch J. A. and Corbett E. S. (1999) A calibration procedure using TOPMODEL to determine suitability for evaluating potential climate change effects on water yield. *Journal of the American Water Resources Association* **35**: 457–468.

Diskin M. and Boneh A. (1973) Determination of optimal kernals for second-order stationary surface runoff systems. *Water Resources Research* **9**: 311–325.

Dixon R. M. and Linden D. R. (1972) Soil air pressure and water infiltration under border irrigation. *Soil Science Society of America Proceedings* **36**: 948–953.

Doe W. W., Saghafian B. and Julien P. Y. (1996) Land-use impact on watershed response: the integration of two-dimensional hydrological modelling and geographical information systems. *Hydrological Processes* **10**: 1503–1511.

Dolciné L. A. H. and French M. N. (1998) Evaluation of a conceptual rainfall forecasting model from observed and simulated rainfall events. *Hydrology and Earth System Science* **2**: 173–182.

Dolman A. J. (1987) Summer and winter rainfall interception in an oak forest. Predictions with an analytical and a numerical simulation model. *Journal of Hydrology* **90**: 1–9.

Donigian A. S. D., Bicknell B. R. and Imhoff J. C. (1995) Hydrological simulation program – Fortran. In Singh V. P. (ed.) *Computer Models of Watershed Hydrology*. Water Resource Publications, Highlands Ranch, CO, pp. 395–442.

Donnelly-Makowecki L. M. and Moore R. D. (1999) Hierarchical testing of three rainfall–runoff models in small forested catchments. *Journal of Hydrology* **219**: 136–152.

Dooge J. C. I. (1957) The rational method for estimating flood peaks. *Engineering* **184**: 311–313; 374–377.

Dooge J. C. I. (1959) A general theory of the unit hydrograph. *Journal of Geophysical Research* **64**: 241–256.

Dooge J. C. I. (1986) Looking for hydrologic laws. *Water Resources Research* **22**(9): 465–585.

Duan J. and Miller N. L. (1997) A generalized power function for the subsurface transmissivity profile in TOPMODEL. *Water Resources Research* **33**: 2559–2562.

Duan Q. S., Sorooshian S. and Gupta V. (1992) Effective and efficient global optimisation for conceptual rainfall–runoff models. *Water Resources Research* **28**: 1015–1031.

Duband D., Obled C. and Rodriguez J.-Y. (1993) Unit hydrograph revisited: an alternative approach to UH and effective precipitation identification. *Journal of Hydrology* **150**: 115–149.

Dubayah R. O., Wood E. F., Engman E. T., Czajkowski K. P., Zion M. and Rhoads J. (1999) Remote sensing in hydrologic modeling. In Schultz G. A. and Engman E. T. (eds) *Remote Sensing in Hydrology and Water Management*. Springer-Verlag, Berlin.

Ducoudre N. I., Laval K. and Perrier A. (1993) SECHIBA: a new set of parameterizations of the hydrological exchanges at the land/atmosphere interface within the LMD atmospheric general circulation model. *Journal of Climate* **6**: 248–273.

Duffy C. (1996) A two-state integral-balance model for soil moisture and groundwater dynamics in complex terrain. *Water Resources Research* **32**: 2421–2434.

Dumenil L. and Todini E. (1992) A rainfall–runoff scheme for use in the Hamburg climate model. In O'Kane J. P. (ed.) *Advances in Theoretical Hydrology, A Tribute to James Dooge*. Elsevier, Amsterdam, pp. 129–157.

Dunn S. M. (1999) Imposing constraints on parameter values of a conceptual hydrological model using baseflow response. *Hydrology and Earth System Science* **3**: 271–284.

Dunn S. M. and Ferrier R. C. (1999) Natural flow in managed catchments: a case study of a modelling approach. *Water Research* **33**: 621–630.

Dunn S. M. and Mackay R. (1996) Modelling the hydrological impacts of open ditch drainage. *Journal of Hydrology* **179**: 37–66.

Dunn S. M., Mackay R., Adams R. and Oglethorpe D. R. (1996) The hydrological component of the NELUP decision-support system: an appraisal. *Journal of Hydrology* **177**: 213–235.

Dunne T. (1978) Field studies of hillslope flow processes. In Kirkby M. J. (ed.) *Hillslope Hydrology*. John Wiley, Chichester, pp. 227–293.

Dunne T. and Black R. D. (1970) An experimental investigation of runoff production in permeable soils. *Water Resources Research* **6**: 478–490.

Dunne T., Price J. G. and Colbeck S. C. (1976) The generation of runoff from subarctic snowpacks. *Water Resources Research* **12**(4): 677–685.

Dunne T., Zhang W. and Aubrey B. F. (1991) Effects of rainfall, vegetation and microtopography on infiltration and runoff. *Water Resources Research* **27**: 2271–2286.

Durand P., Robson A. and Neal C. (1992) Modelling the hydrology of submediterranean montane catchments (Mont Lozère, France), using TOPMODEL: initial results. *Journal of Hydrology* **139**: 1–14.

Durand Y., Brun E., Merindol L., Guymarb'h G., Lesassre B. and Martin E. (1993) A meteorological estimation of relevant parameters for snow models. *Annals of Glaciology* **18**: 65–71.

Duru J. O. and Hjelmfelt, A. T. (1994) Investigating Prediction Capability of HEC-1 and KINEROS Kinematic Wave Runoff Models. *J. Hydrology*, **157**: 87–103, 1994.
Eagleson P. S. (1970) *Dynamic Hydrology*. McGraw-Hill, New York.
Eagleson P. S. (1972) Dynamics of flood frequency. *Water Resources Research* **8**: 878–898.
Editjatno Nascimento N. De O., Yang X., Makhlouf Z. and Michel C. (1999) GR3J: a daily watershed model with three free parameters. *Hydrological Science Journal* **44**: 263–277.
Ehlers W. (1975) Observations on earthworm channels and infiltration on tilled and untilled loess soil. *Soil Science* **119**: 242–249.
Engman E. and Gurney R. (1991) *Remote Sensing in Hydrology*. Chapman and Hall, London.
Engman E. T. and Rogowski A. S. (1974) A partial area model for storm flow synthesis. *Water Resources Research* **10**: 464–472.
Entekhabi D. and Eagleson P. S. (1991) Climate and the Equilibrium State of Land Surface Hydrology Parameterizations. *Surveys in Geophysics*, **12**: 205–220.
Erichsen B. and Myrabø S. (1990) Studies of the relationship between soil moisture and topography in a small catchment. In Gambolati G. (ed.) *Proceedings of the 8th International Conference on Computational Methods in Water Resources*, Venice.
Espino A., Mallants D., Vanclooster M. and Feyen J. (1995) Cautionary notes on the use of pedotransfer functions for estimating soil hydraulic properties. *Agricultural Water Management* **29**: 235–253.
Ewen J. (1997) 'Blueprint' for the UP modelling system for large-scale hydrology. *Hydrology and Earth System Science* **1**: 55–69.
Ewen J. and Parkin G. (1996) Validation of catchment models for predicting land-use and climate change impacts. 1. Method. *Journal of Hydrology* **175**: 583–594.
Ewen J., Sloan W. T., Kilsby C. G. and O'Connell P. E. (1999) UP modelling system for large scale hydrology: deriving large-scale physically-based parameters for the Arkansas–Red River Basin. *Hydrology and Earth System Science*, **3**: 125–136.
Faeh A. O., Scherrer S. and Naef F. (1997) A combined field and numerical approach to investigate flow processes in natural macroporous soils under extreme precipitation. *Hydrology and Earth System Sciences* **1**(4): 787–800.
Fagre D. B., Comanor P. L., White J. D., Hauer F. R. and Running S. W. (1997) Watershed Responses to climate change at Glacier National Park. *Journal of the American Water Resources Association* **33**: 755–765.
Fairfield J. and Leymaire P. (1991) Drainage networks from grid digital elevation models. *Water Resources Research*, **27**: 709–717.
Famiglietti J. S. and Wood E. F. (1991) Evapotranspiration and runoff from large land areas: land surface hydrology for atmospheric general circulation models. *Surveys in Geophysics* **12**: 179–204.
Famiglietti J. S., Wood E. F., Sivapalan M. and Thongs D. J. (1992) A catchment scale water balance model for FIFE. *Journal of Geophysical Research* **97**(D17): 18 997–19 007.
Faurès J.-M., Goodrich D. C., Woolhiser D. A. and Sorooshian S. (1995) Impact of small-scale rainfall variability on runoff modelling. *Journal of Hydrology* **173**: 309–326.
Federer C. A., Vörösmarty C. and Fekete B. (1996) Intercomparison of methods for calculating potential: evaporation in regional and global water balance models. *Water Resources Research* **32**: 2315–2321.
Feldman A. D. (1995) HEC-1 flood hydrograph package. In Singh V. P. (ed.) *Computer Models of Watershed Hydrology*. Water Resource Publications, Highland Ranch, CO, pp. 119–150.
Fernando D. A. K. and Jayawardena A. W. (1998) Runoff forecasting using RBF networks with OLS algorithm. *Journal of Hydrological Engineering ASCE* **3**: 203–209.
Fisher J. and Beven K. J. (1995) Modelling of streamflow at Slapton Wood using TOPMODEL within an uncertainty estimation framework. *Field Studies* **8**: 577–584.
Fleming G. (1975) *Computer Simulation Techniques in Hydrology*. Environmental Science Series, Elsevier.
Fleming G. and Franz D. D. (1971) Flood frequency estimating techniques. *Journal of the Hydraulics Division, ASCE*, **97** (HY9): 1441–1460.

Flügel W.-A. (1995) Delineating hydrological response units by geographical information system analyses for regional hydrological modelling using PRMS/MMS in the drainage basin of the River Bröl, Germany. *Hydrological Processes* **9**: 423–436.

Flury M., Fluhler H., Jury W. A. and Leuenberger J. (1994) Susceptibility of soils to preferential flow of water: a field study. *Water Resources Research* **30**(7): 1945–1954.

Fontaine T. A. (1995) Rainfall–runoff model accuracy for an extreme flood. *Journal of Hydraulic Engineering, ASCE* **121**(4): 365–374.

Forrest S. (1993) Genetic algorithms: principles of natural selection applied to computation. *Science* **261**: 872–878.

Fox D. M., Le Bissonnais Y. and Bruand A. (1998) The effect of ponding depth on infiltration in a crusted surface depression. *Catena* **32**: 87–100.

Franchini M. and Galeati G. (1997) Comparing several genetic algorithm schemes for the calibration of conceptual rainfall–runoff models. *Hydrological Science Journal* **42**: 357–380.

Franchini M. and Pacciani M. (1991) Comparative analysis of several conceptual rainfall–runoff models. *Journal of Hydrology* **122**: 161–219.

Franchini M., Helmlinger K. R., Foufoula-Georgiou E. and Todini E. (1996) Stochastic storm transposition coupled with rainfall–runoff modelling for estimation of exceedence probability of design floods. *Journal of Hydrology* **175**: 511–532.

Franks S. W. and Beven K. J. (1997) Estimation of evapotranspiration at the landscape scale: a fuzzy disaggregation approach. *Water Resources Research* **33**: 2929–2938.

Franks S. W., Beven K. J., Quinn P. F. and Wright I. (1997) On the sensitivity of soil–vegetation–atmosphere transfer (SVAT) schemes: equifinality and the problem of robust calibration. *Agric. Forest Meteorology* **86**: 63–75.

Franks S. W., Gineste Ph., Beven K. J. and Merot P. (1998) On constraining the predictions of a distributed model: the incorporation of fuzzy estimates of saturated areas into the calibration process. *Water Resources Research* **34**: 787–797.

Fread D. L. (1973) Technique for implicit flood routing in rivers with tributaries. *Water Resources Research* **9**: 918–926.

Fread D. (1985) Channel routing. In Anderson M. and Burt T. (eds) *Hydrological Forecasting.* John Wiley, Chichester, pp. 437–503.

Freer J., Beven K. J. and Ambroise B. (1996) Bayesian estimation of uncertainty in runoff prediction and the value of data: an application of the GLUE approach. *Water Resources Research* **32**: 2161–2173.

Freer J., Mc Donnell J., Beven K. J., Brammer D., Burns D., Hooper R. P. and Kendal C. (1997) Topographic controls on subsurface stormflow at the hillslope scale for two hydrologically distinct small catchments. *Hydrological Processes* **11**: 1347–1352.

Freeze R. A. (1972) Role of subsurface flow in generating surface runoff. 2. Upstream source areas. *Water Resources Research* **8**(5): 1272–1283.

Freeze R. A. and Harlan R. L. (1969) Blueprint for a physically-based digitally simulated hydrologic response model. *Journal of Hydrology* **9**: 237–258.

French M. N. and Krajewski W. F. (1994) A model for real-time quantitative rainfall forecasting using remote sensing. 1. Formulation. *Water Resources Research* **30**: 1075–1083.

Gandolfi C., Bishetti G. B. and Whelan M. J. (1999) A simple triangular approximation of the area function for the calculation of network hydrological response. *Hydrological Processes* **13**: 2639–2654.

Gao X., Sorooshian S. and Gupta H. V. (1996) Sensitivity analysis of the biosphere–atmosphere transfer scheme. *Journal of Geophysical Research* **101**(D3): 7279–7289.

Gash J. H. C. (1979) An analytical model of rainfall interception by forests. *Quarterly Journal of the Royal Meteorological Society* **105**: 43–55.

Gash J. H. C. and Morton A. J. (1978) An application of the Rutter model to the estimation of the interception loss from Thetford Forest. *Journal of Hydrology* **38**: 49–58.

Gash J. H. C., Wright I. R. and Lloyd C. R. (1980) Comparative estimates of interception loss from three coniferous forests in Great Britain. *Journal of Hydrology* **48**: 89–105.

Gellens D. (1991) Impact of a CO_2-induced climate change on river flow variability in three rivers in Belgium. *Earth Surface Processes and Landforms* **16**: 619–625.
Genereux D. P., Hemond H. F. and Mulholland P. J. (1993) Spatial and temporal variability in streamflow generation on the west fork of Walker Branch watershed. *Journal of Hydrology* **142**: 137–166.
Germann P. F. (1990) Macropores and hydrologic hillslope processes. In Anderson M. G. and Burt T. P. (eds) *Process Studies in Hillslope Hydrology*. John Wiley, Chichester, pp. 327–364.
Girard G., Ledoux E. and Villeneuve J. P. (1981) Le Modèle Couplé – simulation conjointe des écoulements de surface et des écoulements souterrains sur un système hydrologique. *Cahiers ORSTOM, Sér. Hydrol.* **18**.
Goodrich D. C., Woolhiser D. A. and Keefer T. O. (1991) Kinematic routing using finite elements on a triangular irregular network. *Water Resources Research* **27**: 995–1003.
Goodrich D. C., Schmugge T. J., Jackson T. J., Unkrich C. L., Keefer T. O., Parry R., Bach L. B. and Amer S. A. (1994) Runoff simulation sensitivity to remotely sensed initial soil moisture content. *Water Resources Research* **30**: 1393–1405.
Goodrich D. C., Lane L. J., Shillito R. M., Miller S. N., Syed K. H. and Woolhiser D. A. (1997) Linearity of basin response as a function of scale in a semiarid watershed. *Water Resources Research* **33**: 2951–2965.
Grayson R. B., Moore I. D. and McMahon T. A. (1992a) Physically-based hydrologic modelling. 1. A terrain-based model for investigative purposes. *Water Resources Research* **28**: 2639–2658.
Grayson R. B., Moore I. D. and McMahon T. A. (1992b) Physically-based hydrologic modelling. 2. Is the concept realistic. *Water Resources Research* **28**: 2659.
Grayson R. B., Blöschl G. and Moore I. D. (1995) Distributed parameter hydrologic modelling using vector elevation data: THALES and TAPES-c. In Singh V. P. (ed.) *Computer Models of Watershed Hydrology*. Water Resource Publications, Highlands Ranch, CO, pp. 669–696.
Green W. H. and Ampt G. (1911) Studies of soil physics. Part 1. The flow of air and water through soils. *Journal of the Agricultural Society* **4**: 1–24.
Grove M., Harbor J. and Engel B. (1998) Composite vs. distributed curve numbers: effects on estimates of storm runoff depths. *Journal of the American Water Resources Association* **34**: 1015–1023.
Güntner A., Uhlenbrook S., Seibert J. and Leibundgut C. (1999) Multi-criterial validation of TOPMODEL in a mountainous catchment. *Hydrological Processes* **13**: 1603–1620.
Gupta H. V., Sorooshian S. and Yapo P. O. (1998) Toward improved calibration of hydrologic models: multiple and noncommensurable measures of information. *Water Resources Research* **34**: 751–763.
Gupta H. V., Sorooshian S. and Yapo P. O. (1999) Status of automatic calibration for hydrologic models: comparison with multilevel expert calibration. *Journal of Hydrological Engineering ASCE* **4**: 135–143.
Gupta S. C. and Larson W. E. (1979) Estimating soil water retention characteristics from particle size distribution, organic matter percent and bulk density. *Water Resources Research* **15**: 1633–1635.
Gupta V. K. and Sorooshian S. (1985) The relationship between data and the precision of parameter estimates of hydrologic models. *Journal of Hydrology* **81**: 57–77.
Gurtz J., Baltensweiler A. and Lang H. (1999) Spatially distributed hydrotope-based modelling of evapotranspiration and runoff in mountainous basins. *Hydrological Processes* **13**: 2751–2768.
Hahn G. J. and Meeker W.Q. (1991) *Statistical intervals: a guide for practitioners*. John Wiley, New York.
Hall R. L., Calder I. R., Gunawardena E. R. N. and Rosier P. T. W. (1996) Dependence of rainfall interception on drop size: 3. Implementation and comparative performance of the stochastic model using data from a tropical site in Sri Lanka. *Journal of Hydrology* **185**: 389–407.

Hamon W. R. (1961) Estimating potential evapo-transpiration. *Journal of the Hydraulics Division, ASCE* **87**: 107–120.
Harlin J. and Kung C.-S. (1992) Parameter uncertainty and simulation of design floods in Sweden. *Journal of Hydrology* **137**: 209–230.
Hartman M. D., Baron J. S., Lammers R. B., Cline D. W., Band L. E., Liston G. E. and Tague C. (1999) Simulations of snow distribution and hydrology in a mountain basin. *Water Resources Research* **35**: 1587–1603.
Hashemi A. M., O'Connell P. E., Franchini M. and Cowpertwait P. S. P. (1998) A simulation analysis of the factors controlling the shapes of flood frequency curves. In Wheater H. and Kirby C. (eds) *Hydrology in a Changing Environment*, Volume 1. John Wiley, Chichester, pp. 39–49.
Hassanizadeh S. M. (1986) Derivation of basic equations of mass transport in porous media. Part 2. Generalized Darcy's law and Fick's law. *Advances in Water Resources* **9**: 2–7-222.
Hebson C. and Wood E. F. (1982) A derived flood frequency distribution using Horton order ratios. *Water Resources Research* **18**: 1509–1518.
Henderson D. E., Reeves A. D. and Beven K. (1996) Flow separation in undisturbed soil using multiple anionic tracers (2) steady state core scale rainfall and return flows. *Hydrological Processes* **10**: 1451–1466.
Henderson F. and Wooding R. (1964) Overland flow and groundwater flow from a steady rainfall of finite duration. *Journal of Geophysical Research* **69**: 1531–1540.
Herschy R. W. (1995) *Streamflow Measurement.* Chapman and Hall, London.
Hewlett J. (1974) Comments on letters relating to 'Role of subsurface flow in generating surface runoff. 2. Upstream source areas' by R. Allen Freeze. *Water Resources Research* **10**: 605–607.
Hewlett J. D. (1961) Soil moisture as a source of base flow from steep mountain watersheds. US Department of Agriculture and Forestry Services, Southeastern Forest Experiment Station, Ashville, NC, Station Paper 132.
Hewlett J. D. and Hibbert A. R. (1967) Factors affecting the response of small watersheds to precipitation in humid areas. In Sopper W. E. and Lull H. W. (eds) *Forest Hydrology.* Pergamon Press, Oxford, pp. 275–290.
Hjelmfelt A. T. and Amerman C. R. (1980) The mathematical basin model of Merrill Bernard. IAHS Publication No. 130, Wallingford, UK, pp. 343–349.
Hjelmfelt A. T., Kramer L. A. and Burwell R. (1982) Curve numbers as random variables. In Singh V. (ed.) *Rainfall–Runoff Relationships.* Water Resource Publications, Highlands Ranch, CO, pp. 365–370.
Hollenbeck K. J. and Jensen K. H. (1998) Experimental evidence of randomness and nonuniqueness in unsaturated outflow experiments designed for hydraulic parameter estimation. *Water Resources Research* **34**: 595–602.
Holtan H. N., England C. B., Lawless G. P. and Schumaker G. A. (1968) Moisture-tension data for selected soils on experimental watersheds. Technical Report ARS 41–144, USDA ARS, Agricultural Research Service, Beltsville, MD.
Holwill C. J. and Stewart J. B. (1992) Spatial variability of evapotranspiration derived from aircraft and ground based data. *Journal of Geophysical Research* **97**(D17): 18 673–18 680.
Hooper R. P., Stone A., Christophersen N., de Grosbois E. and Seip H. M. (1986) Assessing the Birkenes model of stream acidification using a multi-signal calibration methodology. *Water Resources Research* **22**: 1444–1454.
Hornberger G. M. and Spear R. C. (1981) An approach to the preliminary analysis of environmental systems. *Journal of Environmental Management* **12**: 7–18.
Hornberger G. M., Beven K. J., Cosby B. J. and Sappington D. E. (1985) Shenandoah watershed study: calibration of a topography-based, variable contributing area hydrological model to a small forested catchment. *Water Resources Research* **21**: 1841–1850.
Horton R. E. (1933) The role of infiltration in the hydrologic cycle. *Transactions of the American Geophysical Union* **14**: 446–460.

Horton R. E. (1936) Maximum ground-water levels. *Transactions of the American Geophysical Union* **17**: 344–357.

Horton R. E. (1940) An approach to the physical interpretation of infiltration capacity. *Soil Science Society of America Proceedings* **5**: 399–417.

Horton R. E. (1942) Remarks on hydrologic terminology. *Transactions of the American Geophysical Union* **23**: 479–482.

Hosking J. R. M. and Clarke R. T. (1990) Rainfall–runoff relations derived from the probability theory of storage. *Water Resources Research* **26**: 1455–1463.

Hottelet C., Braun L., Leibundgut C. and Rieg A. (1993) Simulation of snowpack and discharge in an Alpine karst basin. IAHS Publication No. 218, Wallingford, UK, pp. 249–260.

Houghton-Carr H. A. (1999) Assessment criteria for simple conceptual daily rainfall–runoff models. *Hydrological Science Journal* **44**: 237–263.

Houser P. R., Shuttleworth W. J., Famiglietti J. S., Gupta H. V., Syed K. H. and Goodrich D. C. (1998) Integration of soil moisture remote sensing and hydrologic modelling using data assimilation. *Water Resources Research* **34**: 3405–3420.

Hromadka T. V. II and Whitley R. J. (1994) The rational method for peak flow rate estimation. *Water Resources Bulletin* **30**: 1001.

Huber W. C. (1995) EPA storm water management model – SWMM. In Singh V. P. (ed.) *Computer Models of Watershed Hydrology*. Water Resource Publications, Highlands Ranch, CO, pp. 783–808.

Huff D. D., O'Neill R. V., Emmanuel W. R., Elwood J. W. and Newbold J. D. (1982) Flow variability and hillslope hydrology. *Earth Surface Processes and Landforms* **7**: 91–94.

Huggins L. F. and Monke E. J. (1968) A mathematical model for simulating the hydrologic response of a watershed. *Water Resources Research* **4**: 529–539.

Hulme M. and Jenkins J. (1998) Climate change scenarios for the United Kingdom: scientific report. Technical report, Climatic Research Unit, Norwich, UK.

Hursh C. R. (1936) Storm water and adsorption. *Transactions of the American Geophysical Union* **17**: 301–302.

Hursh C. R. and Brater E. F. (1941) Separating storm-hydrographs from small drainage-areas into surface- and subsurface-flow. *Transactions, American Geophysical Union* **22**: 863–870.

Ibbitt R. P. and O'Donnell T. (1971) Fitting methods for conceptual catchment models. *Journal of Hydraulics Division, ASCE* **97**: 1331–1342.

Ibbitt R. P. and O'Donnell T. (1974) Designing conceptual catchment models for automatic fitting methods. In *Mathematical Models in Hydrology*. IAHS Publication No. 101, Wallingford, UK, pp. 461–475.

IH (1999) *Flood Estimation Handbook* (5 volumes). Centre for Ecology and Hydrology, Wallingford, UK.

Ingber L. (1993) Simulated annealing: practice versus theory. *Statistics and Computing* **11**: 29–57.

Institution of Engineers, Australia (1977) *Australian Rainfall and Runoff*. Institution of Engineers, Sydney, Australia.

Iorgulescu I. and Musy A. (1997) Generalisation of TOPMODEL for a power law transmissivity profile. *Hydrological Processes* **11**: 1353–1355.

Jain S. K., Storm B., Bathurst J. C., Refsgaard J.-C. and Singh R. D. (1992) Application of the SHE to catchments in India: Part 2. Field experiments and simulation studies with the SHE on the Kolar subcatchment of the Narmada River. *Journal of Hydrology* **140**: 25–47.

Jakeman A. and Hornberger G. (1993) How much complexity is warranted in a rainfall–runoff model? *Water Resources Research* **29**(8): 2637–2649.

Jakeman A. J. and Hornberger G. M. (1994) How much complexity is warranted in a rainfall–runoff model? – Reply. *Water Resources Research* **30**(12): 3567.

Jakeman A. J., Littlewood I. G. and Whitehead P. G. (1990) Computation of the instantaneous unit hydrograph and identifiable component flows with application to two small upland catchments. *Journal of Hydrology* **117**: 275–300.

Jakeman A. J., Littlewood I. G. and Whitehead P. G. (1993a) An assessment of the dynamic response characteristics of streamflow in the Balquhidder catchments. *Journal of Hydrology* **145**: 337–355.

Jakeman A. J., Chen T. H., Post D. A., Hornberger G. M., Littlewood I. G. (1993b) Assessing uncertainties in hydrological response to climate at large scale. In Wilkinson W. B. (ed.) *Macroscale Modelling of the Hydrosphere.* IAHS Publication No. 214, Wallingford, UK, pp. 37–47.

Jakeman A. J., Green T. R., Beavis S. G., Zhang L., Dietrich C. R. and Crapper P. F. (1999) Modelling upland and instream erosion, sediment and phosphorus transport in a large catchment. *Hydrological Processes* **13**: 745–752.

Jarvis N. J., Jansson P. E. and Dik P. E. (1991) Modelling water and solute transport in macroporous soil. 1. Model description and sensitivity analysis. *Journal of Soil Science* **42**: 59–70.

Jarvis P. J. (1976) The interpretation of the variation in leaf water potential and stomatal conductance found in canopies in the field. *Philosophical Transactions of the Royal Society, London, Series B* **273**: 593–610.

Jaynes D. B. (1990) *Soil Water Hysteresis: Models and Implications.* John Wiley, Chichester, pp. 93–126.

Jolley T. and Wheater H. (1997) An investigation into the effect of spatial scale on the performance of a one-dimensional water balance model. *Hydrological Processes* **11**: 1927–1944.

Jones N. L., Wright S. G. and Maidment D. R. (1990) Watershed delineation with triangle-based terrain models. *Journal of the Hydraulics Division, ASCE* **110**: 1232–1251.

Jordan J.-P. (1994) Spatial and temporal variability of stormflow generation processes on a Swiss catchment. *Journal of Hydrology* **153**: 357–382.

Karvonen T., Koivusalo H., Jauhiainen M., Palko J. and Weppling K. (1999) A hydrological model for predicting runoff from different land use areas. *Journal of Hydrology* **217**: 253–265.

Keulegan G. (1945) Spatially varied discharge over a sloping plane. *Transactions of the American Geophysical Union* **26**: 821–824.

Kilsby C. G., Fallows C. S. and O'Connell P. E. (1998) Generating rainfall scenarios for hydrological impact modelling. In Wheater H. and Kirby C. (eds) *Hydrology in a Changing Environment.* John Wiley, Chichester, pp. 33–42.

Kilsby C. G., Ewen J., Sloan W. T., Burton A., Fallows C. S. and O'Connell P. E. (1999) The UP modelling system for large scale hydrology: simulation of the Arkansas-Red River basin. *Hydrology and Earth System Science* **3**: 137–149.

Kim S. and Delleur J. W. (1997) Sensitivity analysis of extended TOPMODEL for agricultural watersheds equipped with tile drains. *Hydrological Processes* **11**: 1243–1262.

Kirkby M. (1975) Hydrograph modelling strategies. In Peel R., Chisholm M. and Haggett P. (eds) *Processes in Physical and Human Geography.* Heinemann, London, pp. 69–90.

Kirkby M. (1976) Tests of the random network model and its application to basin hydrology. *Earth Surfaces and Processes* **1**: 197–212.

Kirkby M. J. (1978) Implications for sediment transport. In Kirkby M. J. (ed.) *Hillslope Hydrology.* John Wiley, Chichester.

Kirkby M. J. (1988) Hillslope runoff processes and models. *Journal of Hydrology* **100**: 315–340.

Kirkby M. J. (1997) TOPMODEL: a personal view. *Hydrological Processes* **11**: 1087–1098.

Kite G. W. (1995a) Scaling of input data for hydrologic modelling. *Water Resources Research* **31**: 2769–2781.

Kite G. W. (1995b) The SLURP model. In Singh V. P. (ed.) *Computer Models of Watershed Hydrology.* Water Resource Publications, Highlands Ranch, CO, pp. 521–561.

Kite G. W. and Haberlandt U. (1999) Atmospheric model data for macroscale hydrology. *Journal of Hydrology* **217**: 303–313.

Kite G. and Kouwen N. (1992) Watershed modeling using land classification. *Water Resources Research* **28**: 3193–3200.

Kleissen F. M., Beck M. B. and Wheater H. S. (1990) The identifiability of conceptual hydrochemical models. *Water Resources Research* **26**: 2979–2992.

Klemeš V. (1986) Operational testing of hydrologic simulation models. *Hydrological Science Journal* **31**: 13–24.

Klepper O., Scholten H. and Van de Kamer J. P. G. (1991) Prediction uncertainty in an ecological model of the Oosterschelde estuary. *Journal of Forecasting* **10**: 191–209.

Knight D. W., Yuen K. W. H. and Al-Hamid A. A. I. (1994) Boundary shear stress distributions in open channel flow. In Beven K. J., Chatwin P. C. and Millbank J. (eds) *Mixing and Transport in the Environment.* John Wiley, Chichester, pp. 51–88.

Knisel W. G. and Williams J. R. (1995) Hydrology components of CREAMS and GLEAMS. In Singh V. P. (ed.) *Computer Models of Watershed Hydrology.* Water Resource Publications, Highlands Ranch, CO, pp. 1069–1114.

Konikow L. F. and Bredehoeft J. D. (1992) Groundwater models cannot be validated. *Advances in Water Resources* **15**: 75–83.

Kool J. B., Parker J. C. and Van Genuchten M. Th. (1987) Parameter estimation for unsaturated flow and transport models – a review. *Journal of Hydrology* **91**: 255–293.

Koopmans B. N., Pohl C. and Wang Y. (1995) The 1995 flooding of the Rhine, Wall and Mass rivers in the Netherlands. *Earth Observation Quarterly* **47**: 11–12.

Koster R. D. and Suarez M. J. (1992) A comparative study of two land surface heterogeneity representations. *Journal of Climate* **5**: 1379–1390.

Kothyari U. C. and Singh V. P. (1999) A multiple-input single-output model for flow forecasting. *Journal of Hydrology* **220**: 12–26.

Kuczera G. A. (1987) Prediction of water yield reductions following a bushfire in ash-mixed species eucalypt forest. *Journal of Hydrology* **94**: 215–236.

Kuczera G. (1997) Efficient subspace probabilistic parameter optimisation for catchment models. *Water Resources Research* **33**(1): 177–185.

Kuczera G. and Parent E. (1998) Monte Carlo assessment of parameter uncertainty in conceptual catchment models: the Metropolis algorithm. *Journal of Hydrology* **211**: 69–85.

Kull D. W. and Feldman A. D. (1998) Evolution of Clark's unit graph method to spatially distributed runoff. *Journal of Hydrological Engineering, ASCE* **3**: 9–19.

Kundzewicz Z. W., Szamalek K. and Kawalczak P. (1999) The great flood of 1997 in Poland. *Hydrological Science Journal* **44**: 855–870.

Kuo W.-L., Steenhuis T. S., McCulloch C. E., Mohler C. L., Weinstein D. A., DeGloria S. D. and Swaney D. P. (1999) Effect of grid size on runoff and soil moisture for a variable-source-area hydrology model. *Water Resources Research* **35**: 3419–3428.

Kurothe R. S., Goel N. K. and Mathur B. S. (1997) Derived flood frequency distribution for negatively correlated rainfall intensity and duration. *Water Resources Research* **33**: 2103–2107.

Kustas W. P. and Goodrich D. C. (1994) Preface to the Monsoon'90 issue. *Water Resources Research* **30**: 1211–1225.

Kustas W. P., Rango A. and Uijlenhoet R. (1994) A simple energy budget algorithm for the snowmelt runoff model. *Water Resources Research* **30**: 1515–1527.

Laflen J. M., Lane L. J., and Foster G. R. (1991) WEPP – a new generation of erosion prediction technology. *Journal of Soil Water Conservation*, **46**, 34–38.

Lamb R. (1999) Calibration of a conceptual rainfall–runoff model for flood frequency estimation by continuous simulation. *Water Resources Research* **35**, 3103–3114.

Lamb R., Beven K. J. and Myrabø S. (1997) Discharge and water table predictions using a generalised TOPMODEL formulation. *Hydrological Processes* **11**: 1145–1168.

Lamb R., Beven K. and Myrabø S. (1998a) A generalised topographic-soils hydrological index. In Lane S. N., Richards K. S. and Chandler J. H. (eds) *Landform Monitoring, Modelling and Analysis.* John Wiley, Chichester, pp. 263–278.

Lamb R., Beven K. J. and Myrabø S. (1998b) Use of spatially distributed water table observations to constrain uncertainty in a rainfall–runoff model. *Advances in Water Resources* **22**: 305–317.

Lambert A. O. (1969) A comprehensive rainfall/runoff model for an upland catchment area. *Journal of the Institute of Water Engineering* **23**: 231–238.

Lambert A. O. (1972) Catchment models based on ISO functions. *Journal of the Institute of Water Engineers* **26**: 413–422.

Lane S. N. (1998) Hydraulic modelling in hydrology and geomorphology: a review of high resolution approaches. *Hydrological Processes* **12**: 1131–1150.

Lange J., Leibundgut C., Greenbaum N. and Schick A. P. (1999) A noncalibrated rainfall–runoff model for large, arid catchments. *Water Resources Research* **35**: 2161–2172.

Lavabre J., Sempere-Torres D. and Cernesson F. (1993) Changes in the hydrological response of a small Mediterranean basin a year after a wildfire. *Journal of Hydrology* **142**: 273–299.

Leavesley G. H. and Stannard L. G. (1995) The precipitation–runoff modelling system – PRMS. In Singh V. P. (ed.) *Computer Models of Watershed Hydrology.* Water Resource Publications, Highlands Ranch, CO, pp. 281–310.

LeDrew E. (1979) A diagnostic examination of a complementary relationship between actual and potential evapotranspiration. *Journal of Applied Meteorology* **18**: 495–501.

Lee P. M. (1989) *Bayesian Statistics: An Introduction.* Edward Arnold, London.

Lees M., Young P. C., Ferguson S., Beven K. J. and Burns J. (1994) An adaptive flood warning scheme for the River Nith at Dumfries. In White W. R. and Watts J. (eds) *River Flood Hydraulics.* John Wiley, Chichester, pp. 65–75.

Lek S., Dimopoulos-Derraz M. and El Ghachtoul Y. (1996) Modélisation de la relation pluie-débit à l'aide des réseaux de neurones artificiels. *Revue des Sciences de l'Eau* **3**: 319–331.

Lettenmaier D. P. and Gan T. Y. (1990) Hydrologic sensitivities of the Sacramento-San Joaquin River basin, California, to global warming. *Water Resources Research* **26**: 69–86.

Li R. M., Simons D. B. and Stevens M. A. (1975) Nonlinear kinematic wave approximation for water routing. *Water Resources Research* **11**: 245–252.

Liang G. C., O'Connor K. M. and Kachroo R. K. (1994) A multiple-input, single-output, variable gain-factor model. *Journal of Hydrology* **155**: 185–198.

Liang X., Lettenmaier D. P., Wood E. F. and Burges S. J. (1994) A simple hydrologically-based model of land surface water and energy fluxes for general circulation models. *Journal of Geophysical Research* **99**: 14 415–14 428.

Liang X., Wood E. F. and Lettenmaier D. P. (1996) Surface soil moisture parameterisation of the VIC-2L model: evaluation and modifications. *Global Planetary Change* **13**: 195–206.

Lighthill M. H. and Whitham G. B. (1955) On kinematic waves. 1. Flood movement in long rivers. *Proceedings of the Royal Society London Series A* **229**: 281–316.

Likens G. E., Bormann F. H., Pierce R. S., Eaton J. S. and Johnson N. M. (1977) *Biogeochemistry of a Forested Ecosystem.* Springer-Verlag, New York.

Lin D. S., Wood E. F., Beven K. J. and Saatchi S. (1994) Soil moisture estimation over grass-covered areas using AIRSAR. *International Journal of Remote Sensing* **15**(11): 2323–2343.

Linden D. R. and Dixon R. M. (1973) Infiltration and water table effects of soil air pressure under border irrigation. *Soil Science Society of America Proceedings* **37**: 95–98.

Lindström G., Johansson B., Persson M., Gardelin M. and Bergström S. (1997) Development and test of the distributed HBV-96 hydrological model. *Journal of Hydrology* **201**: 272–288.

Linsley R. K., Kohler M. A. and Paulhus J. L. H. (1949) *Applied Hydrology.* McGraw-Hill, New York.

Liston G. E., Sud Y. C. and Wood E. F. (1994) Evaluating GCM land surface hydrology parameterisations by computing discharges using a runoff routing model: application to the Mississippi Basin. *Journal of Applied Meteorology* **33**: 394–405.

Llorens P., Poch R., Latron J. and Gallart F. (1997) Rainfall interception by a *Pinus sylvestris* forest patch overgrown in a Mediterranean mountainous abandoned area. 1. Monitoring design and results down to the event scale. *Journal of Hydrology* **199**: 331–345.

Lloyd C. R., Gash J. H. C., Shuttleworth W. J. and Marques F., A. de O. (1988) The measurement and modelling of rainfall interception by Amazonian rain forest. *Agric. Forest Meteorology* **43**: 277–294.

Loague K. (1990) R-5 revisited. 2. Reevaluation of a quasi-physically based rainfall–runoff model with supplemental information. *Water Resources Research* **26**: 973–987.

Loague K. (1992a) Soil water content at R-5. Part 1. Spatial and temporal variability. *Journal of Hydrology* **139**: 233–252.

Loague K. (1992b) Soil water content at R-5. Part 2. Impact of antecedent conditions on rainfall–runoff simulations. *Journal of Hydrology* **139**: 253–262.

Loague K. M. and Freeze R. A. (1985) A comparison of rainfall–runoff modelling techniques on small upland catchments. *Water Resources Research* **21**: 229–248.

Loague K. M. and Gander G. A. (1990) R-5 revisited: spatial variability of infiltration on a small rangeland watershed. *Water Resources Research* **26**: 957–971.

Loague K. M. and Kyriakidis P. C. (1997) Spatial and temporal variability in the R-5 infiltration data set: déjà vu and rainfall–runoff simulations. *Water Resources Research* **33**: 2883–2896.

Lohmann D., Raschke E., Nijssen B. and Lettenmaier D. P. (1998a) Regional scale hydrology: I. Formulation of the VIC-2L model coupled to a routing model. *Hydrological Science Journal* **43**: 131–141.

Lohmann D., Raschke E., Nijssen B. and Lettenmaier D. P. (1998b) Regional scale hydrology: II. Application of the VIC-2L model to the Weser River, Germany. *Hydrological Science Journal* **43**: 143–158.

Lohmann D. and 28 others (1998c) The Project for Intercomparison of Land-surface Parameterisation Schemes (PILPS) Phase 2c: Arkansas Red-River basin experiment: 3. Spatial and temporal analysis of water fluxes. *Global Planetary Change* **19**: 161–198.

Luxmoore R. J. and Sharma M. L. (1980) Runoff responses to soil heterogeneity: experimental and simulation comparisons for two contrasting watersheds. *Water Resources Research* **16**: 675–684.

Manguerra H. B. and Engel B. A. (1998) Hydrologic parameterisation of watersheds for runoff prediction using SWAT. *Journal of American Water Resources Association* **34**: 1149–1162.

Manzi A. O. and Planton S. (1994) Implementation of the ISBA parametrization scheme for land surface processes in a GCM – an annual cycle experiment. *Journal of Hydrology* **155**: 353–387.

Marks D. and Dozier J. (1992) Climate and energy exchange at the snow surface in the alpine region of the Sierra Nevada. 2. Snow cover energy balance. *Water Resources Research* **28**: 3043–3054.

Martinec J. and Rango A. (1981) Areal distribution of snow water equivalent evaluated by snow cover monitoring. *Water Resources Research* **17**: 1480–1488.

McAneny K. J., Green A. E. and Astill M. S. (1995) Large aperture scintillometry: the homogeneous case. *Agric. Forest Meteorology* **76**: 149–162.

McCuen R. H. (1973) The role of sensitivity analysis in hydrologic modelling. *Journal of Hydrology* **18**: 37–53.

McCuen R. H. (1982) *A Guide to Hydrologic Analysis Using SCS Methods.* Prentice-Hall, Englewood Cliffs, NJ.

McCulloch J. S. G. and Robinson M. (1993) History of forest hydrology. *Journal of Hydrology* **150**: 189–216.

McDonnell J. J., Freer J., Hooper R., Kendall C., Burns D., Beven K. and Peters J. (1996) New method developed for studying flow on hillslopes. *EOS, Transactions of the AGU* **77**(47): 465–472.

McLauglin D. and Townley R. L. (1996) A reassessment of groundwater inverse problems. *Water Resources Research* **32**: 1131–1161.

Melching C. S. (1995) Reliability estimation. In Singh V. P. (ed.) *Computer Models of Watershed Hydrology.* Water Resource Publications, Highlands Ranch, CO, Chapter 3.

Mesa O. J. and Mifflin E. R. (1986) On the relative role of hillslope and network geometry in hydrologic response. In Gupta V. K., Rodriguez-Iturbe I. and Wood E. F. (eds) *Scale Problems in Hydrology.* 1–17.

Michaud J. and Sorooshian S. (1994) Comparison of simple versus complex distributed runoff models on a midsized semiarid watershed. *Water Resources Research* **30**: 593–606.

Miller E. E. and Miller R. D. (1956) Physical theory for capillary flow phenomena. *Journal of Applied Physics* **27**: 324–332.

Milly P. C. D. (1991) A refinement of the combination equations for evaporation. *Surveys in Geophysics* **12**: 145–154.

Minns A. W. and Hall M. J. (1996) Artificial neural networks as rainfall–runoff models. *Hydrological Science Journal* **41**: 399–417.

Minshall N. E. (1960) Predicting storm runoff on small experimental watersheds. *Journal of the Hydraulics Division, ASCE* **86**(HY8): 17–38.

Mishra S. and Parker J. C. (1989) Parameter estimation for coupled unsaturated flow and transport. *Water Resources Research* **25**: 385–396.

Mishra S. K. and Singh V. P. (1999) Another look at SCS-CN method. *Journal of Hydrological Engineering, ASCE* **4**: 257–264.

Mitchell K. M. and DeWalle D. R. (1998) Application of the snowmelt runoff model using multiple-parameter landscape zones on the Towanda Creek Basin, Pennsylvania. *Journal of American Water Resources Association* **34**: 335–346.

Mockus V. (1949) *Estimation of Total (and Peak Rates of) Surface Runoff for Individual Storms. Exhibit A, Appendix B, Interim Survey Report, Grand (Neosho) River Watershed.* US Department of Agriculture, Washington, DC.

Mohanty B. P. (1999) Scaling hydraulic properties of a macroporous soil. *Water Resources Research* **35**: 1927–1931.

Mohanty B. P., Bowman R. S., Hendrickx J. M. H. and Van Genuchten M. T. (1997) New piecewise – continuous hydraulic functions for modelling preferential flow in an intermittent-flood-irrigated field. *Water Resources Research* **33**: 2049–2064.

Mohanty B. P., Bowman R. S., Hendrickx J. M. H., Simunek J. and Van Genuchten M. T. (1998) Preferential transport of nitrate to a tile drain in an intermittent-flood-irrigated field: model development and experimental evaluation. *Water Resources Research* **34**: 1061–1076.

Monteith J. L. (1965) Evaporation and environment. In *The State and Movement of Water in Living Organisms.* Proceedings of the 19th Symposium, Society of Experimental Biology, Cambridge University Press, London.

Monteith J. (1995a) Accomodation between transpiring vegetation and the convective boundary layer. *Journal of Hydrology* **166**: 251–263.

Monteith J. (1995b) A reinterpretation of stomatal responses to humidity. *Plant, Cell and Environment* **18**: 357–364.

Moore I. D., Mackay S. M., Wallbrink P. J., Burch G. J. and O'Loughlin E. M. (1986) Hydrologic characteristics and modelling of a small forested catchment in southeastern New South Wales. *Journal of Hydrology* **83**: 307–335.

Moore R. J. (1985) The probability-distributed principle and runoff production at point and basin scales. *Hydrological Sciences Journal* **30**: 273–297.

Moore R. J. (1999) Real-time flood forecasting systems: perspectives and prospects. In Casali R. and Margottini C. (eds) *Floods and Landslides: Integral Risk Assessment.* Springer-Verlag, Berlin, pp. 147–189.

Moore R. J. and Clarke R. T. (1981) A distribution function approach to rainfall–runoff modelling. *Water Resources Research* **17**: 1367–1382.

Moore R. J. and Clarke R. T. (1983) A distribution function approach to modelling basin sediment yield. *Journal of Hydrology* **65**: 239–257.

Moore R. J., Jones D. A., Bird P. B. and Cottingham M. C. (1990) A basin-wide flow forecasting system for real-time flood warning, river control and water management. In White W. R. (ed.) *International Conference on River Flood Hydraulics.* John Wiley, Chichester, pp. 21–30.

Moore R. J., Jones D. A., Black K. B., Austin R. M., Carrington D. S., Tinninon M. and Akhondi A. (1994) Rffs and hyrad: Integrated systems for rainfall and river flow forecasting in real-time and their application in Yorkshire. In *Proceedings of the BHS National Meeting on*

Analytical Techniques for the Development and Operations Planning of Water Resource and Supply Systems, BHS Occasional Paper No. 4.
Moore R. J., Bell V. A., Austin R. M. and Harding R. J. (1999) Methods for snowmelt forecasting in upland Britain. *Hydrology and Earth Systems Science* **3**: 233–246.
Morel-Seytoux H. J. and Khanji J. (1974) Derivation of an equation of infiltration. *Water Resources Research* **10**: 795–800.
Morris E. M. (1991) Physics-based models of snow. In Bowles D. S. and O'Connell P. E. (eds) *Recent Advances in the Modelling of Hydrological Systems.* Kluwer Academic, Dordrecht, pp. 85–112.
Morton F. I. (1978) Estimating evapotranspiration from potential evaporation: practicality of an iconoclastic approach. *Journal of Hydrology* **38**: 1–32.
Morton F. I. (1983a) Operational estimates of areal evapotranspiration and their significance to the science and practice of hydrology. *Journal of Hydrology* **66**: 1–76.
Morton F. I. (1983b) Operational estimates of lake evaporation. *Journal of Hydrology* **66**: 77–100.
Mroczkowski M., Raper G. P. and Kuczera G. (1997) The quest for more powerful validation of conceptual catchment models. *Water Resources Research* **33**: 2325–2335.
Mulvaney T. J. (1851) On the use of self-registering rain and flood gauges in making observations of the relations of rainfall and flood discharges in a given catchment. *Proceedings of the Institution of Civil Engineers of Ireland* **4**: 19–31.
Muttiah R. S., Srinivasan R. and Allen P. M. (1997) Prediction of two-year peak stream discharges using neural networks. *Journal of American Water Resources Association* **33**: 625–630.
Myrabø S. (1997) Temporal and spatial scale of response area and groundwater variation in till. *Hydrological Processes* **11**: 1861–1880.
Naden P. S. (1992) Spatial variability in flood estimation for large catchments: the exploitation of channel network structure. *Hydrological Sciences Journal* **37**: 53–71.
Naden P. (1993) A routing model for continental scale hydrology. In Wilkinson W. B. (ed.) *Macroscale Modelling of the Hydrosphere.* IASH Publication No. 214, Wallingford, UK, pp. 67–79.
Naden P. S., Broadhurst P., Tauveron N. and Walker A. (1999) River routing at the continental scale: use of globally-available data and an a priori method of parameter estimation. *Hydrology and Earth System Science* **3**: 109–124.
Naef F. (1981) Can we model the rainfall–runoff process today? *Hydrological Science Bulletin* **26**: 281–289.
Nash J. E. (1959) Systematic determination of unit hydrograph parameters. *Journal of Geophysical Research* **64**: 111–115.
Nash J. E. and Shamseldin A. Y. (1998) The geomorphological unit hydrograph – a critique. *Hydrology and Earth System Science* **2**: 1–8.
Nash J. E. and Sutcliffe J. V. (1970) River flow forecasting through conceptual models 1. A discussion of principles. *Journal of Hydrology* **10**: 282–290.
Natale L. and Todini E. (1977) A constrained parameter estimation technique for linear models in hydrology. In Ciriani T. A. M. U. and Wallis J. R. (eds) *Mathematical Models for Surface Water Hydrology.* John Wiley, Chichester, pp. 109–147.
Nathan R. J., McMahon T. A. and Finlayson B. L. (1988) The impact of the greenhouse effect on catchment hydrology and storage-yield relationships in both winter and summer rainfall zones. In Pearman G. I. (ed.) *Greenhouse, Planning for Climate Change.* East Melbourne, Australia.
Navar J. and Bryan R. B. (1994) Fitting the analytical model of rainfall interception of gash to individual shrubs of semi-arid vegetation in northeastern Mexico. *Agric. Forest. Meteorology* **68**: 133–143.
Neal C., Robson A. J., Shand P., Edmunds W. M., Dixon A. J., Buckley D. K., Hill S., Harrow M., Neal M., Wilkinson, J. and Reynolds B. (1997) The occurrence of groundwater

in the Lower Palaeozoic rocks of upland Central Wales. *Hydrology and Earth System Science* **1**: 3–18.

Nelder J. A. and Mead R. (1965) A simplex method for function minimization. *Computer Journal* **7**: 308–313.

Nelson E. J., Jones N. L. and Berrett R. J. (1999) Adaptive tessellation method for creating TINs from GIS data. *Journal of Hydrological Engineering, ASCE* **4**: 2–9.

NERC (1975) *The Flood Studies Report* (5 volumes). Natural Environment Research Council, Wallingford, UK.

Newson M. D. (1980) The geomorphological effectiveness of floods – a contribution stimulated by two recent events in mid-Wales. *Earth Surface Processes.* **5**: 1–16.

Ng H. Y. F. and Marsalek J. (1992) Sensitivity of streamflow simulation to changes in climatic inputs. *Nordic Hydrology* **23**: 257–272.

Nichols W. E. and Cuenca R. H. (1993) Evaluation of the evaporative fraction for parameterization of the surface energy balance. *Water Resources Research* **29**: 3681–3690.

O'Connell P. E. (1991) A historical perspective. In Bowles D. S. and O'Connell P. E. (eds) *Recent Advances in the Modeling of Hydrologic Systems.* Kluwer, Dordrecht, pp. 3–30.

O'Donnell T. (1966) Methods of computation in hydrograph analysis and synthesis. In *Recent Trends in Hydrograph Synthesis.* Committee for Hydrological Research, Central Organisation for Applied Research in the Netherlands, The Hague.

Olivera F. and Maidment D. (1999) Geographical information systems (GIS)-based spatially distributed model for runoff routing. *Water Resources Research* **35**: 1155–1164.

O'Loughlin E. M. (1981) Saturation regions in catchments and their relation to soil and topographic properties. *Journal of Hydrology* **53**: 229–246.

O'Loughlin E. (1986) Prediction of surface saturation zones in natural catchments by topographic analysis. *Water Resources Research* **22**: 794–804.

Onof C., Faulkner D. and Wheater H. S. (1996) Design rainfall modelling in the Thames catchment. *Hydrological Science Journal* **41**: 715–733.

Oreskes N., Schrader-Frechette K. and Belitz K. (1994) Verification, validation and confirmation of numerical models in the earth sciences. *Science* **263**: 641–646.

Overney O. (1998) *Prédiction Des Crues Par Modélisation Couplé Stochastique et Déterministique: Méthode et Analyse Des Incertitudes.* PhD thesis, Ecole Polutechnique Fédérale de Lausanne, Switzerland

Pachepsky Y. and Timlin D. (1998) Water transport in soils as in fractal media. *Journal of Hydrology* **204**: 98–107.

Palacios-Velez O. L. and Cuevas-Renaud B. (1986) Automated river course, ridge and basin delineation from digital elevation data. *Journal of Hydrology* **86**: 299–314.

Palacios-Velez O. L., Gandoy-Bernasconi W. and Cuevas-Renaud B. (1998) Geometric analysis of surface runoff and the computation order of unit elements in distributed hydrological models. *Journal of Hydrology* **211**: 266–274.

Panagoulia D. (1992) Modelled climatic changes on catchment hydrology. *Hydrological Science Journal* **37**: 141–163.

Paniconi C. and Wood E. F. (1993) A detailed model for simulation of catchment scale subsurface hydrologic processes. *Water Resources Research* **29**: 1601–1620.

Paniconi C., Aldama A. A. and Wood E. F. (1991) Numerical evaluation of iterative and noniterative methods for the solution of the nonlinear Richards equation. *Water Resources Research* **27**: 1147–1163.

Parkin G., O'Donnell G., Ewen J., Bathurst J. C., O'Connell P. E. and Lavabre J. (1996) Validation of catchment models for predicting land-use and climate change impacts. 1. Case study for a Mediterranean catchment. *Journal of Hydrology* **175**: 595–613.

Parlange J.-Y. and Haverkamp R. (1989) Infiltration and ponding time. In Morel-Seytoux H. J. (ed.) *Unsaturated Flow in Hydrologic Modeling.* Kluwer Academic, Dordrecht, pp. 105–126.

Parsons A. J., Wainwright J., Abrahams A. D. and Simanton J. R. (1997) Distributed dynamic modelling of interrill overland flow. *Hydrological Processes* **11**: 1833–1859.

Pereira L. S., Perrier A., Allen R. G. and Alves I. (1999) Evapotranspiration: concepts and future trends. *Journal of Irrigation and Drainage Engineering ASCE* **4**: 45–51.
Peterson J. R. and Hamlett J. M. (1998) Hydrologic calibration of the SWAT model in a watershed containing fragipan soils. *Journal of American Water Resources Association* **34**: 531–544.
Philip J. R. (1957) The theory of infiltration. 1. The infiltration equation and its solution. *Soil Science* **83**: 345–357.
Philip J. R. (1991) Hillslope infiltration: divergent and convergent slopes. *Water Resources Research* **27**: 1035–1040.
Pinder G. F. and Gray W. G. (1977) *Finite Element Simulation in Surface and Subsurface Hydrology*. Academic Press, New York.
Piñol J., Beven K. J. and Freer J. (1997) Modelling the hydrological response of Mediterranean catchments, Prades, Catalonia: the use of distributed models as aids to hypothesis formulation. *Hydrological Processes* **11**: 229–248.
Plate E. J., Ihrringer J. and Lutz W. (1988) Operational models for flood calculations. *Journal of Hydrology* **100**: 489–506.
Poeter E. P. and Hill M. C. (1997) Inverse methods: a necessary next step in ground-water flow modeling. *Ground Water* **35**: 250–260.
Ponce V. M. (1989) *Engineering Hydrology: Principles and Practices.* Prentice-Hall, Englewood Cliffs, NJ.
Ponce V. (1991) The kinematic wave controversy. *Journal of the Hydraulics Division, ASCE* **117**: 511–525.
Post D. and Jakeman A. (1996) Relationships between physical attributes and hydrologic response characteristics in small Australian mountain ash catchments. *Hydrological Processes* **10**: 877–892.
Press W. H., Flannery B. P., Teukolsky S. A. and Vetterling W. T. (1992) *Numerical Recipes: The Art of Scientific Computing,* 2nd edition. Cambridge University Press, Cambridge.
Prevost M., Barry R., Stein J. J. and Plamondon A. (1990) Snowmelt runoff modelling in a balsam fir forest with a variable source area simulator. *Water Resources Research* **26**: 1067–1077.
Puente C. E. (1997) A new approach to hydrologic modeling: derived distributions revisited. *Journal of Hydrology* **187**: 65–80.
Quinn P. F., Beven K. J., Chevallier P. and Planchon O. (1991) The prediction of hillslope flow paths for distributed hydrological modelling using digital terrain models. *Hydrological Processes* **5**: 59–79.
Quinn P. F., Beven K. J. and Culf A. (1995a) The introduction of macroscale hydrological complexity into land surface–atmosphere transfer function models and the effect on planetary boundary layer development. *Journal of Hydrology* **166**: 421–444.
Quinn P. F., Beven K. J. and Lamb R. (1995b) The ln(a/tan β) index: how to calculate it and how to use it in the TOPMODEL framework. *Hydrological Processes* **9**: 161–182.
Quinn P. F., Ostendorf B., Beven K. J. and Tenhunen J. (1998) Spatial and temporal predictions of soil moisture patterns and evaporative losses using TOPMODEL and the GAS-flux model for an Alaskan catchment. *Hydrology and Earth System Sciences* **2**: 41–54.
Rango A. (1995) The snowmelt runoff model (SRM). In Singh V. P. (ed.) *Catchment Models of Watershed Hydrology*. Water Resource Publications, Highlands Ranch, CO, pp. 477–520.
Rango A. and Martinec J. (1995) Revisiting the degree-day method for snowmelt computations. *Water Resources Bulletin* **31**: 657–669.
Rawls W. and Brakensiek D. (1982) Estimating soil water retention from soil properties. *Journal of Irrigation and Drainage ASCE* **108**: 166–171.
Rawls W. J. and Brakensiek D. L. (1989) Estimation of soil water retention and hydraulic properties. In Morel-Seytoux H. J. (ed.) *Unsaturated Flow in Hydrologic Modeling: Theory and Practice.* Kluwer Academic, Dordrecht, pp. 275–300.
Rawls W. J., Brakensiek D. L. and Miller N. (1983) Green–Ampt infiltration parameters from soil data. *Journal of Hydraulics Division, ASCE* **109**: 62–70.

Reed D. W., Johnson P. and Firth J. M. (1975) A non-linear rainfall–runoff model, providing for variable lag time. *Journal of Hydrology* **25**: 295–305.

Refsgaard J.-C. (1997) Parameterisation, calibration and validation of distributed hydrological models. *Journal of Hydrology* **198**: 69–97.

Refsgaard J.-C. and Knudsen J. (1996) Operational validation and intercomparison of different types of hydrological models. *Water Resources Research* **32**: 2189–2202.

Refsgaard J.-C. and Storm B. (1995) MIKE SHE. In Singh V. P. (ed.) *Computer Models of Watershed Hydrology*. Water Resource Publications, Highlands Park, CO, pp. 809–846.

Refsgaard J.-C., Seth S. M., Bathurst J. C., Erlich M., Storm B., Jogenson G. H. and Chandra S. (1992) Application of the SHE to catchments in India. Part 1. General results. *Journal of Hydrology* **140**: 1–23.

Refsgaard J.-C., Storm B. and Abbott M. B. (1996) Comment on 'A discussion of distributed hydrological modelling' by K. Beven. In Abbott M. B. and Refsgaard J.-C. (eds) *Distributed Hydrological Modelling*. Kluwer Academic, Dordrecht, pp. 279–287.

Reggiani P., Hassanizadeh S. M., Sivapalan M. and Gray W. G. (1999) A unifying framework for watershed thermodynamics: constitutive relationships. *Advances in Water Resources* **23**: 15–39.

Richards L. A. (1931) Capillary conduction of liquids through porous mediums. *Physics* **1**: 318–333.

Richards B. D. (1944) *Flood Estimation and Control*. Chapman and Hall, London.

Robinson J. S. and Sivapalan M. (1995) Catchment-scale runoff generation model by aggregation and similarity analyses. *Hydrological Processes* **9**: 555–574.

Robinson J. S. and Sivapalan M. (1997) Temporal scales and hydrological regimes: implications for flood frequency scaling. *Water Resources Research* **33**: 2981–2999.

Robinson J. S., Sivapalan M. and Snell J. D. (1995) On the relative roles of hillslope processes, channel routing, and network geomorphology in the hydrologic response of natural catchments. *Water Resources Research* **31**: 3089–3101.

Robinson M. (1986) Changes in catchment runoff following drainage and afforestation. *Journal of Hydrology* **86**: 71–84.

Robson A. J., Beven K. J. and Neal C. (1992) Towards identifying sources of subsurface flow: a comparison of components identified by a physically based runoff model and those determined by chemical mixing techniques. *Hydrological Processes* **6**: 199–214.

Rodda J. C. and Smith S. W. (1986) The significance of the systematic error in rainfall measurement for assessing wet deposition. *Atmos. Environment* **20**: 1059–1064.

Rodriguez-Iturbe I. (1993) The geomorphological unit hydrograph. In Beven K. J. and Kirkby M. J. (eds) *Channel Network Hydrology*. John Wiley, Chichester, pp. 43–68.

Rodriguez-Iturbe I. and Rinaldo A. (1997) *Fractal River Basins: Chance and Self-organisation*. Cambridge University Press, New York.

Rodriguez-Iturbe I. and Valdes J. (1979) The geomorphic structure of hydrologic response. *Water Resources Research* **15**: 1409–1420.

Romano N. and Santini A. (1997) Effectiveness of using pedo-transfer functions to quantify the spatial variability of soil water retention characteristics. *Journal of Hydrology* **202**: 137–157.

Romanowicz R. (1997) A MATLAB implementation of TOPMODEL. *Hydrological Processes* **11**(9): 1115–1130.

Romanowicz R. and Beven K. J. (1998) Dynamic real-time prediction of flood inundation probabilities. *Hydrological Science Journal* **43**: 181–196.

Romanowicz R., Beven K. J. and Moore R. V. (1993a) TOPMODEL as an application module within WIS. In Kovar K. and Nachtnebel H. P. (eds) *HydroGIS 93: Application of Geographical Information Systems in Hydrology and Water Resources*. IAHS Publication No. 211, Wallingford, UK, pp. 211–226.

Romanowicz R., Beven K. J. and Moore R. V. (1993b) GIS and distributed hydrological modelling. In Mather P. (ed.) *Geographical Information Handling: Research and Applications*. John Wiley, Chichester, pp. 197–206.

Romanowicz R., Beven K. J. and Tawn J. (1994) Evaluation of predictive uncertainty in nonlinear hydrological models using a Bayesian approach. In Barnett V. and Turkman K. F. (eds) *Statistics for the Environment. II. Water Related Issues*. John Wiley, Chichester, pp. 297–317.

Römkens M. J. M., Prasad S. N. and Whisler F. D. (1990) Surface sealing and infiltration. In Anderson M. G. and Burt T. P. (eds) *Process Studies in Hillslope Hydrology*. John Wiley, Chichester, pp. 127–172.

Rose K. A., Smith E. P., Gardner R. H., Brenkert A. L. and Bartell S. M. (1991) Parameter sensitivities, Monte Carlo filtering and model forecasting under uncertainty. *Journal of Forecasting* **10**: 117–134.

Rosenbrock H. H. (1960) An automatic method of finding the greatest or least value of a function. *Computing Journal* **3**: 175–184.

Ross C. N. (1921) The calculation of flood discharge by the use of time contour plan isochrones. *Transactions of the Institute of Engineers, Australia* **2**: 85–92.

Rosso R. (1984) Nash model relation to Horton order ratios. *Water Resources Research* **20**: 914–920.

Rutter A. J., Kershaw K. A., Robins P. C. and Morton A. J. (1971) Predictive model of rainfall interception in forests, 1. Derivation of the model from observations in a plantation of Corsican pine. *Agriculture and Meteorology* **9**: 367–384.

Rutter A. J., Morton D. J. and Robins P. C. (1975) A predictive model of rainfall interception by forests. II. Generalisations of the model and comparisons with observations in some coniferous and hardwood stands. *Journal of Applied Ecology* **12**: 367–380.

Saulnier G.-M., Beven K. J. and Obled C. (1997a) Digital elevation analysis for distributed hydrological modelling: reducing scale dependence in effective hydraulic conductivity values. *Water Resources Research* **33**: 2097–2101.

Saulnier G.-M., Obled Ch. and Beven K. J. (1997b) Analytical compensation between DTM grid resolution and effective values of saturated hydraulic conductivity within the TOPMODEL framework. *Hydrological Processes* **11**: 1331–1346.

Saulnier G.-M., Beven K. J. and Obled Ch. (1998) Including spatially variable soil depths in TOPMODEL. *Journal of Hydrology* **202**: 158–172.

Schaake J. (1990) From climate to flow. In Waggoner P. E. (ed.) *Climate Change and US Water Resources*. John Wiley, Chichester, pp. 177–206.

Schaap M. G. and Bouten W. (1996) Modelling water retention curves of sandy soils using neural networks. *Water Resources Research* **32**: 3033–3040.

Schaap M. G. and Leij F. J. (1998) Database related accuracy and uncertainty of pedotransfer functions. *Soil Science* **163**: 765–779.

Schaap M. G., Leij F. J. and Van Genuchten M. T. (1998) Neural network analysis for hierarchical prediction of soil water retention and saturated hydraulic conductivity. *Soil Science Society of America Journal* **62**: 847–855.

Schmugge T. (1998) Applications of passive microwave observations of surface soil moisture. *Journal of Hydrology* **212/213**: 188–197.

Schmugge T., Jackson T. J., Kustas W. P., Roberts R., Parry R., Goodrich D. C., Amer S. A. and Weltz M. A. (1994) Push broom microwave radiometer observations of surface soil moisture in Monsoon '90. *Water Resources Research* **30**: 1321–1327.

Schnur R. and Lettenmaier D. P. (1998) A case study of statistical downscaling in Australia using weather classification by recursive partitioning. *Journal of Hydrology* **212/213**: 362–379.

Schreider S. Yu., Jakeman A. J., Pittock A. B. and Whetton P. H. (1996) Estimation of possible climate change impacts on water availability, extreme flow events and soil moisture in the Goulburn and Ovens Basins, Victoria. *Climatic Change* **34**: 513–546.

Schreider S. Yu., Whetton P. H., Jakeman A. J. and Pittock A. B. (1997) Runoff modelling for snow-affected catchments in the Australian alpine region, eastern Victoria. *Journal of Hydrology* **200**: 1–23.

Schultz G. A. (1996) Remote sensing applications to hydrology: runoff. *Hydrological Science Journal* **41**: 453–475.
Schultz G. A. (1999) A call for hydrological models based on remote sensing, tracers and other modern hydrometric techniques. In *Integrated Methods in Catchment Hydrology: Tracer, Remote Sensing and New Hydrometric Techniques*. IAHS Publication No. 258, Wallingford, UK, pp. 223–230.
Schumann A. H. and Funke R. (1996) GIS-based components for rainfall–runoff models. In *HydroGIS96: Application of Geographic Information Systems in Hydrology and Water Resources*. IAHS Publication No. 235, Wallingford, UK, pp. 477–484.
Schumm S. A. (1956) Evolution of drainage systems and slopes at Perth Amboy, New Jersey. *Bulletin of the Geological Society of America* **67**: 597–646.
Scoging H. M. and Thornes J. B. (1982) Infiltration Characteristics in a Semiarid Environment. IAHS Publication No. 128, Wallingford, UK, pp. 159–168.
Sefton C. E. M. and Howarth S. M. (1998) Relationships between dynamic response characteristics and physical descriptors of catchments in England and Wales. *Journal of Hydrology* **211**: 1–16.
Seibert J. (1997) Estimation of parameter uncertainty in the HBV model. *Nordic Hydrology* **28**(4/5): 247–262.
Seibert J., Bishop K. H. and Nyberg L. (1997) A test of TOPMODEL's ability to predict spatially distributed groundwater levels. *Hydrological Processes* **11**: 1131–1144.
Sellers P. J. (1985) Canopy reflectance, photosynthesis and transpiration. *International Journal of Remote Sensing* **8**: 1335–1372.
Sellers P. J., Randall D. A., Collatz G. J., Berry J. A., Field C. B., Dzlich D. A., Zhang C., Collela G. D. Bounoua L. (1996) A revised land surface parameterisation (SiB2) for atmospheric GCMs. Part 1. Model formulation. *Journal of Climate* **9**(4): 676–705.
Sempere Torres D., Rodriguez J.-Y. and Obled C. (1992) Using the DPFT approach to improve flash flood forecasting models. *Natural Hazards* **5**: 17–41.
Sen M. and Stoffa P. L. (1995) *Global Optimisation Methods in Geophysical Inversion*. Elsevier, Amsterdam.
Senbeta D. A., Shamseldin A. Y. and O'Connor K. M. (1999) Modification of the probability–distributed interacting storage capacity model. *Journal of Hydrology* **224**: 149–168.
Sharma M. L., Gander G. A. and Hunt C. G. (1980) Spatial variability of infiltration in a watershed. *Journal of Hydrology* **45**: 101–122.
Shaw E. M. (1994) *Hydrology in Practice*, 3rd edition. Chapman and Hall, London.
Sherman L. K. (1932) Streamflow from rainfall by unit-graph method. *Engineering News Record* **108**: 501–505.
Shorter J. A. and Rabitz H. A. (1997) Risk analysis by the guided Monte Carlo technique. *Journal of Statistical Computation and Simulation* **57**: 321–336.
Shouse P. J. and Mohanty B. P. (1998) Scaling of near-saturated hydraulic conductivity measured using disc infiltrometers. *Water Resources Research* **34**: 1195–1205.
Shuttleworth W. J., Gash J. H. C., Lloyd C. R., McNeill D. D., Moore C. J. and Wallace J. S. (1988) An integrated micrometeorological system for evaporation measurement. *Agric. Forest Meteorology* **43**: 295–317.
Silburn D. M. and Connolly R. D. (1995) Distributed parameter hydrology model (answers) applied to a range of catchment scales using rainfall simulator data. I. Infiltration modelling and parameter measurement. *Journal of Hydrology* **172**: 87–104.
Simmons C. S., Nielsen D. R. and Biggar J. W. (1979) Scaling field measured soil water properties (2 parts). *Hilgardia* **47**: 77–173.
Simunek J., Sejna M. and Van Genuchten M. T. (1996) The HYDRUS-2d software package for simulating water flow and solute transport in two dimensional variably saturated media, version 1. Technical Report IGWMC-TP553, International Ground Water Modelling Center, Golden, CO.
Simunek J., Kodesova R., Gribb M. M. and Van Genuchten M. T. (1999) Estimating hysteresis in the soil water retention function from cone permeameter experiments. *Water Resources Research* **35**: 1329–1346.

Singh V. P. (1995) *Computer Models of Watershed Hydrology.* Water Resource Publications, Highlands Ranch, CO.

Singh V. P. (1996) *Kinematic Wave Modelling in Water Resources.* John Wiley, Chichester.

Singh V. P. and Yu C.-Y. (1997) Evaluation and generalisation of 13 mass-transfer equations for determining free water evaporation. *Hydrological Processes* **11**: 311–323.

Singh V. P., Bengtsson L. and Westerstrom G. (1997) Kinematic wave modelling of saturated basal flow in a snowpack. *Hydrological Processes* **11**: 177–187.

Sivapalan M., Wood E. F. and Beven K. J. (1990) On hydrological similarity: 3. A dimensionless flood frequency distribution. *Water Resources Research* **26**: 43–58.

Sklash M. G. (1990) Environmental isotope studies of storm and snowmelt generation. In Anderson M. G. and Burt T. P. (eds) *Process Studies in Hillslope Hydrology.* John Wiley, Chichester, pp. 401–435.

Sklash M. G. and Farvolden R. N. (1979) The role of groundwater in storm runoff. *Journal of Hydrology* **43**: 43–65.

Sklash M. G., Beven K. J., Gilman K. and Darling W. (1996) Isotope studies of pipeflow at Plynlimon, Wales, UK. *Hydrological Processes* **10**: 921–944.

Sloan P. and Moore I. D. (1984) Modelling subsurface stormflow on steeply sloping forested watersheds. *Water Resources Research* **20**(12): 1815–1822.

Slough K. and Kite G. W. (1992) Remote sensing estimates of snow water equivalent for hydrological modelling applications. *Canadian Journal of Water Resources* **17**: 1–8.

Smith J. A., Baeck M. L., Steiner M. and Miller A. J. (1996) Catastrophic rainfall from an upslope thunderstorm in the central Appalachians: the Rapidan storm of June 27th, 1995. *Water Resources Research* **32**(10): 3099–3113.

Smith K. and Ward R. (1998) *Floods: Physical Processes and Human Impacts.* John Wiley, Chichester.

Smith R. E. (1983) Approximate soil water movement by kinematic characteristics. *Soil Science Society of America Journal* **47**: 3–8.

Smith R. E. and Parlange, J.-Y. (1978) A parameter efficient infiltration model. *Water Resources Research* **14**: 533–538.

Smith R. E. and Woolhiser D. A. (1971) Overland flow on an infiltrating surface. *Water Resources Research* **7**(4): 899–913.

Smith R. E., Corradini C. and Melone F. (1993) Modeling infiltration for multistorm runoff events. *Water Resources Research* **29**: 133–144.

Smith R. E., Goodrich D. C., Woolhiser D. A. and Unkrich C. L. (1995) KINEROS – a Kinematic runoff and EROSion model. In Singh V. P. (ed.) *Computer Models of Watershed Hydrology.* Water Resource Publications, Highlands Ranch, CO, pp. 697–732.

Smith R. E., Corradini C. and Melone F. (1999) A conceptual model for infiltration and redistribution in crusted soils. *Water Resources Research* **35**: 1385–1393.

Snell J. D. and Sivapalan M. (1995) Application of the meta-channel concept: construction of the meta-channel hydraulic geometry for a natural channel. *Hydrological Processes* **9**: 485–505.

Sorooshian S. and Gupta V. K. (1995) Model calibration. In Singh V. P. (ed.) *Computer Models of Watershed Hydrology.* Water Resource Publications, Highlands Ranch, CO, pp. 23–68.

Sorooshian S., Gupta V. K. and Fulton J. L. (1983) Evaluation of maximum likelihood parameter estimation techniques for conceptual rainfall–runoff models: influence of calibration data variability and length on model credibility. *Water Resources Research* **19**: 251–259.

Sorooshian S., Duan Q. and Gupta V. K. (1992) Calibration of the SMA-NWSRFS conceptual rainfall–runoff model using global optimisation. *Water Resources Research* **29**: 1185–1194.

Soto B. and Diaz-Fierros F. (1998) Runoff and soil erosion from areas of burnt scrub: comparison of experimental results with those predicted by the WEPP model. *Catena* **31**: 257–270.

Spear R. C., Grieb T. M. and Shang N. (1994) Parameter uncertainty and interaction in complex environmental models. *Water Resources Research* **30**: 3159–3170.

Srinavasan R., Ramanarayanan T. S., Arnold J. G. and Bednarz S. T. (1998) Large area hydrologic modelling and assessment. Part 2. Model application. *Journal of American Water Resources Association* **34**: 91–101.

Stadler D., Wunderli H., Auckenthaler A. and Flühler H. (1996) Measurement of frost-induced snowmelt runoff in a forest soil. *Hydrological Processes* **10**: 1293–1304.

Steel M. E., Black A. R., Werrity A. and Littlewood I. G. (1999) Reassessment of flood risk for Scottish rivers using synthetic runoff data. In *Hydrological Extremes: Understanding, Predicting, Mitigating*. IAHS Publication No. 255, Wallingford, UK, pp. 209–215.

Steenhuis T. S., Winchell M., Rossing J., Zollweg J. A. and Walter M. F. (1995) SCS runoff equation revisited for variable source runoff areas. *Journal of Irrigation and Drainage Engineering, ASCE* **121**: 234–238.

Stephenson G. R. and Freeze R. A. (1974) Mathematical simulation of subsurface flow contributions to snowmelt runoff, Reynolds Creek, Idaho. *Water Resources Research* **10**(2): 284–298.

Stieglitz M., Rind D., Famiglietti J. and Rosenzweig C. (1997) An efficient approach to modelling the topographic control of surface hydrology for regional and global climate modelling. *Journal of Climate* **10**: 118–137.

Stoker J. (1957) *Water Waves*. Interscience, New York.

Surkan A. (1969) Synthetic hydrographs: effects of network geometry. *Water Resources Research* **5**: 115–128.

Swank W. T. and Crossley D. A. (eds) (1988) *Forest Hydrology and Ecology at Coweeta*. Ecological Studies 66, Springer-Verlag, New York.

Tabrizi M. H. N., Said S. E., Badr A. W., Mashor Y. and Billings S. A. (1998) Nonlinear modelling and prediction of a river flow system. *Journal of the American Water Resources Association* **34**: 1333–1339.

Taha A., Gresillon J. M. and Clothier B. E. (1997) Modelling the link between hillslope water movement and stream flow: application to a small Mediterranean forest watershed. *Journal of Hydrology* **203**: 11–20.

Tarantola A. (1987) *Inverse Problem Theory*. Elsevier, New York.

Tarboton D. G. (1997) A new method for the determination of flow directions and upslope areas in grid digital elevation models. *Water Resources Research* **33**: 309–319.

Tardieu F. and Davies W. J. (1993) Integration of hydraulic and chemical signalling in the control of stomatal conductance and water status of droughted plants. *Plane, Cell and Environment* **16**: 341–349.

Tayfur G. and Kavvas M. L. (1998) Areally-averaged overland flow equations at hillslope scale. *Hydrological Science Journal* **43**: 361–378.

Thirumalaiah K. and Deo M. C. (1998) River stage forecasting using artificial neural networks. *Journal of Hydrological Engineering, ASCE* **3**: 26–32.

Thyer M., Kuczera G. and Bates B. C. (1999) Probabilistic optimisation for conceptual rainfall runoff models: a comparison of the shuffled complex evolution and simulated annealing algorithms. *Water Resources Research* **35**: 767–774.

Tietje O. and Tapkenhinrichs M. (1993) Evaluation of pedo-transfer functions. *Soil Science Society of America Journal* **57**: 1088–1095.

Tillotson P. M. and Nielsen D. R. (1984) Scale factors in soil science. *Soil Science Society of America Journal* **48**: 953–959.

Todini E. (1995) New trends in modelling soil processes from hillslope to GCM scales. In Oliver H. R. and Oliver S. A. (eds) *The Role of Water and the Hydrological Cycle in Global Change*. NATA ASI Series, Vol. I 31, pp. 317–347.

Todini E. (1996) The ARNO rainfall–runoff model. *Journal of Hydrology* **175**: 339–382.

Tokar A. S. and Johnson P. A. (1999) Rainfall–runoff modeling using artificial neural networks. *Journal of Hydrological Engineering, ASCE* **4**: 232–239.

Troch P. A., Smith J. A., Wood E. F. and de Troch F. P. (1994) Hydrologic controls of large floods in a small basin: central Appalachian case study. *Journal of Hydrology* **156**: 285–309.

Turner H. M. and Burdoin A. S. (1941) The flood hydrograph. *Journal of the Boston Society of Civil Engineers* **28**: 232–256.

Tyler S. W. and Wheatcraft S. W. (1992) Fractal scaling of soil particle size distributions: analysis and limitations. *Soil Science Society of America Journal* **56**: 362–369.

Uhlenbrook S., Seibert J., Leibundgut C. and Rohde A. (1999) Prediction uncertainty of conceptual rainfall–runoff models caused by problems in identifying parameters and structure. *Hydrological Science Journal* **44**: 779–797.

USDA SCS (1985) *National Engineering Handbook, Supplement A, Section 4*. US Department of Agriculture, Washington, DC, Chapter 10.

USDA SCS (1986) *Urban Hydrology for Small Watersheds, Technical Release 55*. US Department of Agriculture, Washington, DC.

USDA SCS (1992) *STATSGO – State Soils Geographic Data Base*. Soil Conservation Service, Publication No. 1492, Washington, DC.

Van Genuchten M. Th. (1980) A closed-form equation for predicting the hydraulic conductivity of unsaturated soils. *Soil Science Society of America Journal* **44**: 892–898.

Van Genuchten M. Th., Leij F. J., Lond L. J. (eds) (1989) *Indirect Methods for Estimating the Hydraulic Properties of Unsaturated Soils*, Riverside, CA. USDA, Salinity Lab.

Van Genuchten M. Th., Schaap M. G., Mohanty B. P., Simunek J. and Leij F. J. (1999) Modelling flow and transport processes at the local scale. In Feyen J. and Wiyo K. (eds) *Modelling Transport Processes in Soils*. Wageningen Pers, Wageningen, The Netherlands, pp. 23–45.

Van Straten G. and Keesman K. (1991) Uncertainty propagation and speculation in projective forecasts of environmental change: a lake eutrophication example. *Journal of Forecasting* **10**: 163–190.

Venetis C. (1969) The IUH of the Muskingum channel reach. *Journal of Hydrology* **7**: 444–447.

Vereeken H., Maes J., Feyen J. and Darius P. (1989) Estimating the soil moisture retention characteristics from texture, bulk density and carbon content. *Soil Science* **148**: 389–403.

Vereeken H., Maes J. and Feyen J. (1990) Estimating unsaturated hydraulic conductivity from easily measured soil properties. *Soil Science* **149**: 1–11.

Verhoest N. E. C., Troch P. A., Paniconi C. and de Troch F. (1998) Mapping basin scale variable source areas from multitemporal remotely sensed observations of soil moisture behaviour. *Water Resources Research* **34**: 3235–3244.

Vertessy R. A. and Elsenbeer H. (1999) Distributed modelling of storm flow generation in an Amazonian rain forest catchment: effects of model parameterisation. *Water Resources Research* **35**: 2173–2187.

Vertessy R. A., Hatton T. J., O'Shaughnessy P. J. and Jayasuriya M. D. A. (1993) Predicting water yield from a mountain ash forest using a terrain analysis based catchment model. *Journal of Hydrology* **150**: 665–700.

Vogel R. M., Thomas W. O. and McMahon T. A. (1993) Flood-flow frequency model selection in southwestern United States. *Journal of Water Resources Planning and Management* **119**: 353–366.

Wagner B., Tarnawski V., Wessolek G. and Plagge R. (1998) Suitability of models for the estimation of soil hydraulic parameters. *Geoderma* **86**: 229–239.

Walsh R. P. D., Hudson R. N. and Howells K. A. (1982) Changes in the magnitude–frequency of flooding and heavy rainfalls in the Swansea valley since 1875. *Cambria* **9**: 36–60.

Wallace J. S. (1995) Calculating evaporation: resistance to factors. *Agric. Forest Meteorology* **73**: 353–366.

Wang Q. J. (1991) The genetic algorithm and its application to calibrating conceptual rainfall–runoff models. *Water Resources Research* **27**: 2467–2471.

Wang Z., Feyen J., Van Genuchten M. Th. and Nielsen D. R. (1998) Air entrapment effects on infiltration rates and flow instability. *Water Resources Research* **34**: 213–222.

Warrick A. W. and Hussen A. A. (1993) Scaling of Richards equation for infiltration and drainage. *Soil Science Society of America Journal* **57**: 15–18.

Watson F. G. R., Vertessy R. A. and Grayson R. B. (1999) Large-scale modelling of forest hydrological processes and their long-term effect on water yield. *Hydrological Processes* **13**: 689–700.
Weltz M. A., Ritchie J. C. and Fox H. D. (1994) Comparison of laser and field measurements of vegetation height and canopy cover. *Water Resources Research* **30**: 1311–1319.
Western A. W., Grayson R. B., Blöschl G., Willgoose G. and McMahon T. A. (1999) Observed spatial organisation of soil moisture and its relation to terrain indices. *Water Resources Research* **35**: 797–810.
Weyman D. R. (1970) Throughflow on hillslopes and its relation to the stream hydrograph. *Hydrological Science Bulletin* **15**: 25–33.
Wheater H. S., Jakeman A. J. and Beven K. J. (1993) Progress and directions in rainfall–runoff modelling. In Jakeman A. J., Beck M. B. and McAleer M. J. (eds) *Modelling Change in Environmental Systems*. John Wiley, Chichester, pp. 101–132.
Whitehead P. G., Young P. C. and Hornberger G. M. (1979) A systems model of streamflow and water quality in the Bedford–Ouse River. 1. Streamflow modelling. *Water Resources Research* **13**: 1155–1169.
Wigmosta M. and Lettenmaier D. P. (1999) A comparison of simplified methods for routing topographically driven subsurface flow. *Water Resources Research* **35**: 255–264.
Wigmosta M. S., Vail L. W. and Lettenmaier D. P. (1994) A distributed hydrology–vegetation model for complex terrain. *Water Resources Research* **30**(6): 1665–1679.
Wilby R. L. and Wrigley T. M. L. (1997) Downscaling general circulation model output: a review of methods and limitations. *Progress in Physical Geography* **21**: 530–548.
Wilby R. L., Hassan H. and Hanaki K. (1998) Statistical downscaling of hydrometeorological variables using general circulation model output. *Journal of Hydrology* **205**: 1–19.
Willgoose G. and Kuczera G. (1995) Estimation of subgrid scale kinematic wave parameters for hillslopes. *Hydrological Processes* **9**: 469–482.
Williams J. R. (1995) The EPIC model. In Singh V. P. (ed.) *Computer Models of Watershed Hydrology*. Water Resource Publications, Highlands Ranch, CO, pp. 909–1000.
WMO (1964) Guide for hydrometeorological practices. Technical report, World Meteorological Organisation, Geneva, p. 24.
WMO (1975) Intercomparison of conceptual models used in hydrological forecastings. Operational Hydrology Technical Report No 7, World Meteorological Organisation, Geneva.
WMO (1986) Intercomparison of models of snowmelt runoff. Operational Hydrology Technical Report No. 23, World Meteorological Organisation, Geneva.
Wolock D. M. (1993) Simulating the variable source area concept of streamflow generation with the watershed model. Water-Resources Investigations Report 93-4124. Technical report, US Geological Survey, Lawrence, Kansas.
Wolock D. M. and Hornberger G. M. (1991) Hydrological effects of changes in levels of atmospheric carbon dioxide. *Journal of Forecasting* **10**: 105–116.
Wong T. H. F. and Laurenson E. M. (1983) Wave speed – discharge relations in natural channels. *Water Resources Research* **19**: 701–706.
Wood E. F., Sivapalan M. and Beven K. J. (1990) Similarity and scale in catchment storm response. *Reviews in Geophysics* **28**: 1–18.
Wood E. F., Lettenmaier D. P. and Zatarian V. G. (1992) A land-surface hydrology parametersiation with sub-grid variability for general circulation models. *Journal of Geophysical Research* **97**(D3): 2717–2728.
Wood E. F., Lin D.-S., Mancini M., Thongs D., Troch P. A., Jackson T.J., Famiglietti J. S. and Engman E. T. (1993) Intercomparisons between passive and active microwave remote sensing and hydrological modeling for soil moisture. *Advances in Space Research* **13**: 167–176.
Woolhiser D. A. and Goodrich D. C. (1988) Effect of storm intensity patterns on surface runoff. *Journal of Hydrology* **102**: 335–354.
Wösten J. H. M. (1999) The HYPRES database of hydraulic properties of European soils. In Feyen J. and Wiyo K. (eds) *Modelling Transport Processes in Soils*. Wageningen Pers, Wageningen, The Netherlands, pp. 675–681.

Wright T . R., Gash J. H. G., Da Rocha H. R., Shuttleworth W. J., Nobre C. A., Maitelli G. T., Zamperoni C. A. G. D. and Carvalho P. R. A. (1992) Dry season micrometeorology of central Amazonian ranchland. *Quarterly Journal of the Royal Meteorology Society* **118**: 1083–1099.
Xevi E., Christiaens K., Espino A., Sewnandan W., Mallants D., Sorensen H. and Feyen J. (1997) Calibration, validation and sensitivity analysis of the MIKE SHE model using the Neuenkirchen catchment as case study. *Water Resources Management* **11**: 219–242.
Xinmei H., Lyons T. J., Smith R. C. G. and Hacker J. M. (1995) Estimation of land surface parameters using satellite data. *Hydrological Processes* **9**: 631–643.
Yapo P. O., Gupta H. V. and Sorooshian S. (1996) Calibration of conceptual rainfall–runoff models: sensitivity to calibration data. *Journal of Hydrology* **181**: 23–48.
Yapo P. O., Gupta H. and Sorooshian S. (1998) Multi-objective global optimization for hydrologic models. *Journal of Hydrology* **204**: 83–97.
Young P. C. (1975) Recursive approaches to time series analysis. *Bulletin of the Institute of Math. Appl.* **10**: 209–224.
Young P. C. (1983) The validity and credibility of models for badly defined systems. In Beck M. B. and Van Straten G. (eds) *Uncertainty and Forecasting of Water Quality*. Springer-Verlag, New York, pp. 69–98.
Young P. C. (1984) *Recursive Estimation and Time Series Analysis*. Springer-Verlag, Berlin.
Young P. C. (1992) Parallel processes in hydrology and water-quality – a unified time-series approach. *Journal of the Institution of Water and Environmental Management* **6**(5): 598–612.
Young P. C. (1993) Time variable and state dependent parameter modelling of nonstationary and nonlinear time series. In Subba Rao T. (ed.) *Developments in Time Series*. Chapman and Hall, London, pp. 374–413.
Young P. C. (2001) Data-base mechanistic modelling and validation of rainfall-flow processes. In Anderson M. G. and Bates P. D. (eds) *Model Validation: Perspectives in hydrological Science*. John Wiley, Chichester, (in press).
Young P. C. and Beven K. J. (1991) Computation of the instantaneous unit-hydrograph and identifiable component flows with application to two small upland catchments – comment. *Journal of Hydrology* **129**(1–4): 389–396.
Young P. C. and Beven K. J. (1994) Data-based mechanistic modelling and the rainfall–flow nonlinearity. *Environmetrics* **5**(3): 335–363.
Young P. C. and Wallis S. G. (1985) Recursive estimation: a unified approach to the identification, estimation and forecasting of hydrological systems. *Applied Maths and Computation* **17**: 299–334.
Young P. C., Jakeman A. J. and Post D. A. (1997) Recent advances in the data-based modelling and analysis of hydrological systems. *Water Science Technology* **36**: 99–116.
Young R. A., Onstad C. A. and Bosch D. D. (1995) AGNPS: An agricultural nonpoint source model. In Singh V. P. (ed.) *Computer Models of Watershed Hydrology*. Water Resource Publications, Highlands Ranch, CO, pp. 1001–1020.
Yu B. (1998) Theoretical justification of SCS method for runoff estimation. *Journal of Irrigation and Drainage Engineering, ASCE* **124**: 306–309.
Zhang W. and Montgomery D. R. (1994) Digital elevation model grid size, landscape representation and hydrologic simulations. *Water Resources Research* **30**: 1019–1028.
Zhang L., Dawes W. R., Hatton T. J., Reece P. H., Beale G. T. H. and Packer Z. (1999) Estimation of soil moisture and groundwater recharge using the TOPOG–IRM model. *Water Resources Research* **35**: 149–161.
Zhao R. J. (1992) The Xinanjiang model applied in China. *Journal of Hydrology* **135**: 371–381.
Zhao R. J. and Liu X.-R. (1995) The Xinanjiang model. In Singh V. P. (ed.) *Computer Models of Watershed Hydrology*. Water Resource Publications, Highlands Ranch, CO, pp. 215–232.
Zhao R. J., Zhuang Y.-L., Fang L.-R., Liu X.-R. and Zhang Q.-S. (1980) The Xinanjiang model. In *Hydrological Forecasting. IAHS Publication No. 129*. Wallingford, UK, pp. 351–356.
Zoch R. T. (1934) On the relation between rainfall and streamflow. *Monthly Weather Review* **62**(9): 315–322.

Zoppou C. and O'Neill I. (1982) Criteria for the choice of flood routing methods in natural channels. In *Hydrology and Water Resources Symposium*. Institution of Engineers of Australia, pp. 75–81.

Zuidema P. K. (1985) Hydraulik der Abflussbildung Wahrend Starniederschlagen. PhD thesis, Versuchanstalt fur Wasserbau, Hydrologie und Glaziologie, ETH, Zurich.

Index